# Fundamentals of Data Science

## Part II: Statistical Modeling

Jared M. Maruskin

# Fundamentals of Data Science

Part II: Statistical Modeling

Cayenne Canyon Press

Dr. Jared M. Maruskin

Published by Cayenne Canyon Press
San José, California

ISBN: 978-1-941043-12-7  (softcover)

10  9  8  7  6  5  4  3  2  1

*Imagination is more important than knowledge.*
—A. Einstein

# Preface

The goal of this project is to lay out a mathematical foundation of data science, from basic statistics through machine learning. In my experience, many data science books largely focus on data-science applications and programming packages, and not the nitty gritty of the underlying mathematical theory, which is largely buried across numerous graduate texts in statistics and machine learning. In addition to laying out the mathematics, I also complement the reading with numerous examples in Python, and strive (especially as this series progresses) to follow an object-oriented approach to programming.

Due to the encyclopedic scope of this project, I made certain necessary trade offs in order to make a condensed version of the topics presented throughout these pages, while covering as much material as possible in a reasonable amount of time. In particular, I favored clear explanation of the mathematical principles, supplemented with working Python code, over a parade of interesting examples, which I find to be commonly available in many other wonderful texts. Most chapters are themselves condensed versions of entire textbooks, which I refer to for additional discussion and depth throughout.

This work is intended for students of data science with a prerequisite of mathematical maturity, PhDs from the broader panoply of mathematical sciences with a weak statistics background who are interested into transitioning into the field, and as a reference source for practicing data scientists. I personally belonged to this second camp: my own PhD was in dynamical systems theory, with applications to aerospace engineering (tracking space debris; see Maruskin, *et al.* [2012]) and physics (geometric theory of the variational principles of nonholonomic systems; see Maruskin [2018]), whereas my knowledge of statistics, modeling, and machine learning, on the other hand, I picked up on the job after transitioning into the data science field. Since my understanding of many of these topics was superficial or tangential, I found myself one day wishing that I had a masters degree in statistics and machine learning and figured, *no worries, I used to be a professor, I can do that myself.* So I sat down and outlined my own personal curriculum and divided it among three volumes—statistics and experiment, statistical modeling, and machine learning—which form the basis of this series. In short: my goal was to lay a foundation of expertise in statistical modeling and machine learning for myself, while leaving a foundational path for others to follow.

I am very excited to present the second volume of the series, which covers principles of validation, linear and logistic regression, generalized linear models, causality, time series, and Bayesian statistics. This project has been a labor of love, and I've dedicated many weekends and early mornings to read, learn, solve and write the content of these pages. I hope the result benefits your learning and understanding and proficiency of the topics as it has mine.

Coding examples throughout are written in Python. I have found that writing simple simulations is an effective tool in learning theory. It also serves as a tool to validate our theoretical formulas. Several times in the process of writing this text, I developed a formula, ran a simulation, and found that it didn't work the way I expected. This led me to discover my errors and, when the simulation finally worked, served to validate that I had arrived at the correct conclusions.

## Python Distribution

I recommend obtaining the free Anaconda Python 3 distribution, available from `anaconda.com`. It comes with all of the scientific packages pre-installed. Anaconda is a bundle of applications, so you can use it to launch R-Studio, Jupyter Notebook, or Spyder. I write my code in Spyder. The Sypder interface gives you access to a code editor, an iPython console, and help docs, all in one screen. I find it extremely useful running code in the console while I develop source code in the editor. I have also have several friends who use PyCharm from `jetbrains.com` as an alternative.

Many of the examples throughout this text will make use of various packages. The import statements that are used throughout are collected in Code Block 1. I will further assume the reader to have familiarity with Python. For those starting out, however, I recommend McKinney [2017] as an excellent place to start. For additional references on Python for machine learning, featuring sci-kit learn and tensorflow, see Gèron [2019].

San José, California

*Jared M. Maruskin*
December 2021

```python
1  import numpy as np
2  import pandas as pd
3  import sklearn, scipy
4  import time, datetime
5  import matplotlib.pyplot as plt
6  import pymc3 as pm
7  from mpl_toolkits.mplot3d import Axes3D # For 3D
8  from pandas import DataFrame, Series
9  from abc import ABC, abstractmethod
10 from queue import Queue
11 from collections import Counter
12 from scipy.special import comb
13 from scipy.special import beta as BETA
14 from scipy.special import gamma as GAMMA
15 from sklearn.preprocessing import OrdinalEncoder, OneHotEncoder,
       KBinsDiscretizer, KBinsDiscretizer, StandardScaler
16 from sklearn.impute import SimpleImputer
17 from sklearn.pipeline import Pipeline
18 from sklearn.compose import ColumnTransformer
19 from sklearn.model_selection import train_test_split
20 from sklearn.metrics import r2_score, roc_curve, auc,
       confusion_matrix
21 from sklearn.calibration import calibration_curve
22 from statsmodels.tsa.stattools import acf, pacf
23 from tqdm import tqdm
```

Code Block 1: Common package imports.

# Contents

## Part II  Statistical Modeling

# Part II

# Statistical Modeling

# 6

# Modeling Runway

In this chapter, we lay some of the groundwork for the models we will discuss throughout the rest of this book. We begin with a discussion of basic nomenclature, and then discuss common approaches for model validation. We then discuss aspects of preprocessing that are common to most data science applications. We conclude the chapter with a discussion on engineering aspects of data science; in particular, we discuss how to take an object-oriented approach to data-science projects.

## 6.1 Modeling and Validation

In this section we lay out some key definitions of modeling and machine learning. We then describe various metrics and methods for model validation that will be used extensively throughout the remainder of this text.

### 6.1.1 Models and Model Learning

We begin by laying some terminology for modeling and machine learning.

**Definition 6.1.** A model *is a mathematical representation of a system or a process.*

A predictive model *is a model for an output (or* target*) variable as it depends on a prescribed set of inputs (or* features*).*

A statistical model *is a predictive model whose output is a probability distribution for the target variable.*

Statistical models are often expressed in the form

$$Y = f(X) + \epsilon,$$

where $f$ is a predictive model and $\epsilon$ is an *error term*, with $\mathbb{E}[\epsilon] = 0$, so that $\mathbb{E}[Y|X] = f(x)$. Some additional assumptions regarding the noise $\epsilon$

are typically included in the model, such as independence and assumptions about its variance or distribution. Sometimes, for brevity, we will refer to a predictive model by its deterministic component $f : X \rightarrow \hat{Y}$, where $\hat{Y} = \mathbb{E}[Y|X]$.

**Definition 6.2.** *A* machine learning algorithm *is any computer algorithm that is able to automatically improve its performance of a task through experience.*

Supervised learning *refers to any machine learning algorithm that learns a predictive model $f : X \rightarrow \hat{Y}$ from a set of labeled data $\mathcal{D} = \{(x_i, y_i)\}_{i=1}^n$. The input $X$ is often referred to as a* feature vector, *and its components as a set of* features. *The output $Y$ is often referred to as the* target variable. *Each datum $(x_i, y_i)$ is referred to as an* instance.

Unsupervised learning *is any machine learning algorithm that learns patterns in unlabeled data.*

Reinforcement learning *is any machine learning algorithm in which an* agent *takes a sequence of* actions *in an* environment *in order to maximize a cumulative* reward.

In Part II of this text, we will predominantly focus on predictive models. In Part III, we will explore machine learning algorithms more closely. As a quick illustration of the distinction laid out in Definition 6.2, consider the following.

*Example 6.1.* A computer algorithm is trained with a set of images as inputs, each input having a label "cat" or "dog." It is tested on its ability to recognize whether or not a new image is a cat or dog. This is an example of supervised learning, as the labels ("cat" and "dog") were provided with the training set.

A different algorithm is provided the same set of input images, except without being explicitly told what is a cat and what is a dog. It is asked to seek patterns in the data. Here, the algorithm might come up with patterns that are unexpected. For example, the algorithm might classify images of the side view versus images of a front view; or it might classify partially hidden versus fully visible animals. Or it might classify head shots versus body shots. Or it might classify based on whether or not the pet is wearing a collar. Or it might come up with its own concept, something that distinguishes the images that is present in the data but would not normally be recognized by the human eye. This is an example of unsupervised learning, as the algorithm is left to its own devices to infer patterns from the data.

Finally, our third algorithm has learned how to play the game *Go.* Here, the environment is the board. The actions is the set of permissible (legal) Go moves. And the reward is issued at the end of each game as +1 win or 0 for loss. The computer can be trained with actual Go games, and it will continue to learn based on its experience. That is, as it plays, it becomes better at playing. In fact, two computers can be set up to play each other, with entire games occurring at lightning speed, while the algorithm learns which strategies work best.                                                    ▷

Predictive models themselves can be grouped into two different types, based on the type of target variable they are predicting.

**Definition 6.3.** *A* classification algorithm *is any supervised machine learning algorithm that learns a predictive model with a categorical output variable. The enumeration of possible outputs are referred to as* labels.

*A* regression algorithm *is any supervised machine learning algorithm that learns a predictive model with a continuous real output variable.*

In Example 6.1, the supervised algorithm is a classification algorithm, as it learned whether each picture represented a dog or a cat. An example of a regression algorithm is an algorithm that predicts home price based on a set of features (e.g., square footage, city, number of beds / baths, school district, etc.).

In general, the feature vector of a model will consist of a mix of continuous, categorical, and ordinal features. (An ordinal feature is similar to a categorical feature in the sense that it is finite and discrete, but it differs as it constitutes an ordered collection; e.g., *small*, *medium*, and *large*.) We will explore various requirements for data preprocessing for each of these feature types in Section 6.2.

**Definition 6.4.** *Given a predictive model $Y = f(X) + \epsilon$, with error term $\mathbb{E}[\epsilon] = 0$ and $\mathbb{V}(\epsilon) = \sigma^2$, an instance $(x, y)$ is said to have a* weight *(or* prior weight*) $w > 0$ if the variance of its error $\epsilon$ is reduced by a factor $w$; i.e., if $y_i \sim f_Y$, then $\mathbb{E}[Y] = f(x_i)$ and $\mathbb{V}(Y - f(x_i)) = \sigma^2/w$.*

*A* weighted data set *$\mathcal{D} = \{(x_i, y_i, w_i)\}_{i=1}^n$ is a data set with a prescribed (known) set of prior weights, so that if $y_i \sim f_{Y_i}$, then $\mathbb{E}[Y_i] = f(x_i)$ and $\mathbb{V}(Y_i - f(x_i)) = \sigma^2/w_i$. The* total weight *$\omega$ is the sum of the individual weights; i.e., $\omega = \sum_{i=1}^n w_i$.*

An instance $(x, y)$ with positive integer weight $w \in \mathbb{Z}^+$ is equivalent to a set of $w$ identical instances $\{(x, y)\}_{i=1}^w$, so that weights are a convenient encapsulation for multiplicity.

We conclude this section with a more detailed definition on what constitutes a machine learning model.

**Definition 6.5.** *A* (supervised) machine learning model *over a feature space $\mathbb{F}$ into a target space $\mathbb{T}$ is a mapping*

$$f : \mathbb{D} \to \mathscr{F}(\mathbb{F}, \mathbb{T}), \tag{6.1}$$

*where $\mathbb{D} = \mathscr{P}_{\text{fin}}(\mathbb{F} \times \mathbb{T})$ is the set of all possible finite subsets (or samples) of the data, and where $\mathscr{F}(\mathbb{F}, \mathbb{T})$ is the set of functions from $\mathbb{F}$ into $\mathbb{T}$.*

Given a sample of data $\mathcal{D} \in \mathbb{D}$, *known as* training data, *the function* $f_{\mathcal{D}} : \mathbb{F} \rightarrow \mathbb{T}$ *(known as the* trained model*) is a predictive model for the target variable as it depends on the feature vector.*

*If the target space is a finite set* $\mathbb{T} = \mathcal{C}$ *of categories, we say that the model is a* classification *model. Otherwise, if the target space is a subset of the reals,* $\mathbb{T} \subset \mathbb{R}$*, we say that the model is a* regression *model.*

*An* unsupervised machine learning model *is similar, except that the set of data* $\mathbb{D} = \mathscr{P}_{\mathrm{fin}}(\mathbb{F})$ *consists only of the feature data.*

Another aspect of learning algorithms is that the mapping shown in Equation (6.1) need not be performed in a single step; i.e., the model can continuously *improve* as data are added. In this way, we can view a machine learning algorithm as a mapping

$$f : \mathbb{D} \times \mathscr{F}(\mathbb{F}, \mathbb{T}) \rightarrow \mathscr{F}(\mathbb{F}, \mathbb{T}),$$

which symbolizes the important aspect that the model can improve (or *learn*) with time.

### 6.1.2 Validation Metrics for Regression

In this section, we will consider a predictive model of the form $\mathbb{E}[Y|X] = f(X)$ that was trained using a regression algorithm. We will consider various metrics that can be used to measure the performance of the model.

**Squared Errors**

Consider a data set $\mathcal{D} = \{(x_i, y_i)\}$ and a model $f$ with predictive values $\hat{y}_i = f(x_i)$. Note that the data set $\mathcal{D}$ is not necessarily the same data that was used to generate the model. We will discuss this in more detail in Section 6.1.5. We consider the following sum-of-squares statistics.

**Definition 6.6.** *Given a weighted data set* $\mathcal{D} = \{(x_i, y_i, w_i)\}_{i=1}^{n}$ *and model predictions* $\hat{y}_i = f(x_i)$*, we define the* sum of squared errors (SSE)*, the* regression sum of squares (SSR)*, and the* total sum of squares (SST) *as*

$$\mathrm{SSE} = \sum_{i=1}^{n} w_i (y_i - \hat{y}_i)^2, \tag{6.2}$$

$$\mathrm{SSR} = \sum_{i=1}^{n} w_i (\hat{y}_i - \overline{y})^2, \tag{6.3}$$

$$\mathrm{SST} = \sum_{i=1}^{n} w_i (y_i - \overline{y})^2, \tag{6.4}$$

*respectively.*

*Note 6.1.* The difference $y_i - \hat{y}_i$ between the true and predicted value is referred to as a *residual*. Therefore, many authors, especially in machine learning, refer to the sum of squared errors as the *residual sum of squares*, or RSS. We will use the phrases *sum of squared errors* and *residual sum of squares* interchangeably, however we favor the notation SSE as it is consistent with other sum-of-squares notation and the *mean-squared error* MSE, defined as MSE = SSE/$n$ or, for a weighted data set, as MSE = SSE/$\omega$.    ▷

One might be tempted to assume that the residual sum of squares and the regression sum of squares sum to the total sum of squares. This assumption, however, is not true in general. In fact, it is only true if the cross terms

$$\sum_{i=1}^{n} (y_i - \hat{y}_i)(\hat{y}_i - \overline{y})$$

sum to zero. Whether or not these terms vanish clearly depends on the model. For instance, consider the model that predicts $f(x) = 0$, for all $x$. Then the cross terms do not cancel, but instead sum to $-n\overline{y}^2$.

A common metric used in model evaluation is the coefficient of determination, which is often denoted simply and erroneously as $R^2$, which we define as follows.

**Definition 6.7.** *The* coefficient of determination $(R^2)$ *is defined as*

$$R^2 = 1 - \frac{\text{SSE}}{\text{SST}},\tag{6.5}$$

*where* SSE *and* SST *are defined as in Definition 6.6.*

This "$R$-squared" statistic really isn't the square of anything, other than its own square root, which need not be real. In other words, the $R^2$ statistic in certain cases can be negative. It cannot, however, be larger than 1. In fact, a value of $R^2 = 1$ would be a perfect model, as this is only achieved whenever $\hat{y}_i = y_i$, for all $i = 1, \ldots, n$. Hence we have the inequality

$$R^2 \leq 1.$$

In fact, the $R^2$ statistic *will* be negative whenever our predictions have a greater total error than had we simply predicted the average value $f(x) = \overline{y}$ for the entire data set. (See Exercise 6.1.) So a *positive* $R^2$ indicates we are doing better than no model at all (where, in absence of a model, we can simply use the average value as the value of all of our predictions).

The $R^2$ statistic is commonly interpreted as the *fraction of variance explained* by the model. To see this, note that the total sum of squares is essentially (a multiple of) the sample variance of the data. This represents the variance observed in the data set itself. The residual sum of squares, on the other hand, is how much variance is still *unexplained* by our model, as the residual sum of squares measures the variance between our predictions and the true values of the data.

The $R^2$ statistic is good in that the residual sum of squares is scaled by the total sample variance. Whereas interpretation of the MSE of a model completely relies on domain expertise, the $R^2$ statistic is a normalized metric that has a standard meaning.

A pitfall of the $R^2$ statistic, however, is that there is no universally adopted notion of which values of $R^2$ are acceptable. This is due, in part, to the fact that $R^2$ depends on the variance in the data. If we hold the mse fixed, the larger the variance in the data, the higher the value of $R^2$. In one context, an $R^2 = 0.20$ might be acceptable, but in another, an $R^2 = 0.90$ might not. For example, if we are modeling housing prices, which show a significant variability, a value of $R^2 = 0.90$ might translate to a RMSE in the tens of thousands of dollars, which would probably not be considered acceptable to the given application (example from Kuhn and Johnson [2013]).

The $R^2$ statistic is, however, useful in comparing different models. But even then there is an important caveat: the $R^2$ statistic will increase as one adds explanatory variables in the model. This leads to the erroneous conclusion that one should keep adding features to the model: everything except the kitchen sink. One important modification that addresses this is the following.

**Definition 6.8.** *The* adjusted coefficient of determination $\bar{R}^2$ *is defined as*

$$\bar{R}^2 = 1 - \frac{\text{SSE}/\text{df}_e}{\text{SST}/\text{df}_t}, \tag{6.6}$$

*where $df_t = n - 1$ is the number of degrees of freedom of the estimate of the population variance and $df_e = n - p$ is the number of degrees of freedom of the estimate of the residual variance, where $p$ is the total number of independent modeling parameters of the model.*

Unlike regular $R^2$, the adjusted $\bar{R}^2$ will increase with the inclusion of additional features only if the the model improvement is more than one would expect to see by chance.

An alternate assessment metric is given as follows.

**Definition 6.9.** *The* mean absolute error (MAE) *of a model is given by*

$$MAE = \frac{1}{n} \sum_{i=1}^{n} |\hat{y}_i - y_i|. \tag{6.7}$$

*Similarly, the* mean absolute percentage error (MAPE) *of a model is given by*

$$MAPE = \frac{1}{n} \sum_{i=1}^{n} \left| \frac{\hat{y}_i - y_i}{y_i} \right|. \tag{6.8}$$

Essentially, MAPE is a measure of how close the predictions are to the true values. A MAPE of 10% indicates that the predictions are typically about $\pm 10\%$ of the true values. This metric also has various flaws: it cannot be used if there are any zeros in the data, and there is an asymmetry between under-predictions (which cannot exceed 100%) and over-predictions (which are unlimited).

**Bias–Variance Tradeoff**

We will discuss model selection in Section 6.1.5. Presently, however, we shall content ourselves with laying out some of the theoretical groundwork. We begin with two definitions.

**Definition 6.10.** *Given a normal predictive model $Y = f(X) + \epsilon$, for $\epsilon \sim N(0, \sigma^2)$, where $f(X)$ is an unknown function of the input $X$. Let $\hat{f}(X) = \hat{f}(X; \mathcal{D})$ be an estimate for the unknown function $f(X)$ that is learned on a set of data $\mathcal{D} = \{(X_i, Y_i)\}_{i=1}^n$. Then we define the* bias *of the model $\hat{f}$ at a point $X = x$ as*

$$\text{bias}\left[\hat{f}(x)\right] = \mathbb{E}\left[\hat{f}(x; \mathcal{D})\right] - f(x). \tag{6.9}$$

*Similarly, we define the model variance at a point $X = x$ as*

$$\mathbb{V}\left[\hat{f}(x)\right] = \mathbb{E}\left[\left(\hat{f}(x; \mathcal{D}) - \mathbb{E}[\hat{f}(x; \mathcal{D})]\right)^2\right]. \tag{6.10}$$

*In Equations (6.9) and (6.10), all expectations are taken relative to the data $\mathcal{D} \sim f_{X,Y}^n$.*

Definitions in hand, we are now ready to state our main result.

**Theorem 6.1 (Bias–Variance tradeoff).** *Let $Y = f(X) + \epsilon$, where $\epsilon \sim N(0, \sigma^2)$, and $\hat{f}(X) = \hat{f}(X; \mathcal{D})$ be as in Definition 6.10. Then the expected squared residual at a point $X = x$ is given by*

$$\mathbb{E}\left[\left(Y - \hat{f}(x; \mathcal{D})\right)^2\right] = \sigma^2 + \text{bias}\left[\hat{f}(x)\right]^2 + \mathbb{V}\left[\hat{f}(x)\right], \tag{6.11}$$

*where the expectation on the left-hand side is taken over the joint distribution for $Y$ and $\mathcal{D}$.*

*Note 6.2.* The term $\sigma^2$ is often referred to as the *irreducible error*, as it is due to the white noise $\epsilon \sim N(0, \sigma^2)$ in the random variable $Y$; i.e., it is inherent uncertainty that cannot be remedied by modeling.     ▷

*Proof.* We begin by expanding the residual for fixed $X = x$ as

$$Y - \hat{f}(x; \mathcal{D}) = (Y - f(x)) + \left(f(x) - \mathbb{E}\left[\hat{f}(x; \mathcal{D})\right]\right) + \left(\mathbb{E}\left[\hat{f}(x; \mathcal{D})\right] - \hat{f}(x; \mathcal{D})\right).$$

Now, the first term depends on the random variable $Y$, whereas the second and third terms depend on the random variable $\mathcal{D}$. Therefore, the first term is independent with the second and third terms, and we conclude that

$$\mathbb{E}\left[(Y - f(x))\left(f(x) - \mathbb{E}\left[\hat{f}(x; \mathcal{D})\right]\right)\right] = 0$$

and

$$\mathbb{E}\left[(Y - f(x))\left(\mathbb{E}\left[\hat{f}(x; \mathcal{D})\right] - \hat{f}(x; \mathcal{D})\right)\right] = 0.$$

The second term is actually independent of $\mathcal{D}$ itself, so a similar result holds for the penultimate and final terms

$$\mathbb{E}\left[\left(f(x) - \mathbb{E}\left[\hat{f}(x; \mathcal{D})\right]\right)\left(\mathbb{E}\left[\hat{f}(x; \mathcal{D})\right] - \hat{f}(x; \mathcal{D})\right)\right] = 0.$$

By squaring and taking expectation over our expression for the residual, and by recalling that $Y - f(x) = \epsilon$, we therefore arrive at the expression

$$\mathbb{E}\left[\left(Y - \hat{f}(x; \mathcal{D})\right)^2\right] = \mathbb{E}[\epsilon^2] + \left(\mathbb{E}\left[\hat{f}(x; \mathcal{D})\right] - f(x)\right)^2$$
$$+ \mathbb{E}\left[\left(\hat{f}(x; \mathcal{D}) - \mathbb{E}\left[\hat{f}(x; \mathcal{D})\right]\right)^2\right].$$

The result follows.     □

*Note 6.3.* The *mean-squared error* MSE of a model can be defined by taking the expectation of Equation (6.11) over the random variable $X$.     ▷

We will see illustrations of this result as we progress through the chapter. Note that the bias-variance theorem is inherently concerned with regression problems, as it is a statement concerning the our expectation (literally) for the residual sum of squares.

### 6.1.3 Validation Metrics for Classification

In classification problems, the target variable $Y$ constitutes a categorical response, taking one of $c$ values in a set $\mathcal{C}$, whose elements can be labeled as $1, \ldots, c$. Oftentimes we will directly model a set of functions $f_i(X)$, for $i = 1, \ldots, c$, that are required to satisfy the constraints that $0 < f_i(X) < 1$, for all $i = 1, \ldots, c$ and $X \in \mathcal{X}$, and the normalization condition

$$\sum_{i=1}^{c} f_i(X) = 1.$$

Moreover, these functions should satisfy the property that the larger the value of $f_i(X)$, for a given class label $i$, the more likely it is that $Y = i$. They can be thought of as probabilities, but these outputs typically do not correspond to the actual real probability. When the raw functions $f_i'(X)$ do not satisfy the normalization conditions, we can always apply the *softmax transformation*

$$f_i(X) = \frac{e^{f_i'(X)}}{\sum_{i=1}^{c} e^{f_i'(X)}}. \tag{6.12}$$

(See Exercise 6.2.) However, our goal is not to predict a number between zero and one, but to predict class labels for a new instance $X$. A common method is to simply select the class with the largest value of $f_i$:

$$\hat{Y}(X; \mathcal{D}) = \arg\max_{i=1,\ldots,c} f_i(X; \mathcal{D}). \tag{6.13}$$

However, in many cases, we will be specifically considering classification algorithms for the purpose of predicting a *binary response*, so that $Y \in \{0, 1\}$. In this case, we only need to specify a single function $f(X) = f_1(X)$. In this case, it is more common to use a threshold $t$ to determine the final classification:

$$\hat{Y}_t(X; \mathcal{D}) = \mathbb{I}\left[f(X; \mathcal{D}) > t\right]. \tag{6.14}$$

Here, the same model will yield a different set of predictions for different thresholds. We will discuss this in more depth at the end of this subsection.

*Note 6.4.* We should further note that many classification algorithms *do not model an actual probability*! That is, the function $f(X)$ that is learned by the classification algorithm will satisfy the properties of a probability, but will not correspond to a probability itself. A value of $f(X) = 0.5$ means that the response $Y = 1$ is more likely than it would have been if $f(X) = 0.4$, but it does *not* mean that there is a 50% *probability* that the response is $Y = 1$. We will discuss this concern, and its remedy, in Section 6.1.4.    ▷

### Standard Classification Metrics

For the remainder of this section, we shall consider the common case of binary classification problems. In this context, the response $Y = 1$ is referred to as "the event," and it typically corresponds to the target of our prediction task. Common applications include a diagnostic for a lab test and a user in a freemium mobile app making a purchase. This distinction is made without loss of any generality, as $Y = 0$ and $Y = 1$ could just as easily correspond to dogs and cats. However, for the purpose of the following, we will consider the event $Y = 1$ to be the target of our prediction problem.

**Definition 6.11.** *Given a classification problem and a set of true values $Y_i$ and predictions $\hat{Y}_i$, for $i = 1, \ldots, n$, where each $Y_i$ may take one of $m$ distinct values, the* confusion matrix $C$ *is the $c \times c$ matrix of counts, where*

$$C_{ij} = \sum_{k=1}^{n} \mathbb{I}\left[\hat{Y}_k = i \text{ and } Y_k = j\right].$$

Thus, in a confusion matrix, the rows correspond to the predicted classes, and the columns correspond to the actual (true) classes. For the case of binary classification, it is conventional (for some reason) to write the rows and columns of the confusion matrix in *descending* order[1]. A typ-

|           | $Y = 1$ | $Y = 0$ |
|-----------|---------|---------|
| $\hat{Y} = 1$ | TP      | FP      |
| $\hat{Y} = 0$ | FN      | TN      |

Table 6.1: Confusion matrix for binary classification.

ical confusion matrix is shown in Table 6.1. The elements of the matrix are labeled TP, FP, FN, and TN. These correspond to true/false positive/negative, as defined below.

**Definition 6.12.** *In a binary classification problem, the elements of the confusion matrix are referred to as*

1. *True Positive (TP)—the number of correct positive predictions ($\hat{Y} = 1$ and $Y = 1$),*
2. *False Positive (FP)—the number of incorrect positive predictions ($\hat{Y} = 1$ and $Y = 0$),*
3. *False Negative (FN)—the number of incorrect negative predictions ($\hat{Y} = 0$ and $Y = 1$), and*
4. *True Negative (TN)—the number of correct negative predictions ($\hat{Y} = 0$ and $Y = 0$).*

*Note 6.5.* In Definition 6.12, think of the terminology in the context of a laboratory test for a medical condition: a positive result means you tested positive for the condition and a negative result means you tested negative for the condition. The True/False indicates whether or not the test result (the predicted value) was correct.　　　　　　　　　　　　　　　　　　　　▷

The first validation metric for a classifier is accuracy.

**Definition 6.13.** *The accuracy of a classifier is the ratio of the number of correct predictions to the number of overall predictions, i.e.,*

$$accuracy = \frac{TP + TN}{TP + FP + FN + TN}.$$

However, considering accuracy alone can be misleading.

---

[1] I suppose otherwise there wouldn't be anything confusing about it.

*Example 6.2.* Consider a binary classification problem for a rare event. Now consider the classifier that classifies *everything* as $\hat{Y} = 0$. The confusion matrix for 100 samples is given in Table 6.2. The accuracy of this model on

|  | $Y = 1$ | $Y = 0$ |
|---|---|---|
| $\hat{Y} = 1$ | 0 | 0 |
| $\hat{Y} = 0$ | 1 | 99 |

Table 6.2: Confusion matrix for Example 6.2; 99% accuracy.

this data set is 99%, even though the model (by design) doesn't do anything at all.                                                                        ▷

To remedy this, statisticians typically refer to two quantities known as sensitivity and specificity, defined below.

**Definition 6.14.** *The* sensitivity *and* specificity *of a binary classifier are the ratios of the number of true positives or negatives, respectively, to the total number of positives or negatives, respectively; i.e.,*

$$sensitivity = \frac{true\ positives}{total\ actual\ positives} = \frac{TP}{TP + FN} = TPR,$$

$$specificity = \frac{true\ negatives}{total\ actual\ negatives} = \frac{TN}{FP + TN} = TNR = 1 - FPR.$$

*The sensitivity is equivalent to the* true positive rate *(TPR) and the specificity is equivalent to the* true negative rate *(TNR), which is equivalent to one minus the* false positive rate *(FPR).*

It is desirable to have high sensitivity and specificity, and a low false positive rate. However, these two metrics still do not capture the full picture, as illustrated in our next example.

*Example 6.3.* Consider a binary classification problem with confusion matrix given in Table 6.3. The overall accuracy ($190/210 \approx 90.47\%$) is good.

|  | $Y = 1$ | $Y = 0$ |
|---|---|---|
| $\hat{Y} = 1$ | 10 | 20 |
| $\hat{Y} = 0$ | 0 | 180 |

Table 6.3: Confusion matrix for Example 6.3; 100% sensitivity, 90% specificity.

Similarly, the sensitivity and specificity both look healthy:

$$\text{sensitivity} = \frac{10}{10} = 100\%,$$

$$\text{specificity} = \frac{180}{200} = 90\%.$$

However, what does a positive test result $\hat{Y} = 1$ mean? Given a positive test result, we have to look at the first row of the confusion matrix. And we see that we only have a 33% probability of actually being a true positive. This discrepancy is not captured at all when looking at sensitivity and specificity alone.    ▷

As a result of Example 6.3, in the field of machine learning, it is more common to analyze the following pair of metrics.

**Definition 6.15.** *The* precision *and* recall *of a binary classifier are defined as*

$$precision = \frac{true\ positives}{total\ predicted\ positives} = \frac{TP}{TP + FP},$$

$$recall = \frac{true\ positives}{total\ actual\ positives} = \frac{TP}{TP + FN}.$$

*Recall is the same as sensitivity.*

In Example 6.3, the precision is 33% and the recall is 100%. Thus, a positive example has a 100% chance of being identified as positive; however, an example with a positive prediction only has a 33% chance of being an actual positive.

If a combined metric is required, one can use the $F_1$-score, which is the harmonic mean of precision and recall. More generally, we can use $F_\beta$, for $\beta > 0$, which is defined as

$$F_\beta = (1 + \beta^2) \frac{\text{precision} \cdot \text{recall}}{\beta^2 \text{precision} + \text{recall}}.$$

For the case $\beta = 1$, we recover the harmonic mean of the two metrics.

### Receiver-operator Characteristic (ROC) Curves

As stated above, the decision of whether or not to classify the output $f(X)$ of a binary classifier as an event is typically determined by setting a threshold $t \in [0, 1]$ and applying Equation (6.14). All of our validation metrics thus far apply to a single predictive model, i.e., following the selection of an appropriate threshold. But how should we compare different classifiers without regard to the particular value of the threshold parameter? The answer lies in the following.

**Definition 6.16.** *The* receiver–operator characteristic (ROC) *curve of a predictive model $f(X; \mathcal{D})$, with classification function*

$$\hat{Y}_t(X; \mathcal{D}) = \mathbb{I}\left[f(X; \mathcal{D}) > t\right],$$

*for $t \in [0, 1]$, is the parametric curve $\mathbf{r} : [0, 1] \to [0, 1]^2$ defined by*

$$\mathbf{r}(t) = \langle FPR(t), TPR(t) \rangle,$$

*where $FPR(t)$ and $TPR(t)$ are the false- and true-positive rates as a function of the threshold parameter $t$.*

Typically, the $x$-label of the ROC curve is represented as "1 - specificity," which, of course, is equivalent to the false-positive rate. All ROC curves satisfy the endpoint conditions. For an example ROC curve, see Figure 7.3.

**Proposition 6.1.** *Let $\mathbf{r} : [0, 1] \to [0, 1]^2$ be an ROC curve. Then*

$$\mathbf{r}(0) = \langle 1, 1 \rangle \qquad and \qquad \mathbf{r}(1) = \langle 0, 0 \rangle.$$

*Moreover, the TPR is a nondecreasing function of the FPR.*

*Proof.* We begin by considering the threshold $t = 0$. In this case, $\hat{Y}_0(X) = 1$, for all $X$, since the function $f \in (0, 1)$. Therefore

$$TPR(0) = FPR(0) = 1.$$

Similarly, at the endpoint $t = 1$, we have $\hat{Y}_1(X) = 0$, for all $X$, so that

$$TPR(1) = FPR(1) = 0.$$

This proves the two endpoint conditions; i.e., the ROC curve connects the endpoints $\langle 0, 0 \rangle$ and $\langle 1, 1 \rangle$.

Next, let us examine the monotonicity of the true-positive rate, when cast as a function of the false-positive rate. Consider any two $t_1, t_2$, with $0 \le t_1 < t_2 \le 1$. By increasing the threshold from $t_1$ to $t_2$, any instance $X$ with $f(X; \mathcal{D}) \in (t_1, t_2)$ will no longer be classified as a positive example. The total of each column of the confusion matrix must remain constant; however, counts will drip from row $i = 1$ into row $i = 0$. Thus, we have

$$TPR(t_2; x) \le TPR(t_1; x) \qquad and \qquad FPR(t_2; x) \le FPR(t_1; x).$$

The result follows.                                                                  $\square$

There are two interesting points to note. The first is that the line $TPR = FPR$ represents a model that classifies at random. That is, if the ranking of the model outputs $f(X_1) < f(X_2) < \cdots < f(X_n)$ is random, then the probability of classifying an instance as a positive is independent of whether or not the instance actually is positive. Thus, we should expect that the actual positive instances are classified as positive (TPR) at the same rate as are the actual negative instances (FPR). This gives some guidance on interpreting the ROC curve. Curves that fall below the diagonal "$y = x$" line are fairing worse than had you just assigned probabilities at random. Curves at are entirely above the diagonal are performing better than average.

The second point is that a perfect model would have $TPR = 1$ and $FPR = 0$. Thus, the closer the curve gets to the point $\langle 1, 0 \rangle$, typically the better the model. Stated differently: the area under the curve (which is bound between 0 and 1) is a measure of predictive performance for the model. We therefore defining the following.

**Definition 6.17.** *The* area under the curve (AUC) *of a classifier is defined as the area bounded by the ROC curve and the lines $TPR = 0$ and $FPR = 1$.*

Clearly, it follows that $0 < AUC < 1$, for any model. Moreover, an $AUC = 1/2$ is a model that performs as good as random. And a model with an AUC close to 1 is a superior model.

### 6.1.4 Predicting Class Probabilites

As mentioned in Note 6.4, the function $f(X; \mathcal{D})$, produced by a binary classifier, is not a true probability, though it possesses the basic properties of probabilities. We now turn to the case where the goal of our classifier is actually to determine an accurate *probability*, as opposed to an actual label.

### Calibration

A classification model is said to be *well calibrated* if its output values $f(X)$ represent the probability that $Y = 1$; i.e., a value $f(X) = 0.3$ actually means that the given instance has a 30% probability of being positive. In order to visualize how well calibrated a model is, we rely on the following.

**Definition 6.18.** *Given a set of data $\{X_i, Y_i\}_{i=1}^{n}$ and prediction values $f(X_i)$ of a classification algorithm. The instances are ordered relative to their prediction values, so that $f(X_1) < f(X_2) < \cdots < f(X_n)$. Then they are divided into $k$ buckets or bins, of approximately equal size $n/k$. Next, let $f_i$ represent the average prediction value of the instances in the ith bucket, and let $p_i$ represent the fraction of positive instances within the ith bucket. Then the curve connecting the points $\langle f_i, p_i \rangle$ constitutes a* calibration curve *for the model.*

*Note 6.6.* The diagonal "$y = x$" line represents a perfectly calibrated model, as the prediction values are equivalent to the true probabilities.                    ▷

*Note 6.7.* An alternative formulation of the calibration curve plots the bin number on the $x$-axis and concurrently plots two separate curves on the $y$-axis: one for the predicted values and one for the true values of each bin. This format is suitable for viewing the "calibration" of a regression model as well, and can serve as a visual diagnostic as to whether a model is well calibrated. By comparing the true and predicted curves concurrently, it is also more interpretable to business stakeholders.                    ▷

Typically, a calibration curve will have a sigmoidal shape. We can therefore adjust the prediction values by fitting the calibration curve to the sigmoid function

$$p_i = \frac{1}{1 + \exp(-\beta_0 - \beta_1 f_i)}.$$

Applying the same curve to each instances prediction value $f(X)$ therefore returns a prediction value that is better calibrated as a probability. This method was introduced by Platt [2000]. This approach and an alternative approach, isotonic regression, are discussed in Niculescu-Mizil and Caruana [2005]. Isotonic regression is especially suitable when the calibration curve does not have a sigmoidal shape.

## Brier Score and Cohort Probabilities

We next discuss a metric used to assess probability models, due to Brier [1950].

**Definition 6.19.** *The* Brier score *for a binary classifier is defined by*

$$\mathrm{BS} = \frac{1}{n} \sum_{i=1}^{n} (y_i - p(X_i; \mathcal{D}))^2, \tag{6.15}$$

*where $y_i$ is the actual value and $p(X_i; \mathcal{D})$ is the predicted probability for of the $i$th instance.*

The Brier score differs from the residual sum of squares, as the residual sum of squares would be defined based on the final classification—and not the predicted probability—of the each instance. The prediction probabilities are typically generated from the prediction values $f(X_i; \mathcal{D})$ of a classifier by implementing either Platt scaling or isotonic regression, as discussed in the preceding paragraph.

The Brier score has an interesting decomposition when the prediction probabilities are made for fixed *cohorts*, or groups with similar characteristics, as opposed to at the individual level. This is common when there is a finite and manageable number of permutations of the feature set.

**Proposition 6.2.** *Suppose a unique probability is provided for each of k cohorts. Then the Brier score is equivalent to the following two-component decomposition*

$$\text{BS} = \frac{1}{n} \sum_{i=1}^{k} n_i \left(p_i - \overline{y}_i\right)^2 + \frac{1}{n} \sum_{i=1}^{k} n_i \overline{y}_i \left(1 - \overline{y}_i\right), \qquad (6.16)$$

*where $n_i$ is the count of instances in the ith cohort, and $\overline{y}_i$ is the observed event probability in the ith cohort.*

The first term in Equation (6.16) is related to the calibration: how well the cohort predicted probabilities align with the observed true values. The second term is an expression of the average inherent uncertainty within each cohort. This second term represents an irreducible error, as it cannot be changed with model improvements, for a fixed set of cohorts. We leave the proof to the reader. (See Exercise 6.3.)

### 6.1.5 Validation Methodology

Over the preceding pages, we were a bit nonchalant regarding to the data to which each of the formulas should be applied. We now seek to remedy that coolness by discussing this issue in greater depth. For additional references, see Hastie, *et al.* [2009] and Kuhn and Johnson [2013].

### Loss Functions

We begin with a discussion on *loss functions*. This will allow us to speak generally in our conversation on error, without regard to the particular type of predictive model we are addressing.

**Definition 6.20.** *In a predictive learning model, any function of the form $L : \mathbb{R}^2 \rightarrow [0, \infty)$ is called a* loss function *if it has the properties that*

$$L(y, y) = 0$$

*and $L(Y, \hat{Y}(X)) \rightarrow 0$ as $\hat{Y}(X) \rightarrow Y$, for any instance $(X, Y) \sim f_{X,Y}$.*

*For regression problems, common loss functions include squared error and absolute error,*

$$L(y, \hat{y}) = (y - \hat{y})^2 \qquad (6.17)$$
$$L(y, \hat{y}) = |y - \hat{y}|, \qquad (6.18)$$

*respectively.*

*For a c-class classification problems, where our random target variable $Y \in \mathcal{C}$, common loss functions include binary loss and log-loss,*

$$L(y, \hat{y}) = \mathbb{I}[y \neq \hat{y}] \qquad (6.19)$$

$$L(y, p) = -\sum_{k=1}^{c} \mathbb{I}[y = k] \log p_k = -\log p_y, \qquad (6.20)$$

*respectively.*

Unlike Equations (6.17)–(6.19), Equation (6.20) is a function of the predicted probability of a classification model, not the class predictions itself. Here, $p_k$ is the probability that a given instance belongs to class $k$, for $k = 1, \ldots, c$. Suppose that the instance does indeed belong to class $k$, for some particular $k \in \{1, \ldots, m\}$. Then the larger the value of $p_k$, the better the performance of the model. Recall from basic logarithm properties, that $-\log p_k > 0$ and $-\log p_k \to 0$ as $p_k \to 1$. The smaller the predicted value $p_k$, the worse the model did (since $y = k$ is correct), and the larger the value of $-\log p_k$.

The loss function Equation (6.17) corresponds to the residual sum of squares from Equation (6.2), whereas the loss function Equation (6.18) corresponds to the MAE from Equation (6.7). Similarly, the loss function Equation (6.19) is related to accuracy; in fact, it is 1 minus the accuracy.

Finally, we note that, for binary classification problems, if we let $p$ represent the predicted probability of the positive label (typically, the minority label), then the log loss of Equation (6.20) simplifies as

$$L(y, p) = -y \log p - (1 - y) \log(1 - p). \tag{6.21}$$

## Training, Test, and Prediction Errors

When applying any of the above metrics (e.g., Equations (6.2) and (6.15)) to the training set $\mathcal{D}$ itself, the result is referred to as the *training error*

$$\mathrm{ERR}_\mathcal{D} = \sum_{i=1}^{n} L(y_i, f(x_i, \mathcal{D})). \tag{6.22}$$

This, however, is not a good metric with which to assess a model, because the model was able to train on the results that we are testing its performance on. This can lead to *overfitting* of a model, which occurs when a model has a low training error but generalizes poorly to new data. To remedy this, we next define several types of errors, that are determined based on how they are applied, but not the individual context to which they are applied. Thus, we will develop the following for a generic loss function, which will depend on the particular context.

**Definition 6.21.** *In a predictive learning model $f(X; \mathcal{D})$, learned from a data set $\mathcal{D}$, with loss function $L$, the* test error *or* generalization error *is the expected error over an independent sample*

$$\mathrm{ERR}_\mathcal{D} = \mathbb{E}\left[L(Y, f(X; \mathcal{D}))\right]. \tag{6.23}$$

*Here, the expectation is with respect to the random variable $(X, Y) \sim f_{X,Y}$.*

The test error is thus the expected loss on an *independent sample*. Note that the test error is for a particular trained model $f(\cdot; \mathcal{D})$. The predictive model, and thus the test error, might have come out differently had the algorithm learned from an alternate training set $\mathcal{D}$. Thus, measuring test error doesn't go far enough to properly assess our model: what if our model had been learned from a different training set?

**Definition 6.22.** *In a predictive learning model $f(X; \mathcal{D})$, which can be learned from any data set $\mathcal{D}$, with loss function $L$, the* prediction error *is the expected test error*

$$\text{ERR} = \mathbb{E}\left[\text{ERR}_{\mathcal{D}}\right], \tag{6.24}$$

*where, the expectation is with respect to the training set $\mathcal{D}$.*

Thus, the test error is the expected loss given our particular training set, and the prediction error is the expected test error of a model, without regard to the particular training set deployed. Clear as pudding[2].

### Model Selection and Assessment

A predictive model does not occur in a vacuum. Commonly, we are never interested in a single model, but a family of related models that follow a similar methodology. That is to say, predictive models are typically dependent on a number of *tuning parameters* or *hyperparameters* that specify the model or how it operates. The tuning parameters may vary the complexity of the model, or otherwise determine which features should be used to train the model. Once we have determined (what we believe to be) the optimal model, we then want to determine an accurate measure of its expected performance.

We therefore find ourselves faced with two distinct tasks:

- *Model selection*: estimate the performance of the various models under consideration in order to select the best one;
- *Model assessment*: evaluate the performance of the selected model.

At first pass, the task of model assessment might seem redundant; after all, have we not already estimated each model's performance before making our selection? Such an approach, however, often leads to folly, as it fails to account for any *selection bias* incurred during the model selection phase.

*Example 6.4.* A set of ten models are trained on a training set $\mathcal{D}$, yielding $f_i(X)$ for $i = 1, \ldots, 10$. The models are then applied to a separate test set $\mathcal{T}$, for which the test error is estimated using the mean-squared error ($\text{MSE} = \text{SSE}/df$).

Now, suppose that each model is exactly equivalent, except that they may perform differently on different data sets. That is, suppose that the

---

[2] Perhaps I'll start a new trend of ending mathematical statements with CAP, like proofs are ended with QED.

prediction error for each of our ten models is exactly $\text{ERR} = 10$, and the model variances are all $\mathbb{V}(f_i(x)) = 9$, for each $i = 1, \ldots, 10$. We can simulate such a scenario by drawing a random sample of ten numbers from $N(10, 9)$, obtaining the values

$$8.72, \ 9.44, \ 10.13, \ 6.75, \ 10.59, \ 11.67, \ 8.61, \ 6.35, \ 11.81, \ 2.76.$$

Upon seeing these results, we have a clear winner: model 10. (Even though, in reality, the models are exactly identical.)

Now, if we stop there, and report that our model has an expected generalization error of 2.76, we have a problem. This is called selection bias, and it illustrates the importance of the model assessment task as a separate task from the model selection process.                                        ▷

To address the issue of selection bias in model selection and assessment, we commonly divide our data set into three subsets:

1. *A training set* $\mathcal{D}$: the data set used to train the models;
2. *A validation set* $\mathcal{V}$: the data set used to estimate the test error of the various models, used for model selection;
3. *A test set* $\mathcal{T}$: the data set used to estimate the generalization error for the final model.

A typical split may be 50–25–25, but this in reality depends on the application and the size of data. Several tools that can be used for data sets that are two small to be amenable to such a split are discussed in Hastie, *et al.* [2009].

## Cross Validation

Another approach often used to estimate the prediction error of a model is $k$-fold cross validation. The idea is that we divide our data set randomly into $k$ partitions (or folds). We then proceed to train our model $k$ times, each time we leave out one of the folds as a test set. This method therefore yields $k$ separate estimates of the test error, which can be averaged together to estimate the expected test error.

Let $\mathcal{D} = \{(X_i, Y_i)\}_{i=1}^{n}$ be our data set, and let $\kappa : \{1, \ldots, n\} \to \{1, \ldots, k\}$ be a function that randomly assigns each data point into one of $k$ bins (or folds), and let $\kappa^{-1}(j) = \{i : \kappa(i) = j\}$. Now define

$$\mathcal{D}_j = \{(X_i, Y_i)\}_{\kappa(i) \neq j} \quad \text{and} \quad \mathcal{T}_j = \{(X_i, Y_i)\}_{\kappa(i) = j}$$

be the set with the $j$th fold removed. We want to train our model on $\mathcal{D}_j$ and test our model on $\mathcal{T}_j$, for $j = 1, \ldots, k$. We may estimate the test error for the model trained on the set $\mathcal{D}_j$ as

$$\hat{\text{ERR}}_{\mathcal{D}_j} = \frac{1}{|\kappa^{-1}(j)|} \sum_{i \in \kappa^{-1}(j)} L(Y_i, f(X_i; \mathcal{D}_j)), \tag{6.25}$$

for $j = 1, \ldots, k$. By averaging these results, we thus arrive at an estimate for the prediction error

$$\hat{\text{ERR}} = \frac{1}{k} \sum_{j=1}^{k} \hat{\text{ERR}}_{\mathcal{D}_j}. \tag{6.26}$$

Similarly, we can compute the sample variance of our $k$ estimated test errors to estimate the model's variance. An illustration of five-fold cross validation is shown in Figure 6.1.

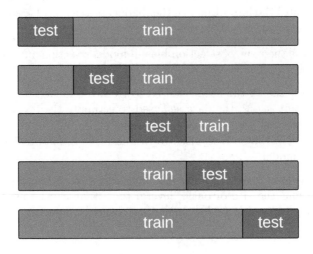

Fig. 6.1: An illustration of five-fold cross validation.

A typical choice of $k$ may be $k = 5$ or $k = 10$. The case $k = n$ is referred to as *leave-one-out cross validation* or the *jackknife*.

### 6.1.6 Validation on Temporal Datasets

In Section 5.4, we discussed various kinds of stochastic processes that generate data over time. In practice, we must take extra care to perform validation on data that has an explicit time component. There are two potential problems that can arise. First, even when traditional machine-learning validation methodologies are available (e.g., cross validation), which requires that we have a well defined target variable in our training data, user behavior can still vary over time, so that a model that is valid today might not have been valid yesterday. Second, we might encounter data that is not *fully baked*, so that there is no well defined target variable. This occurs with survival processes (Section 5.3), as well as in customer lifetime value models (Section 10.3). In these models, we are learning from behavioral data we have *to date*, in order to predict future behavior. In both cases, it is therefore important to perform a historical validation over a stretch of time.

Before jumping into how we are going to validate such a model, let us first take a moment to examine how our model will be deployed in production. In order to run our model in production, we will require a well defined *training window* and a *prediction range*. In addition, we often desire to buffer the immediately trailing data from our training window. Formally, we define these as follows.

**Definition 6.23.** *When deploying a model over a* temporal dataset*, or a dataset with an explicit time component, the* training window *is the date range for the training set, and the* prediction range *is the date range for the prediction set.*

*These are commonly specified relative to a* point in time*, which defaults to the present date, and three configuration parameters*

1. delay*: the number of days between the end of the training set and the point in time;*
2. window*: the length (in days) of the training set;*
3. range*: the length (in days) of the prediction set.*

Given these definitions, our training window and prediction range can be expressed as

$$\text{training window} = [\texttt{point\_in\_time} - \texttt{delay} - \texttt{window}, \texttt{point\_in\_time} - \texttt{delay})$$
$$\text{prediction range} = [\texttt{point\_in\_time} - \texttt{range}, \texttt{point\_in\_time}).$$

This is shown schematically in Figure 6.2. Here, the training set consists of the green data, whereas the prediction set is the blue data. In the figure, the delay and range are shown as equal, though this is not a requirement. In production, the "point in time" is always taken to be the present date.

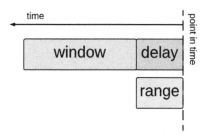

Fig. 6.2: Training (green) and prediction (blue) data for temporal datasets. In production, *point in time* is the current date.

So why don't we just say "now" instead of "point in time"? The beauty of this design, is that we are treating "now" as a parameter. This means that we can pass a "different now" into our runtime job in order to see what our model results would have been had we run our model, with the same configuration, at some point in time in the past.

We can take advantage of this configuration to perform a *backtesting validation* (or, simply, *backtest*), which consists of running our model using historical data at various points in time in the past, in order to write predictions over a period of time. Not only can we run our standard validation metrics to determine how well our model would have performed, but we can also see how our performance varied over our stretch of time. For example, we can determine how much our model predictions might fluctuate in time. A backtesting validation with four jobs is shown in Figure 6.3.

Note that each job is staggered by the range, so that we have full coverage of our predictions over a substantial period of time. In other words,

$$\texttt{point\_in\_time(job\_i)} = \texttt{point\_in\_time} - \texttt{range} \times i,$$

where $i \in 0, \ldots, n - 1$, where $n$ is the number of backtest jobs deployed, and `point_in_time` is the ending point in time.

Backtesting is especially critical for online process data, which is commonly not fully baked at the time we write our predictions (e.g., survival processes). For such a scenario, we can set the point in time for `job-0` as the latest point in time for which fully baked data is available. The model still trains using only the data it would have had available at that point in time, but we can use the fully baked data from those cohorts in order to validate our predictions. In such a scenario, cross validation is not necessary. We are training our model over a historical backtesting period in the exact same way that we would train the model during production. We can therefore see exactly how well our model would have fared over our historical backtesting period, had it been live and in production at the time.

## 6.2 Preprocessing

It is seldom the case that one can use the features (independent variables) to train a model in their raw form. Rather, transformations are commonly applied prior to feeding the features into the model. Now, what is good for the goose, is good for the gander: transformations applied to the training set should equally be applied to the validation and test sets prior to computing the model's predictions.

Most of this section will rely heavily on the scikit-learn package. For an introduction to scikit-learn in machine learning, see Pedregosa, *et al.* [2011]. For a discussion of the design principles of its API, see Buitinck, *et al.* [2013]. And, of course, for the latest and up-to-date changes on functionality, visit the scikit-learn user guide at

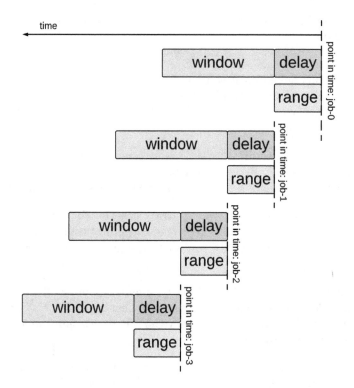

Fig. 6.3: Backtesting validation for temporal datasets.

https://scikit-learn.org/stable/user_guide.html.

### 6.2.1 Categorical Features

A *categorical variable* is a variable that can take on any value from a finite set $\mathcal{C}$. We will consider two types of categorical features: ordinal and nominal. To differentiate between these two, we introduce the following.

**Definition 6.24.** *Let $\mathcal{S}$ be a set and "$\leq$" a binary relation over $\mathcal{S}$[3]. Then the binary relation is a* total order *if it satisfies*

1. Connexity: $x \leq y$ or $y \leq x$, for all $x, y \in \mathcal{S}$;
2. Reflexivity: $x \leq x$, for all $x \in \mathcal{S}$;
3. Transitivity: $x \leq y$ and $y \leq z$ implies $x \leq z$, for all $x, y, z \in \mathcal{S}$;
4. Antisymmetry: $x \leq y$ and $y \leq x$ implies $x = y$, for all $x, y \in \mathcal{S}$.

---

[3] A binary relation of a set $\mathcal{S}$ is a subset of the Cartesian product $\mathcal{S} \times \mathcal{S}$. Connexity simply means that all pairs are in the binary relation, one way or the other.

*Given a total order, a* strict total order[4] *is the binary relation "<" defined by the condition*

$$x < y \text{ if } x \leq y \text{ and } x \neq y.$$

Gloss over this definition if you will, a total order and strict total order are simply a generalization of the *less than or equals to* and *less than* operators that act over real numbers ($\pi^2 < 10$), except now they act over elements of a set (dog < cat).

**Definition 6.25.** *A* categorical variable *is a variable that can take on any value from a finite set $\mathcal{C}$. A categorical variable is said to be* ordinal *if its underlying set has a natural total order, otherwise it is said to be* nominal.

*Note 6.8.* We can always contrive an arbitrary ordering for any set, which is why we require ordinal variables to possess a *natural* total order; i.e., a total order that has meaning with respect to the meaning of the variable.
▷

*Note 6.9.* An ordinal feature differs from a numeric feature as it possesses no sense of scale. For example, we might know that $A < B < C$, but the nature of an ordinal variable is that we have no way to determine *how much greater* one variable is to another.                                        ▷

Thus, an ordinal feature is simply a discrete feature that has a natural ordering. Examples include T-shirt size ($XS < S < M < L < XL$) and responses on a survey ("very dissatisfied" < "dissatisfied" < "neutral" < "satisfied" < "very satisfied"). Similarly, a nominal feature does not have such an intrinsic ordering. Examples of nominal features include color (red, blue, yellow) and country (UK, CA, JP, AU).

The presence or lack of ordering affects how we preprocess a discrete feature.

**Ordinal Features**

In order to make ordinal features suitable for learning algorithms, we use a technique referred to as *label encoding*.

**Definition 6.26.** *Let $X \in \mathcal{C}$ be an ordinal variable. Then the mapping $\iota : \mathcal{C} \to \{1, \ldots, |\mathcal{C}|\}$ is an* ordinal encoding *if it is invariant with respect to the strict total ordering on $\mathcal{C}$; i.e., if*

$$x < y \text{ implies } \iota(x) < \iota(y), \text{ for all } x, y \in \mathcal{C}.$$

---

[4] A strict total order satisfies the same set of axioms as a total order, except instead of reflexivity it is irreflexive, $x \not< x$ for all $x \in \mathcal{S}$, and instead of antisymmetry it is asymmetric, $x < y$ implies $y \not< x$ for all $x, y \in \mathcal{S}$.

*Note 6.10.* The comparison $x < y$ in Definition 6.26 is with respect to the ordering of the set $\mathcal{C}$, whereas the comparison $\iota(x) < \iota(y)$ is with respect to the standard Cartesian ordering of the integers.    ▷

*Example 6.5.* Fortunately, the scikit-learn (sklearn) package[5] has a built-in ordinal encoder, ready for use. In Code Block 6.1, we define three ordinal variables: survey, t-shirt, and color. We then instantiate an instance of the OrdinalEncoder class on line 7, declaring the required ordering of each feature as well as how to handle unknown values. The encoder fits the data frame (which is required, despite being, in this case, redundant from the object initialization) on line 17 and transforms it into an encoded matrix on line 18. Note that since we are using an ordinal encoding, we are actually saying that red < yellow < blue, at least for the purpose of the current illustration.

```
1   from sklearn.preprocessing import OrdinalEncoder
2
3   survey = ['very dissatisfied', 'dissatisfied', 'neutral',
          'satisfied', 'very satisfied']
4   t_shirt = ['XS', 'S', 'M', 'L', 'XL']
5   color = ['red', 'yellow', 'blue']
6
7   enc = OrdinalEncoder(
8          categories = [survey, t_shirt, color],
9          handle_unknown='use_encoded_value',
10         unknown_value=-1)
11
12  df = DataFrame({
13      'survey': np.random.choice(survey, size=10),
14      't_shirt': np.random.choice(t_shirt, size=10),
15      'color': np.random.choice(color, size=10)})
16
17  enc.fit(df)
18  X = enc.transform(df)
19  enc.categories_
```

Code Block 6.1: Syntax for sklearn's ordinal encoder

For this example, we generated our data frame by randomly choosing (with replacement) from our possible classes for each feature. A randomized data frame and its encoding is given in Table 6.4.    ▷

---

[5] For an up-to-date guide on using the scikit-learn package, see https://scikit-learn.org/stable/user_guide.html.

| row | survey | t-shirt | color | X[:, 0] | X[:, 1] | X[:, 2] |
|-----|--------|---------|-------|---------|---------|---------|
| 0 | satisfied | L | red | 3 | 3 | 0 |
| 1 | very dissatisfied | XS | red | 0 | 0 | 0 |
| 2 | very dissatisfied | M | red | 0 | 2 | 0 |
| 3 | satisfied | L | yellow | 3 | 3 | 1 |
| 4 | satisfied | S | blue | 3 | 1 | 2 |
| 5 | satisfied | L | blue | 3 | 3 | 2 |
| 6 | neutral | L | red | 2 | 3 | 0 |
| 7 | very satisfied | S | yellow | 4 | 1 | 1 |
| 8 | very dissatisfied | XL | yellow | 0 | 4 | 1 |
| 9 | very satisfied | M | yellow | 4 | 2 | 1 |

Table 6.4: Example of an ordinal encoding.

*Note 6.11.* In a multi-class classification problem, one must further encode the target variable. This is achieved with a *label encoder*, available
in sklearn's `LabelEncoder` class. The label encoder functions almost like
the ordinal encoder, with a few exceptions:

- The label encoder does not accept any arguments into its constructor
  (except for, optionally, specifying the number of classes),
- The fit method may only be applied to a one-dimensional array or a
  pandas Series (equivalently, a single column of a data frame).

The label encoder is specifically used for encoding a *dependent* (i.e., target)
variable.                                                                      ▷

## Nominal Features

Similar to how we encoded ordinal features into integers, we encode nominal
categorical features into binary vectors.

**Definition 6.27.** *Let $X \in C$ be a nominal categorical variable. Then the
mapping $\iota : C \to \mathbb{B}^{|C|}$ is a one-hot encoding if it is an bijection (i.e., if it is
one-to-one and onto) and if it satisfies the normalization condition*

$$||\iota(x)||_1 = 1, \text{ for all } x \in C,$$

*where $||\cdot||_1$ is the $\ell_1$ norm[6] and the set $\mathbb{B} = \{0,1\}$ is the set of binary digits
(bits).*

*Note 6.12.* The normalization condition in Definition 6.27 implies that the
encoding $\iota(x)$ is a binary vector consisting of all zeros except for a solitary
one.                                                                           ▷

---

[6] Here, the quantity $||x||_1 = \sum_{i=1}^{n} |x_i|$ represents the $\ell_1$ *norm*, or *Manhattan
norm*, of the vector $x \in \mathbb{R}^n$.

*Example 6.6.* Consider the set of colors $\mathcal{C} = \{R, B, Y\}$. The mapping defined by

$$\iota(R) = \begin{bmatrix} 1 \\ 0 \\ 0 \end{bmatrix}, \qquad \iota(B) = \begin{bmatrix} 0 \\ 1 \\ 0 \end{bmatrix}, \qquad \iota(Y) = \begin{bmatrix} 0 \\ 0 \\ 1 \end{bmatrix},$$

constitutes a one-hot encoding.                                                   ▷

*Example 6.7.* Let's return to Example 6.5, but treat the variables as nominal, and implement a one-hot encoding. This is achieved in Code Block 6.2 (using the dataframe definition from Code Block 6.1). Here, we squash the fit and transform methods into a single command `fit_transform`, which is available to both encoders. Note, we pass in `sparse=False` if we actually want to visually see or print the output array X. Otherwise, the transform method will return a sparse array object, which is computationally more efficient for large data sets.

```
from sklearn.preprocessing import OneHotEncoder

enc = OneHotEncoder(sparse=False)
X = enc.fit_transform(df)
enc.categories_
```

Code Block 6.2: Syntax for sklearn's one-hot encoder

| neut. | sat. | vy. dissat. | vy. sat. | L | M | S | XL | XS | blue | red | yellow |
|-------|------|-------------|----------|---|---|---|----|----|------|-----|--------|
| 0 | 1 | 0 | 0 | 1 | 0 | 0 | 0 | 0 | 0 | 1 | 0 |
| 0 | 0 | 1 | 0 | 0 | 0 | 0 | 0 | 1 | 0 | 1 | 0 |
| 0 | 0 | 1 | 0 | 0 | 1 | 0 | 0 | 0 | 0 | 1 | 0 |
| 0 | 1 | 0 | 0 | 1 | 0 | 0 | 0 | 0 | 0 | 0 | 1 |
| 0 | 1 | 0 | 0 | 0 | 0 | 1 | 0 | 0 | 1 | 0 | 0 |
| 0 | 1 | 0 | 0 | 1 | 0 | 0 | 0 | 0 | 1 | 0 | 0 |
| 1 | 0 | 0 | 0 | 1 | 0 | 0 | 0 | 0 | 0 | 1 | 0 |
| 0 | 0 | 0 | 1 | 0 | 0 | 1 | 0 | 0 | 0 | 0 | 1 |
| 0 | 0 | 1 | 0 | 0 | 0 | 0 | 1 | 0 | 0 | 0 | 1 |
| 0 | 0 | 0 | 1 | 0 | 1 | 0 | 0 | 0 | 0 | 0 | 1 |

Table 6.5: Example of a one-hot encoding.

The resultant matrix is shown in Table 6.5. Now, the actual output matrix X is just a two-dimensional numpy array. In order to determine the column headings, we have to look at the attribute `enc.categories_`.

Notice that the columns do not follow the same ordering that we specified in Example 6.5. We could have passed in a `categories` argument into the constructor to have specified the ordering of the encoding. However, this is tantamount to shuffling the columns and does not typically make a difference in practice, as it does with ordinal encoding, where the relative size matters.

Notice, also, that there is not a column for *dissatisfied*, as that category does not occur in our example data frame, from Table 6.4. The `fit` method, therefore, does not create space for this feature. To include *dissatisfied*, the categories must be passed into the constructor, as in Code Block 6.1.    ▷

Even though a one-hot encoding returns a much larger matrix, which requires a separate column for each category within each nominal feature, it is necessary when encoding nominal data, i.e., data without any intrinsic ordering. Use of an ordinal encoder on nominal data, as so often with youth, leads to folly, as a learning algorithm will treat the encoded data as ordered.

### 6.2.2 Continuous Features

Continuous features are features which can take any real value. Unlike categorical features, they are already numeric. However, they often require their own form of manipulation prior to feeding into a learning algorithm. As was the case with categorical features, many of those transformations are built in to scikit-learn.

### Discretization

Our first method for continuous features is a method that can be used to convert a continuous feature into a categorical feature. The quintessential example is age range: the raw age of a customer is converted into an ordinal age-range feature: 18–14, 25–34, 35–44, 45–54, 55–64, 65+. Instead of working with a customer's raw age, which, as an integer, is equivalent to dealing with an ordinal feature with $O(100)$ values, we work with the age-range, which is a more manageable set of seven buckets. In addition, there is usually an eighth bucket for *unknown* values.

**Definition 6.28.** *A mapping $\iota : \mathbb{R} \to \mathcal{C}$ shall constitute a* discretization *or* binning *of the real numbers if $\mathcal{C}$ is a finite, totally ordered set and if the mapping $\iota$ is invariant with respect to the total ordering, i.e., if*

$$x \leq y \text{ if and only if } \iota(x) \leq \iota(y).$$

*A* discretization *is an* ordinal discretization *if the set $\mathcal{C}$ is ordinally encoded, and it is a* one-hot discretization *if the set $\mathcal{C}$ is one-hot encoded.*

Typically, when we speak about discretization, we will implicitly refer to ordinal discretization, as the real numbers have a natural ordering.

The `sklearn.preprocessing` package has a built-in class, `KBinsDiscretizer`, that can handle most of our discretization needs for us. The constructor of this class has the following keyword arguments:

1. *n_bins* (int, default=5): The number of bins to use, must be an integer greater than 1.
2. *encode* ('onehot' (default), 'onehot-dense', 'ordinal'): Determines whether to return an ordinal encoding, or a one-hot encoding, which can be either dense or sparse.
3. *strategy* ('uniform', 'quantile' (default), 'kmeans'): Determines how the widths of the bins are calculated: uniform strategy will produce bins with equal widths; a quantile strategy will bin the data points (approximately) equally into bins; and a kmeans strategy will implement a clustering technique known as *k*-means, which we will not discuss at the present.

As with the OrdinalEncoder, LabelEncoder, and OneHotEncoder classes, the KBinsDiscretizer has a `fit`, `transform`, and a `fit_transform` method. The KBinsDiscretizer can discretize multiple numeric features at once; in this case, an object from this class should be instantiated with an array of integers passed in for the `n_bins` argument. Attributes of this class include `n_bins_`, an array with the number of bins for each feature, and `bin_edges_`, an array of arrays that determines the edges of the bins for each feature.

*Example 6.8.* Consider student data generated from a large university course: the student's age, exam score, and attendance percentage. The data are generated in Code Block 6.3. Age is then encoded into seven bins, exam score into twelve bins, and attendance into five bins. A quantile strategy is used, so that approximately an equivalent number of data points should appear in each bucket.

We can view the bin edges and the first few rows of our data frame and output matrix, as shown in lines 14–15.                                    ▷

## Scaling Numeric Features

Many machine learning algorithms are highly sensitive to the scaling of numeric features and perform much better when the feature set is appropriately scaled. For example, if an algorithm were to use the Euclidean distance between two numeric features for the purpose of predicting whether or not a home was in a good school district, we should expect things to break down if one feature is the number of bedrooms and another is the home price: these two features live on radically different scales.

The `sklearn.preprocessing` package has three main types of built-in scalers that represent the three most commonly used forms of scaling.

```
1   from sklearn.preprocessing import KBinsDiscretizer
2
3   n_rows = 100
4   df = DataFrame({
5           'age': 18 + np.random.exponential(10,
                  size=n_rows).astype(int),
6           'score': np.fmin(100, np.random.normal(loc=70, scale=20,
                  size=n_rows).astype(int)),
7           'attendance': np.random.beta(5, 1, size=n_rows)
8           })
9
10  K = KBinsDiscretizer(n_bins=[7, 12, 5], encode='ordinal',
        strategy='quantile')
11  X = K.fit_transform(df)
12  # K.fit(df)
13  # X = K.transform(df)
14  print(K.bin_edges_)
15  print(df.head(10), X[:10, :])
```

Code Block 6.3: Syntax for sklearn's discretizer

**Definition 6.29.** *Let* $X_1, \ldots, X_n \sim f_X$ *be a sample of* IID *data. Then the transformed data set*

$$Z_i = \frac{X_i - \overline{X}_n}{S_n},$$

*where* $\overline{X}_n$ *is the sample mean and* $S_n^2$ *is the sample variance, is said to be a z-score normalized or* standardized *data set. Similarly, the transformed data set*

$$Y_i = \frac{X_i - \min(X_i)}{\max(X_i) - \min(X_i)}$$

*is said to be a* min-max normalized *data set. Finally, the transformed data set*

$$W_i = \frac{X_i - Q_{0.5}}{Q_{1-\alpha} - Q_\alpha},$$

*where* $\alpha \in (0, 1/2)$*, typically* $\alpha = 0.2$ *or* $\alpha = 0.25$*, and* $Q_\alpha$ *is the* $\alpha$*th quantile of the sample, is said to be a* quantile normalized *data set. The value* $Q_{0.5}$ *is the median of the original data set. For the case* $\alpha = 0.25$*, the denominator* $Q_{0.75} - Q_{0.25}$ *is referred to as the* interquartile range.

Common methods available in the **sklearn.preprocessing** module are

- **StandardScaler**: standardization produces scaled data with zero mean and unit variance;
- **MinMaxScaler**: min-max normalization produces scaled data on range $[0, 1]$;

- MaxAbsScaler: divides by the largest absolute value of the data, without shifting; scaled data on range $[-1, 1]$. Can be used with sparse data.
- RobustScaler: quantile normalization does not limit range.

The StandardScaler takes arguments with_mean and with_std (default True). If used with sparse data, with_mean should be set to False. The Min-MaxScalar takes arguments feature_range (default $(0,1)$) and clip (default False). If clip is set to True, test data (not used during the original fit) is clipped to the same range $[0,1]$ as the input data. Finally, the RobustScaler has arguments with_centering (default True), with_scaling (default True), and quantile_range (default $(0.25, 0.75)$).

Once an object of any of these classes is constructed, the standard fit, transform, and fit_transform methods are available to transform a data set.

In addition to the aforementioned scalers, scikit-learn has two nonlinear transformation methods, QuantileTransformer and PowerTransformer, which transform the data set into a uniform distribution or a Gaussian distribution, respectively.

### 6.2.3 Data Pipelines

In building a model with a large feature set, one typically encounters multiple transformations prior to training a model. This process will be simplified with a structure known as a pipeline. But first, we will consider two additional preprocessing tasks: handling missing values and randomizing the data into training and test sets.

### Imputing missing values

Data sets often contain missing values. Instead of deleting an entire datum, we can define a strategy for handling missing data. To achieve this, we will use the SimpleImputer from the sklearn.impute package. The constructor has an argument strategy, which can take a value from 'mean,' 'median,' 'most_frequent,' and 'constant.' If 'constant' is selected, the argument fill_value should also be specified.

In Code Block 6.4, we define an outer join of two data frames, which results in two columns (b and c) with missing values. We then demonstrate imputing with the median and most frequent strategies. Note that the *mean* and *median* strategy are not available for categorical features.

Often, in practice, it is useful to use the constant strategy, in order to give a unique and separate value to the missing data.

### Building a Pipeline

So far, we have reviewed many types of transformers: encoders, discretizers, scalers, and imputers. Each of these classes has a fit and a transform

```
1  from sklearn.impute import SimpleImputer
2
3  df_1 = DataFrame({'a':[1,2,5,6], 'b':['x','y','z','z']})
4  df_2 = DataFrame({'a':[0, 2, 3, 4, 6], 'c':[4, 8, 15, 16, 23]})
5  df = pd.merge(df_1, df_2, how='outer')
6
7  imp = SimpleImputer(strategy='median')
8  X_num = imp.fit_transform(df[['a', 'c']])
9  imp = SimpleImputer(strategy='most_frequent')
10 X_cat = imp.fit_transform(df[['b']])
11 imp = SimpleImputer(strategy='constant', fill_value='MISSING')
12 X_cat = imp.fit_transform(df[['b']])
```

Code Block 6.4: Imputing missing values

method, along with, for convenience a `fit_transform` method, which is equivalent to applying both methods to the same data set in order. The actual machine learning machinery lives in classes as well, a collection of classes known as *estimators*. Estimators typically have both a *fit* and a *predict* method, which operate separately: `fit` is used with the training data, and `predict` is used with the test data.

A *pipeline* is a series of transformers with an optional estimator in its last position. The `fit` method only need be called once on a pipeline object: The dataset is passed through the `fit_transform` methods of each transformer in series. The pipeline object will have all the functionality as its last estimator or transformer.

A pipeline is constructed by passing a list of key-value pairs: the key being an arbitrary (but useful) name for each step of the pipeline, and the value being the transformer or estimator object that is to be called at that step.

A simple pipeline is constructed in Code Block 6.5. The `mask` method on line 6 randomly (with 10% probability) *masks* or removes each value of the data frame. The pipeline consists of two transformers: impute missing values with the median, and then perform standardization.

A similar pipeline for categorical data is constructed in Code Block 6.6. Here, the pipeline consists of three steps: impute missing values, apply an ordinal encoder, apply a one-hot encoding. The application of a one-hot encoding following an ordinal encoding is completely redundant, and could have (and should have) been accomplished with the one-hot encoding alone. The redundancy was only for the purpose of illustration.

In practice, the pipelines in Code Blocks 6.5 and 6.6 would contain an estimator at the end of the pipeline, so that a model could have been built. We will see examples of this soon enough.

```
1  from sklearn.pipeline import Pipeline
2  from sklearn.preprocessing import StandardScaler
3
4  X = np.random.randint(0, 20, size=(10, 4))
5  df = DataFrame(X, columns=['a','b','c','d'])
6  df = df.mask(np.random.random(df.shape) < .1)
7
8  transformers = [
9          ('impute', SimpleImputer(strategy='median')),
10         ('standardize', StandardScaler())]
11
12 pipe = Pipeline(transformers)
13 pipe.fit_transform(df)
```

Code Block 6.5: A simple pipeline for continuous features

```
1  s = """Our revels now are ended. These our actors,
2  As I foretold you, were all spirits and
3  Are melted into air, into thin air"""
4
5  s = s.replace(",", "").replace(".","").replace("\n"," ").split(' ')
6  df = DataFrame(np.random.choice(s, size=(10,4)), columns=['a', 'b',
       'c', 'd'])
7  df = df.mask(np.random.random(df.shape) < .1)
8
9  transformers = [
10         ('impute', SimpleImputer(strategy='constant',
               fill_value='UNKNOWN')),
11         ('ordinal',
               OrdinalEncoder(handle_unknown='use_encoded_value',
               unknown_value=-1)),
12         ('ohe', OneHotEncoder(sparse=False,
               handle_unknown='ignore'))]
13
14 pipe = Pipeline(transformers)
15 pipe.fit_transform(df)
```

Code Block 6.6: A simple pipeline for categorical features

The astute reader may be wondering how to handle a feature set that contains mixed data types. This, too, is easily accomplished with the construction of a `ColumnTransformer` object, which can then be inserted as a pipe in the pipeline. The column transformer simply specifies which transformations to apply to which columns. Finally, a keyword argument `remainder` tells the transformer what to do with columns not specified: 'drop' (default) or 'passthrough'. An example pipeline using a column transformer is constructed in Code Block 6.7.

```python
from sklearn.compose import ColumnTransformer

class_levels = ['freshman', 'sophomore', 'junior', 'senior']
majors = ['science', 'engineering', 'math', 'computer science']

n_rows = 100
df = DataFrame({
        'level': np.random.choice(class_levels, size=n_rows),
        'major': np.random.choice(majors, size=n_rows),
        'age': 18 + np.random.exponential(10,
            size=n_rows).astype(int),
        'score': np.fmin(100, np.random.normal(loc=70, scale=20,
            size=n_rows).astype(int)),
        'attendance': np.random.beta(5, 1, size=n_rows)})

column_transformer = ColumnTransformer(
        [('ordinal', OrdinalEncoder(categories=[class_levels],
            handle_unknown='use_encoded_value', unknown_value=-1),
            ['level']),
        ('ohe', OneHotEncoder(sparse=False,
            handle_unknown='ignore'), ['major']),
        ('bins', KBinsDiscretizer(n_bins=8, encode='ordinal',
            strategy='quantile'), ['age', 'score'])],
        remainder='passthrough')

column_transformer.fit(df)
```

Code Block 6.7: A column transformer can handle mixed data types

Now, ideally, you get some ideas. For example, following the illustration in Code Block 6.7, we might think to ourselves that we should construct two column transformers—one to impute missing values and one to apply the various encodings—and connect them together in a pipeline. A problem with this is that the `transform` method outputs an *array*, not a pandas data frame.

To remedy this, I devised a clever workaround. The idea is we can define our own subclass of `ColumnTransformer` that returns a pandas data frame from its `transform` and `fit_transform` methods, instead of an array. The code is given in Code Block 6.8. A final caveat: the `OrdinalEncoder` does not seem to actually handle unknowns as advertised, which requires us to redefine the `class_levels` list to include our unknown value. (It is important to actually redefine line 22 with the updated list of `class_levels`.) Finally, this method only works if *every* column is explicitly handled in the column imputer; hence the use of `remainder='drop'`.

```
1   class ColumnImputer(ColumnTransformer):
2       def transform(self, X):
3           X = ColumnTransformer.transform(self, X)
4           cols = []
5           for t in self.transformers_:
6               cols += t[2]
7           return DataFrame(X, columns=cols)
8
9       def fit_transform(self, X, y=None):
10          X = ColumnTransformer.fit_transform(self, X, y)
11          cols = []
12          for t in self.transformers_:
13              cols += t[2]
14          return DataFrame(X, columns=cols)
15
16  column_imputer = ColumnImputer([
17          ('impute_cat', SimpleImputer(strategy='constant',
                  fill_value='UNKNOWN'), ['level', 'major']),
18          ('impute_num', SimpleImputer(strategy='median'), ['age',
                  'score', 'attendance']),
19          remainder='drop'])
20
21  class_levels = ['UNKNOWN', 'freshman', 'sophomore', 'junior',
            'senior']
22  column_transformer = ColumnTransformer( # ... Same as before.
23
24  pipe = Pipeline([
25          ('impute', column_imputer),
26          ('encode', column_transformer)])
27
28  pipe.fit_transform(df)
```

Code Block 6.8: Multiple column transformers in a pipeline; continuation of Code Block 6.7.

**Training and Test Sets**

Naturally, scikit-learn also has a built-in tool for handling the randomization of a data set (features and labels) into a training and test set. The syntax is shown in Code Block 6.9.

```
from sklearn.model_selection import train_test_split
from sklearn.metrics import r2_score

X = np.random.randint(0, 20, size=(100, 4))
y = np.random.randint(0, 2, size=100)
X_train, X_test, y_train, y_test = train_test_split(X, y,
    test_size=0.20, random_state=42)

pipe = Pipeline() #.....
pipe.fit(X_train, y_train)
y_pred = pipe.predict(X_test)
print(f"R2 score: {r2_score(y_test, y_pred)}")
```

Code Block 6.9: Train-test split

*Note 6.13.* In Code Block 6.9, notice that `train_test_split` is used directly without being instantiated. This is because it is a function, not a class.  ▷

Here, we suppose that we have built a pipeline with the various preprocessing operations, topped with an actual model, or estimator, with a **predict** method. The code shows how we can automatically create a train-test split, train the model on the training data, and use the holdout data for our validation step.

## 6.3 Object-Oriented Data Science

*Object-oriented programming* (OOP) is a software-engineering paradigm centered around the use of modular, reusable code. OOP has allowed engineers to develop sophisticated software architectures and platforms that would otherwise be intractable. Though much of this text is focused around the statistical aspects of data science, we devote this section to some of the engineering aspects. We advocate for object-oriented design in the development of data-science projects in order to both better leverage the reusability aspects and to better manage sophisticated projects that are developed across a team. We conclude this section with a discussion of *agile*, which is a process for managing engineering projects that was developed around a set of principles known as the *agile manifesto*.

For more in depth introduction to data structures in Python, see Lambert [2014] and Lee and Hubbard [2015]. For more on data structures and algorithms, see Lafore [2003] (JAVA) or Cormon *et al.* [2009].

### 6.3.1 Classes and Objects

Object-oriented programming is focused on the use of *classes* and their specific realizations, *objects*. The class encapsulates a set of instructions for creating (or instantiating) individual objects. The class is the cookie cutter and the objects are the cookies.

**Definition 6.30.** *In* object-oriented programming, *a* class *is a code block that consists of a number of functions, called* methods, *and variables, called* attributes, *that provides instructions for how to construct, or* instantiate *any number of* objects. *An* object (or, instance of a class) *is a variable that is created from a class, that stores all of the class's methods and attributes. Two instances (objects) from the same class may have different values for their attributes, but maintain the same structure, as defined by the underlying class.*

In Python, the instructions for instantiating an object from a class are given by the __init__ method, which is reserved for such use. This method is not called directly, rather we use the name of the class like a function that takes in the prescribed input parameters and returns an object of that class. In addition, all methods of a class must take at least one argument, called **self**, which represents the object itself. It is through this argument that any method from a class can access the values stored as attributes and any of the other methods for any particular object. To illustrate, consider the following.

*Example 6.9.* Consider the class defined in Code Block 6.10, which defines a concept of a *car*.

Each "car object" that is constructed from this class has six attributes: name, color, max speed, position, velocity, and time. The values of those attributes will differ from car to car, though the structure itself remains constant. In addition, each object will possess four methods: speed up, time lapse, stop, and repaint.

Individual cars may be instantiated using the __init__ method, though this method is not called directly. Two cars are instantiated in Code Block 6.11. In addition, we provide a method called __str__ which returns a string representation of an individual object[7]. This is used with the **print** command, as shown in Code Block 6.11.

In Code Block 6.11, the objects are represented by the variables car_1 and car_2. In addition to printing the object, as defined by the internal

---

[7] It is not common to define a __str__ method for data-science applications, but it seemed helpful for this example

```
 1  class Car:
 2      def __init__(self, name, color, max_speed=100):
 3          self.name = name
 4          self.color = color
 5          self.max_speed = max_speed
 6
 7          self.position = 0
 8          self.velocity = 0
 9          self.time = 0
10
11      def __str__(self):
12          s  = f"name: {self.name}\n"
13          s += f"color: {self.color}\n"
14          s += f"max speed: {self.max_speed}\n"
15          s += f"position: {self.position}\n"
16          s += f"velocity: {self.velocity}\n"
17          s += f"time: {self.time}\n"
18          return s
19
20      def speedUp(self, delta_v=0):
21          self.velocity = min(self.max_speed, self.velocity + delta_v)
22
23      def timeLapse(self, t):
24          self.time += t
25          self.position += t * self.velocity
26
27      def stop(self):
28          self.velocity = 0
29
30      def repaint(self, new_color):
31          self.color = new_color
```

Code Block 6.10: Car class for Example 6.9

__str__ method defined in the class, we may also print specific attributes using, for example, print(car_1.speed). Finally, we may modify any of the attributes directly, using, for example, car_1.speed += 200.    ▷

### 6.3.2 Principles of OOP

Object-oriented programming is built around the following four principles:

1. *Inheritance*: We may define new classes ("children") from old ("parents") by updating only the new aspects, whereas all functionality not specifically modified is *inherited* from the parent class;
2. *Encapsulation*: An object only exposes certain methods and attributes to the outside world, while keeping unnecessary details hidden;

```
1  # Instantiate two car objects
2  car_1 = Car('1GAT123', 'blue', max_speed=140)
3  car_2 = Car('EIPI+1', 'red', max_speed=160)
4
5  car_1.speedUp(100)
6  car_1.timeLapse(3)
7  car_2.speedUp(80)
8  car_2.timeLapse(3)
9  car_1.stop()
10 car_1.repaint('black')
11 print(car_1) # invokes __string__ method
12 print(car_2) # invokes __string__ method
13 print(car_1.speed) # access particular attribute
14 car_1.speed += 200 # modify attribute directly
```

Code Block 6.11: Creating objects from the Car class

3. *Abstraction*: The operations of an object can be defined without reference to its internal implementation;

4. *Polymorphism*: Code can be agnostic with respect to which class it is operating on, as long as it is operating from a given inheritance hierarchy; similarly, two classes defined within an inheritance may have different internal definitions for the same method.

While encapsulation and abstraction both hide data, they differ in a crucial manner: encapsulation hides data at an implementation level, whereas abstraction hides data at a design level. We discuss each of these four principles in turn, focusing on how they are implemented in Python.

### Inheritance

*Inheritance* allows us to define new classes from old, changing only what is needed. In this context, the new class is called a *child* of the old class, which is called the *parent*. For example, our Car class of Example 6.9 might be a subclass from a Vehicle class. The Vehicle class would define functionality and attributes common to all kinds of vehicles: bicycles, trains, boats, cars, planes, and so forth. The Car class could then overwrite—or freshly define—those elements that are unique to automobiles. This process can continue: the Car class might then be a parent to a number of subclasses representing different makes of cars (BMW, Subaru, Ford, etc.). Perhaps the Car class has a engine method, which could be uniquely defined for each of its subclasses.

In addition, we can allow for *multiple inheritance*, in which a new class is defined from an ordered sequence of parents. This can lead to ambiguity if one is not careful, as demonstrated by the *diamond problem* of Figure 6.4.

In this figure, class $D$ inherits from both classes $B$ and $C$, which are each,

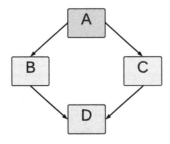

Fig. 6.4: Diamond problem: Class $D$ inherits from both $B$ and $C$.

in turn, children of the class $A$. In Python 3, the method resolution order is $D \to B \to C \to A$. (This differs from Python 2, which used a DLR—depth-first left-to-right—method, which would have favored $A$ over $C$.)

In other words, suppose that class $A$ has a method, which is overwritten by class $C$, but not by class $B$ or by class $D$. Objects of type $D$ would then follow the overwritten method as defined by class $C$.

*Example 6.10.* We can construct a subclass from the `Car` class (Code Block 6.10) that provides for a more realistic `speedUp` method, in the sense that it takes the vehicle's acceleration into account, so that the velocity jump is not instantaneous. The result is shown in Code Block 6.12.

```python
class RealCar(Car):
    def __init__(self, name, color, max_speed=100, acc=5):

        self.acc = acc
        super().__init__(name, color, max_speed=max_speed)

    def speedUp(self, delta_v=0):
        delta_v = min(delta_v, self.max_speed - self.velocity)
        delta_v = max(-self.velocity, delta_v)
        delta_t = abs(delta_v) / self.acc
        self.velocity += delta_v
        self.timeLapse(delta_t)
```

Code Block 6.12: Subclass of `Car`

Notice that we indicate the parent class in line 1. All methods not explicitly redefined within this code are inherited from the parent class, so that we may still access the `timeLapse`, `stop`, and `repaint` methods.

In our `RealCar` class, we update the `__init__` method, to allow for an additional input parameter, representing the vehicle's acceleration. We access the parent class using Python's built-in `super()` method. Finally, we update the `speedUp` method of the parent class with the code shown on lines 7–12.                                                                          ▷

## Encapsulation

The concept of *encapsulation* refers to the art of only providing access to certain functionality to the outside world. Many programming languages have a concept of *private* and *public* attributes and methods: public attributes and methods are exposed to the outside world, whereas private attributes and methods are only available from within the class itself. For example, when building a car, there are many operations that are required for it to run. Only a handful of these methods are exposed to the driver: gas pedal, brakes, steering wheel. Even though a car requires many other operations to run—piston and cylinders, spark ignition, diverting energy to battery, and so forth—those functions are not directly exposed to the driver.

Python does not offer private attributes and methods. Instead, we may follow the convention that attributes and methods not intended to be exposed to the outside world be preceded by an underscore. For instance, we could signify our intent of keeping `timeLapse` as an internal, private method by renaming it as `_timeLapse`.

For stronger protection of private attributes and methods, we may instead use a double underscore; for example, `__timeLapse`. Attributes and methods with names beginning with a double underscore are not *directly* available from the outside, but need to be preceded by an additional underscore and the class name. For instance, to access `__timeLapse` from *outside* of the class, we first instantiate an object `my_car = Car(..)`, and then we call `my_car._Car__timeLapse(5)`. This process is known as *name mangling*, and is a layer of protection Python offers for variables intended to be kept private.

When defining classes within an inheritance hierarchy, we take the following approach: for private attributes and methods meant to be accessible to (or overwritten by) the various children and descendants, we use the single underscore, whereas private attributes and methods that are meant to be attached to a single class, we use the double underscore. The double underscore mangles the variable name when referred to from any subclass, making it more difficult to access (or overwrite) the original functionality.

## Abstraction

Whereas encapsulation hides specific chunks of code or data from the outside world, at the implementation level, abstraction hides the details of an implementation at the design level. In other words, abstraction is used to define the inputs and outputs without any reference to any specific implementation. In this way, abstraction defines the interface that any user of a class must follow. Such an interface is often referred to as an *application user interface*, or *API*.

In practice, abstraction is used whenever we want to define a family of classes that follow a common API. We achieve this by constructing an *abstract class*, which contains one or more *abstract methods*, which are simply methods that have not been implemented. This defines a set of methods, along with their inputs and outputs, that must be defined for any subclass of the abstract class. Abstract classes can be written in Python using the abc (abstract base class) package, as shown in Code Block 6.13.

```python
from abc import ABC, abstractmethod

class AbstractVehicle(ABC):

    @abstractmethod
    def speedUp(self, delta_v: int) -> None:
        pass

    @abstractmethod
    def timeLapse(self, t: float) -> None:
        pass

    @abstractmethod
    def stop(self) -> None:
        pass
```

Code Block 6.13: Abstract classes and methods in Python.

Thus, to create an abstract base class in Python, we simply subclass the ABC class from the abc package. To define abstract methods, we *decorate* the method by adding the text @abstractmethod, which is referred to as a *decorator*, to the line prior to the method definition. Decorators are also known as *wrappers*, as they can be used to wrap our functions in a blanket of code. We will discuss wrapper functions and decorators shortly, but for now, think of the extra line as a magic spell that prevents the class, or any of its subclasses, from instantiating objects unless they have overridden the abstract methods with specific implementations.

Code Block 6.13 defines three methods that must be common to all vehicles: speedUp, timeLapse, and stop. Since they are abstract methods, Python will throw an error if we try to instantiate an object from the AbstractVehicle class. However, when we defined Car, in Code Block 6.10, we could replace the first line with

```
class Car(AbstractVehicle):
```

to ensure that our Car class satisfies the API defined by our concept of an abstract vehicle. We have also established a lengthy hierarchy of subclasses:

$$\text{ABC} \rightarrow \text{AbstractVehicle} \rightarrow \text{Car} \rightarrow \text{RealCar}.$$

It's not hard to imagine a hierarchical tree structure: we might have many types of vehicles (cars, boats, airplanes, submarines), which each might have many types of manifestations (car alone might branch to BMW, Mercedes, Subaru, Ford, Ferrari, Toyota, Trabant, etc.).

Finally, we used *type hinting* to specify the types of inputs and outputs we should expect from our abstract methods. Type hinting is not limited to abstract classes; it can be used when defining methods for any class. We typically omit type hinting from our examples to save space, though it is typically beneficial to use in practice.

## Polymorphism

Our final principle of OOP is polymorphism, which simply means that objects from different classes can be designed to share behaviors. For example, all vehicles descendant from the AbstractVehicle method defined in Code Block 6.13 must have an implementation for speedUp, timeLapse, and stop. Their internal workings might vary dramatically: how an airplane *speeds up* is quite different than how a skateboarder speeds up. Polymorphism allows us to instantiate a fleet of vehicles from our AbstractVehicle family—cars, bicycles, trains, planes, ships, submarines, and skateboarders—and use them agnostically in regards to their internal workings. We will return to this concept in our discussion of factory methods, and again when we discuss how the object-oriented paradigm can be used in a data-science context.

### 6.3.3 Generators, Wrappers, and Factories

### Special Methods

Certain methods of a class have names that are both preceded and followed by double underscore, such as __init__ and __str__. These are referred to as *special methods*, and each have specific meaning related to how objects can be used in a Python environment. Commonly used special methods are shown in Tables 6.6 and 6.7. For a more comprehensive list, see docs.python.org[8]. Though many of these special methods are not used in practice, it is useful to be aware of them.

---

[8] https://docs.python.org/3/reference/datamodel.html#special-method-names

| method | usage | explanation |
| --- | --- | --- |
| __init__(self,...) | x=X(...) | instantiates an object from class X |
| __str__(self) | print(x) | returns a string representation of object |
| __len__(self) | len(x) | returns the length of object |
| __eq__(self, other) | x==y | compares two objects for equality |
| __getitem__(self, key) | x[key] | gets the value associated with a key |
| __setitem__(self, key, value) | x[key] = value | sets value of a key |
| __contains__(self, item) | item in x | determine if item in object |
| __iter__(self) | iter(x) | returns an iterator object |
| __next__(self) | next(x) | returns next item from container |

Table 6.6: Special methods in Python for class X with objects x and y.

| method | usage | explanation |
|---|---|---|
| __add__(self, other) | x + y | add |
| __sub__(self, other) | x - y | subtract |
| __mul__(self, other) | x * y | multiply |
| __truediv__(self, other) | x / y | divide |
| __pow__(self, other) | x ** y | power |
| __lt__(self, other) | x < y | less than |
| __le__(self, other) | x <= y | less than or equal to |
| __eq__(self, other) | x == y | equals to |
| __ne__(self, other) | x != y | not equals to |
| __gt__(self, other) | x > y | greater than |
| __ge__(self, other) | x >= y | greater than or equal to |

Table 6.7: Math-type special methods in Python for class X with objects x and y.

## Iterators and Generator Functions

In python, an *iterable object* is simply an object that is an object that is capable of returning its data, one element at a time. Iterables are often used in the context of a `for` loop. An iterable object must implement two special methods, _iter_ and _next_, collectively known as the *iterator protocol*. Lists, tuples, and dictionaries are all *iterable* objects.

An *iterator* is an object that represents a specific stream of data. The _iter_ method of an iterable object must return an iterator. Once an iterator has been instantiated, repeated calls to its _next_ method return successive items from the stream, until a `StopIteration` exception is raised.

For example, if we define a list `my_list = [1,2,3,4,5]`, calling `next(my_list)` throws an error, as `my_list` is not itself an iterator. Instead, if we call `my_iter = iter(my_list)`, we can then call `next(my_iter)` multiple times, receiving, in turn, the values 1–5, and the `StopIteration` exception thereafter.

*Example 6.11.* An iterable that represents the Fibonacci sequence is shown in Code Block 6.14. Once an object has been instantiated, we can iterate

```
1   class Fibonacci:
2       def __init__(self, max_length=10):
3           self.max_length = max_length
4       def __iter__(self):
5           self.position = 0
6           self.lag1 = 0
7           self.lag2 = 0
8           return self
9       def __next__(self):
10          self.position += 1
11          if self.position > self.max_length:
12              raise StopIteration
13
14          next_item = 1 if self.position == 1 else self.lag1 +
                  self.lag2
15          self.lag1, self.lag2 = next_item, self.lag1
16          return next_item
17
18  fib = Fibonacci()
19  for x in fib:
20      print(x) # prints 1, 1, 2, 3, 5, 8, 13, 21, 34, 55
```

Code Block 6.14: Defining an iterable in Python.

over it using a for-loop, as shown on lines 18–20.                    ▷

The code required to define an iterator is, however, a bit verbose. A commonly used simplification is that of a *generator function*, which creates an iterator without the need to explicitly invoke the iterator protocol. Generator functions are similar to regular Python functions, except they have one or more `yield` statements, instead of the typical `return` statement.

*Example 6.12.* The Fibonacci numbers can be generated using the generator function of Code Block 6.15, achieving the same result as Code Block 6.14.

```python
def fibonacci(max_length=10):
    position = 0
    lag1 = 0
    lag2 = 0
    while position < max_length:
        position += 1
        next_item = 1 if position == 1 else lag1 + lag2
        lag1, lag2 = next_item, lag1
        yield next_item

for x in fibonacci():
    print(x) # prints 1, 1, 1, 3, 5, 8, 13, 21, 34, 55

f = fibonacci()
next(f) # 1
next(f) # 1
next(f) # 2
# .... 3, 5, and so forth...

sum(fibonacci()) # 143
```

Code Block 6.15: Fibonacci numbers using generators

The generator function `fibonacci` defined in Code Block 6.15 can be used in the context of a for loop, as shown on lines 11–12. The function itself returns a fresh iterator object, which can be used to get successive values through the `next` method, as shown on lines 14–17. Additionally, we can simply sum the iterator object returned by a call to `fibonacci()`, as shown on line 19.                                                             ▷

## Wrapper Functions and Decorators

In Python, a *wrapper function* is a special type of function that takes a function as input[9] and returns a new function as output. They can be

---

[9] In Python, functions are objects, and therefore they may be passed into other functions as inputs.

used to "wrap" around other functions by adding a decorator—a line of code with the @ symbol followed by the name of the wrapper function— to the line *preceding* the function definition, in the same way we used the @abstractmethod command to decorate various functions in Code Block 6.13. An example of a simple wrapper is shown in Code Block 6.16.

```
1   def greetings(func):
2
3       def wrapper(*args, **kwargs):
4           print("Hello, there!")
5           result = func(*args, **kwargs)
6           print("Googdbye!")
7           return result
8
9       return wrapper
10
11  @greetings
12  def pow2(x, power=2):
13      return x**power
14
15  t = pow2(5, power=3) # Prints Hello / Goodbye; sets t = 125
```

Code Block 6.16: A simple wrapper.

Wrappers are often used when defining hierarchies. For example, there might be a number of assertions, or data checks, that many methods of various subclasses share. Instead of writing these assertions into each individual method, we can use a single wrapper function in the parent class that is then inherited to each subclass. For example, we can define a wrapper in the AbstractVehicle abstract class defined in Code Block 6.13. The code is shown for defining the wrapper function is shown in lines 5–13 of Code Block 6.17.

The _timeWrap decorator can be used to wrap any of the methods from the Car class, which is subclassed from AbstractVehicle. Proper usage of the decorate is shown on line 19. Notice that in order to access the decorator, we must make reference to the class in which the method is defined. (Technically, the method _timeWrap is implemented as a static method, which we discuss next.)

**Static Methods and Class Methods**

Sometimes it is useful to define methods, within a class, that can be called *without reference to any particular object*. This scenario occurs with two distinct flavors: static methods and class methods. *Static methods* are simply

```
 1   class AbstractVehicle(ABC):
 2
 3       # .... Methods from AbstractVehicle
 4       .
 5       def _timeWrap(func):
 6
 7           def wrapper(*args, **kwargs):
 8               t = time.time()
 9               results = func(*args, **kwargs)
10               print(f"Time to execute {func.__name__} is: {time.time()
                   - t}")
11               return results
12
13           return wrapper
14
15   class Car(AbstractVehicle):
16
17       # ....
18
19       @AbstractVehicle._timeWrap
20       def speedUp(self, delta_v=0):
21           # ....
```

Code Block 6.17: Adding a wrapper to the `AbstractVehicle` class.

methods attached to a class that can be used without instantiating an object from that class. *Class methods* are similar, except that a class method is allowed to modify any of the *class attributes*, or variables attached to the class itself, as opposed to individual objects. Class attributes are uniform and accessible across all objects from the class; this is a particularly handy way to store constant that do not vary object-to-object. Class methods must take the variable `cls`, instead of `self`, as their first argument, which refers to the class itself, and not the object. Static methods should have neither `cls` nor `self` as arguments, as they are pure functions attached to the class, and can modify neither object nor class.

Class methods should be preceded by the `@classmethod` decorator. Static methods should be preceded by the `@staticmethod` operator. The exception to the latter rule is when defining static decorators, as in lines 5– 13 of Code Block 6.17. Though `_timeWrap` is a static method, the absence of the `@staticmethod` decorator simply means it is not available when attached to any object of the class.

**Factories**

Static methods are commonly used to define *factories*, which are certain methods used to generate new objects from a particular family of classes.

*Example 6.13.* Suppose that we have defined a variety of subclasses from the `AbstractVehicle` class, including `Bmw`, `Submarine`, and `Airplane`. A simple factory method and its usage is shown in Code Block 6.18.

```
class VehicleFactory:

    @staticmethod
    def get(name, *args, **kwargs):
        lookup = {'bmw': Bmw,
                  'sub': Submarine,
                  'airplane': Airplane}

        return lookup[name](*args, **kwargs)

c = VehicleFactory.get('bmw', 'my sports car', 'red')
```

Code Block 6.18: Factory method to generate various objects from the `AbstractVehicle` family.

Note that, unlike other classes we've studied, the `VehicleFactory` never gets instantiated. We *directly* invoke the `get` method of the `VehicleFactory` class, without reference to any specific instance of the class. This method takes as argument a name, used as the key to a lookup map, and returns an instantiated object of the appropriate type.

We could, alternatively, define the lookup map *outside* of the `get` method, making it a class attribute. If we did so, we would instead define `get` as a class method, and reference `cls.lookup`.

Another variation is to use a factory method to *dynamically* generate new Python classes on the fly, using the built-in `type` function. Suppose, in addition to our `AbstractVehicle` class, we also had an `AbstractEngine` class, with various subclasses, such as `Diesel`, `Turbine`, `Combustion`, and `Rocket`. Further, suppose that any engine can be used for any type of vehicle, as they each follow a consistent API. By following the convention `vehicle_engine`, we can dynamically generate an object that inherits the vehicle components from the `vehicle` class and the engine components from the `engine` class, as shown in Code Block 6.19.

This factory returns an instantiated object from the new class, inheriting properties from both the vehicle as well as the engine. This is a convenient way to handle the case in which a class has multiple components,

```
1   class VehicleFactory:
2
3       # class variables
4       vehicles = {'bmw': BMW,
5                   'sub': Submarine,
6                   'airplane': Airplane}
7
8       engines = {'diesel': Diesel,
9                  'turbine': Turbine,
10                 'combustion': Combustion,
11                 'rocket': Rocket}
12
13      @classmethod
14      def get(cls, name, *args, **kwargs):
15          assert '_' in name
16          vehicle, engine = name.split('_')
17
18          return type(name, (cls.vehicles[vehicle],
                cls.engines[engine]), {})(*args, **kwargs)
19
20  c = VehicleFactory.get('bmw_rocket', 'my rocket car', 'red')
21  type(c) # prints: __main__.bmw_rocket
```

Code Block 6.19: Dynamic class creation in a factory method

and each component can have multiple variations. Instead of defining each permutation as an individual subclass, we can define an abstract class for each component, and dynamically cast one subclass from each component together to generate a new mix-and-match object.                    ▷

### 6.3.4 OOP in Data Science

We next consider how we can deploy the principles of object-oriented programming to data science practice. In doing so, we seek to benefit from the various benefits offered by an object-oriented mindset, including use of modular, reusable code as well as a structure for building data-science products in teams.

By leveraging the four principles of object-oriented design, we can deploy greatly simplified runtime scripts, as shown in Code Blocks 6.20 and 6.21. Py3912 shows the overall structure of the file, whereas Code Block 6.20 defines the main function. Notice that all of the complexity is abstracted within three classes: our database class D, which can read and write to our database, our schema class S, which defines the table structure (e.g., column types, partitions, etc.) for each table we wish to write to, and our model class M, which encapsulates all of the implementation details of our

```
1    def main(project, model, mode, db_src, exec_id, **config):
2
3        D = DataBaseFactory.get(db_src)
4        S = SchemaFactory.get(project)
5
6        S.set('run_config')
7        df_config = DataFrame({'project': project,
8                               'model': model,
9                               'mode': mode,
10                              'exec_id': exec_id,
11                              'config': str(config)}, index=[0])
12       D.put(S, df_config)
13
14       input_table = config.pop('input_table')
15       target_table = config.pop('target_table')
16       cols = config.pop('cols')
17       target_cols = config.pop('target_cols')
18       time_col = config.pop('time_col')
19       point_in_time = config.pop('point_in_time',
                datetime.datetime.utcnow().strftime('%Y-%m-%d'))
20       delay = config.pop('delay', 30)
21       window = config.pop('window': 30)
22       range_ = config.pop('range_': 90)
23
24       df_train = D.get(table=input_table,
25                        cols=cols,
26                        time_col=time_col,
27                        time=[point_in_time, delay, window])
28
29       df_labels = D.get(table=target_table,
30                         cols=cols,
31                         time_col=time_col,
32                         time=[point_in_time, delay, window])
33
34       df_predict = D.get(table=input_table,
35                          cols=target_cols,
36                          time_col=time_col,
37                          time=[point_in_time, 0, range_])
38
39       M = ModelFactory.get(project, model, **config)
40       M.train(df_train, df_labels)
41       df_out = M.predict(df_predict)
42       df_out['exec_id'] = exec_id
43
44       S.set('run_output')
45       D.put(S, df_out)
```

Code Block 6.20: Example of a runtime script.

```
 1  import datetime, argparse
 2  import SchemaFactory, DataBaseFactory, ModelFactory
 3
 4  def main(project, model, mode, db_src, exec_id, **config):
 5      # ...
 6      # main script here....
 7
 8  if __name__ == '__main__':
 9      parser = argparse.ArgumentParser(description='Inputs for
            Project.')
10      parser.add_argument('--project', type=str, required=True)
11      parser.add_argument('--model', type=str, required=True)
12      parser.add_argument('--mode', type=str, default='TEST')
13      parser.add_argument('--db_src', type=str, default='hive')
14      parser.add_argument('--exec_id', type=int, required=True)
15      # .... additional model configuration parameters
16
17      args = parser.parse_args()
18      main(**args)
```

Code Block 6.21: Example of a runtime script (continued)

model. Additionally, by defining an abstract Model class, as shown in Code Block 6.22, we can ensure that whatever model we deploy will be compatible with our main script, as long as it is subclassed from Model with an implementation of the train and predict methods.

Moreover, by leveraging the principle of polymorphism, we construct our three classes using factory methods, so that our code is agnostic as to which particular subclass of each we are using, as long as that subclass follows the API laid out by its abstract base class. In particular, note that a single script can be leveraged across multiple projects, and many models of the same kind within a project. The model, itself, is simply passed as a parameter into the script, and the factory method delivers the appropriate subclass corresponding to that model. Moreover, we parameterized the point_in_time, so that the same runtime script can be reused in the backtesting framework, as discussed in Section 6.1.6. Another unexpected advantage occurred for one data science team, when their organization migrated to Hive, as we see in our next example.

*Example 6.14.* A large organization is planning on migrating all its data from Vertica to Hive. One data science team in the organization consists of a number of data scientists who operate mostly independently, writing their own scripts and methods for accessing the database. Moreover, many projects that have been finished have the database queries and connections hard-coded into dusty scripts that are scattered about. The team spends

```
1  from abc import ABC, abstractmethod
2  class Model:
3
4      def __init__(self, **kwargs):
5          # override to provide for default parameter values
6          self.params = kwargs
7
8      @abstractmethod
9      def train(self,
10                 df: DataFrame,
11                 y: np.array,
12                 weights: np.array=None):
13          return
14
15      @abstractmethod
16      def predict(self,
17                 df: DataFrame):
18          return
```

Code Block 6.22: Abstract Model class

several months determining all of their production dependencies and overwriting the code in order to point to the new data source. They also have to determine how, exactly, they are going to switch over to Hive, given that the rollout occurs over time and not all of the data is available yet, though they are still constrained to hit the deprecation timeline. The team loses an entire quarter of productivity handling the migration.

Another team, one which embraced the principles of object-oriented design from the ground up, has a much simpler task ahead of it. First, it must write a new subclass from following their AbstractDataBase API, replacing the functionality of DataBase_Vertica with DataBase_Hive. In particular, they must connect to a new data source and update their queries, as there are slight syntactical differences between the two languages. Next, they add one line to their DataBaseFactory lookup map:

'hive' : DataBaseHive.

Finally, they update a single config file that stores db_src for all production jobs, changing it to point from vertica to hive. And Voila! With a few simple changes, the team has migrated all jobs across all projects to the new database. Moreover, they only required help from a single member of their team to write the new database subclass. By using the principles of inheritance, encapsulation, abstraction, and polymorphism, the actual runtime scripts don't need to be changed at all, as they are agnostic with regard to the actual database engine that delivers its data.    ▷

In addition to the migration problem of Example 6.14, there are two other benefits of generating a database class from a factory method. The first is that it might not be possible to connect to Hive from local laptops. Instead, the team might connect to Presto. The factory method allows them to continue to run their jobs locally, by simply passing `db_src='presto'` instead of `db_src='hive'`. The second advantage is when writing unit tests for the code. Sample data can be stored in the form of `.csv` files in a data folder within the project. A new database subclass, `DataBase_Csv`, can then be created to read from the sample datasets in the local folder, requiring no connection to any database. The testing scripts can then pass `db_src='csv'` into the runtime script, so that unit tests will always have access to sample data, without wasting time creating database connections or managing the different connections required based on whether the tests are run locally or on a virtual workstation.

The final advantage of an object-oriented design is that it lubricates collaboration within teams. Since everything follows the same API, different members can easily spin up different models, or try different approaches, in a seamless fashion. One can easily spin up a new model subclass, add the new subclass to the model factory, and then deploy a backtest to see how well the new model would have fared compared to the current production model. Moreover, the modular, a la carte nature of the object-oriented design allows anyone on the team to run any model from anyone else. Such a system is therefore more manageable and maintainable as individual contributors on the team migrate switch roles or migrate to other teams.

## Problems

**6.1.** Construct an counterexample to the claim that $R^2 \geq 0$; i.e., construct a data set and a predictive model with a negative coefficient of determination.

**6.2.** Show that the softmax transformation defined in Equation (6.12) has the properties $f_i(X) \in (0, 1)$ and $\sum_{i=1}^{m} f_i(X) = 1$.

**6.3.** Prove Proposition 6.2.

**6.4.** Determine which of the following discrete variables should be ordinal versus categorical:

1. Age range: $\{18\text{--}24, 25\text{--}34, 35\text{--}44, 45\text{--}54, 55\text{-}64, 65+\}$;
2. Gender: $\{$male, female$\}$;
3. Year in college: $\{$freshman, sophomore, junior, senior$\}$;
4. Brand: $\{$Nike, Adidas, Asics$\}$;
5. Compass heading $\{$north, south, east, west$\}$;
6. Blood type $\{$A, B, AB, O$\}$.

**6.5.** Prove that every label encoder is invertible.

**6.6.** Write your own transformer class in Python that handles one-hot encoding of a categorical variable.

**6.7.** Write an abstract class for each of the three classes required for the runtime script Code Block 6.20: `AbstractSchema`, `AbstractDataBase`, `AbstractModel`. *Hint*: The SQL syntax for `SELECT`, `CREATE TABLE`, and `INSERT` might vary from database to database, so that `AbstractDataBase` should house the abstract methods to generate the given SQL strings.

# To Regression and Beyond

## 7.1 Linear Regression

And now for some mathematics.

### 7.1.1 The Model

Linear regression is the cornerstone of statistical analysis. We present the basic analysis here; for more details see Dunn and Smyth [2018], Hastie, *et al.* [2009], or James, *et al.* [2013]. For details of many of the related proofs, see Olive [2017] and Seber and Lee [2003].

### Regression on Random Variables

**Definition 7.1.** *A* linear regression *of a random variable $Y$ over a set of random variables $\{X_1, \ldots, X_p\}$, all defined over a common probability space, is the random variable*

$$\hat{Y} = \beta_0 + \sum_{j=1}^{p} X_j \beta_j, \tag{7.1}$$

*where the regression coefficients $\beta_0, \ldots, \beta_p$ are obtained by minimizing the expected squared error*

$$\text{MSE} = \mathbb{E}\left[\left(Y - \beta_0 - \sum_{j=1}^{p} X_j \beta_j\right)^2\right]. \tag{7.2}$$

*In a linear regression, the random variables*

$$\epsilon = Y - \hat{Y}$$

*are referred to as the* residuals *or* residual errors.

A *linear regression model is* simple *if if consists of a single explanatory variable (p = 1). Otherwise, it is said to be a* multiple linear regression model.

*Note 7.1.* In a linear regression, the regressors $X_1, \ldots, X_p$ are also referred to as *explanatory variables* or *features*. A linear regression model with $p$ explanatory variables has a total of $p+1$ unknown model parameters. The parameter $\beta_0$ is referred to as the *intercept*, whereas the parameters $\beta_1, \ldots, \beta_p$ are referred to as *slopes*.                                                            ▷

**Proposition 7.1.** *The regression coefficients $\beta_0, \ldots, \beta_p$ to a linear regression of $Y$ on $\{X_1, \ldots, X_p\}$ are the solutions to the system of equations*

$$\mathbb{E}\left[\left(Y - \beta_0 - \sum_{j=1}^{p} \beta_j X_j\right) X_k\right] = 0, \tag{7.3}$$

*for $k = 0, \ldots, p$, where we take $X_0 = 1$.*

*Proof.* Equation (7.3) is obtained by differentiating Equation (7.2) with respect to the parameters $\beta$ and setting the result equal to zero.    □

Note that the system of equations Equation (7.3) is equivalent to

$$\beta_0 + \sum_{j=1}^{p} \beta_j \mathbb{E}[X_j] = \mathbb{E}[Y] \tag{7.4}$$

$$\beta_0 \mathbb{E}[X_k] + \sum_{j=1}^{p} \beta_j \mathbb{E}[X_j X_k] = \mathbb{E}[Y X_k], \tag{7.5}$$

for $k = 1, \ldots, p$. This can be expressed in matrix form as

$$\begin{bmatrix} 1 & \mathbb{E}[X_1] & \cdots & \mathbb{E}[X_p] \\ \mathbb{E}[X_1] & \mathbb{E}[X_1^2] & \cdots & \mathbb{E}[X_1 X_p] \\ \vdots & & \ddots & \vdots \\ \mathbb{E}[X_p] & \mathbb{E}[X_p X_1] & \cdots & \mathbb{E}[X_p^2] \end{bmatrix} \cdot \begin{bmatrix} \beta_0 \\ \beta_1 \\ \vdots \\ \beta_p \end{bmatrix} = \begin{bmatrix} \mathbb{E}[Y] \\ \mathbb{E}[Y X_1] \\ \vdots \\ \mathbb{E}[Y X_p] \end{bmatrix}. \tag{7.6}$$

The matrix on the left can be inverted to solve for the coefficients $\beta_0, \ldots, \beta_p$.

Note that by applying the law of total expectation (Equation (1.19)) to the first equation in our system Equation (7.3), we can alternatively this equation as

$$\mathbb{E}[\mathbb{E}[Y|X] - \hat{Y}] = 0. \tag{7.7}$$

When the values of $X$ are regarded as fixed, as opposed to random, we can drop the outer expectation and express this as the class *regression equation*

$$\hat{Y} = \mathbb{E}[Y|X]. \tag{7.8}$$

However, when $X$ is regarded as random, the regression equation is only true *on average* (i.e., Equation (7.7)) or in special cases (e.g., the joint distribution is Gaussian).

### Conditional Covariance and Regression

We next discuss a useful result that links regression with conditional covariance. First, we define conditional covariance as a straightforward generalization of Definition 1.17.

**Definition 7.2.** *The* conditional covariance *of random variables $X$ and $Y$, conditioned on the random variable $Z$, is defined as*

$$\text{COV}(X, Y|Z) = \mathbb{E}\left[(X - \mathbb{E}[X|Z])(Y - \mathbb{E}[Y|Z])|\, Z\right]. \tag{7.9}$$

It is easy to show that the conditional covariance is equivalent to

$$\text{COV}(X, Y|Z) = \mathbb{E}[XY|Z] - \mathbb{E}[X|Z]\mathbb{E}[Y|Z], \tag{7.10}$$

in analogy to Equation (1.13).

**Proposition 7.2.** *Given random variables $X$, $Y$, and $Z$. Then the expected conditional covariance is given by*

$$\mathbb{E}\left[\text{COV}(X, Y|Z)\right] = \text{COV}(X - \mathbb{E}[X|Z], Y - \mathbb{E}[Y|Z]). \tag{7.11}$$

Note that the expectation on the left-hand side of Equation (7.11) is with respect to the random variable $Z$.

*Proof.* First, note that

$$\mathbb{E}[X - \mathbb{E}[X|Z]] = 0 \qquad \text{and} \qquad \mathbb{E}[Y - \mathbb{E}[Y|Z]] = 0,$$

from the law of total expectation. Therefore, the covariance of $X - \mathbb{E}[X|Z]$ and $Y - \mathbb{E}[Y|Z]$ is given by the expectation of the product

$$\begin{aligned}
\text{COV}(X - \mathbb{E}[X|Z], Y - \mathbb{E}[Y|Z]) &= \mathbb{E}[(X - \mathbb{E}[X|Z])(Y - \mathbb{E}[Y|Z])] \\
&= \mathbb{E}[XY - X\mathbb{E}[Y|Z] - Y\mathbb{E}[X|Z] + \mathbb{E}[X|Z]\mathbb{E}[Y|Z]].
\end{aligned}$$

Now, by applying the law of total expectation, and noting that both $\mathbb{E}[X|Z]$ and $\mathbb{E}[Y|Z]$ are functions of $Z$, we obtain

$$\text{COV}(X - \mathbb{E}[X|Z], Y - \mathbb{E}[Y|Z]) = \mathbb{E}\left[\mathbb{E}[XY|Z] - \mathbb{E}[X|Z]\mathbb{E}[Y|Z]\right].$$

Finally, we recognize the right-hand side as the expected conditional covariance expressed in the form of Equation (7.10), thereby completing the result. □

**Regression on Data**

Oftentimes we are interested in estimating the regression coefficients from a set of data. We can do this by replacing the expected squared error Equation (7.2) with the sum of squared errors, and then solving for the coefficients using elementary calculus. The result of such a procedure is defined as follows.

**Definition 7.3.** *Given a weighted data set* $\{(x_{i1}, \ldots, x_{ip}, y_i, w_i)\}_{i=1}^n$, *the least-squares solution is an estimate for the linear regression coefficients that consists of the unique set of values* $\hat{\beta}_0, \ldots, \hat{\beta}_p$ *that minimizes the sum of squared errors*

$$\text{SSE}(\beta) = \sum_{i=1}^n w_i \left( y_i - \beta_0 - \sum_{j=1}^p x_{ij}\beta_j \right)^2. \tag{7.12}$$

Note that we can recast Equation (7.12) in matrix form as

$$\text{SSE}(\beta) = (\mathbf{y} - \mathbf{X}\beta)^T \mathbf{W}(\mathbf{y} - \mathbf{X}\beta), \tag{7.13}$$

where we define the matrix $\mathbf{X} \in \mathbb{R}^{n \times (p+1)}$ by $\mathbf{X}_{i0} = 1$ and $\mathbf{X}_{ij} = x_{ij}$, for $j \neq 0$, known as the *design matrix*, and the $n \times n$ diagonal matrix $W = \text{diag}(w_1, \ldots, w_n)$. Naturally, $\mathbf{y} \in \mathbb{R}^n$ and $\beta \in \mathbb{R}^{p+1}$.

It is a straightforward exercise to show the following.

**Proposition 7.3.** *The estimate* $\hat{\beta}$ *is a global minimum of Equation* (7.13) *if and only if*

$$\left( \mathbf{X}^T \mathbf{W} \mathbf{X} \right) \hat{\beta} = \mathbf{X}^T \mathbf{W} \mathbf{y}. \tag{7.14}$$

*This equation is known as the* normal equation

We will leave the proof as an exercise for the reader (Exercise 7.1). It follows from the normal equation that the least-squares solution to the linear regression problem is given by

$$\hat{\beta} = \left( \mathbf{X}^T \mathbf{W} \mathbf{X} \right)^{-1} \mathbf{X}^T \mathbf{W} \mathbf{y}. \tag{7.15}$$

*Note 7.2.* It is more computationally efficient to solve the system of equations given by Equation (7.14) than it is to solve Equation (7.15) directly. Python built-in solvers are, however, more efficient still. The scikit-learn method `sklearn.linear_model.LinearRegression` is actually a wrapper around the least-squares solver `scipy.linalg.lstsq`, which is based on an efficient form of the singular-value decomposition.                    ▷

**Corollary 7.1.** *For a simple linear regression problem, the least-squares coefficients are given by*

$$\hat{\beta}_0 = \bar{y} - \hat{\beta}_1 \bar{x},$$

$$\hat{\beta}_1 = \frac{\mathrm{SS}_{xy}}{\mathrm{SS}_x} = \frac{\sum_{i=1}^n w_i (x_i - \bar{x}_i) y_i}{\sum_{i=1}^n w_i (x_i - \bar{x}_i)^2},$$

*where $\bar{x}$ and $\bar{y}$ are the* weighted sample means

$$\bar{x} = \frac{1}{\omega} \sum_{i=1}^n w_i x_i \qquad and \qquad \bar{y} = \frac{1}{\omega} \sum_{i=1}^n w_i y_i,$$

*where $\omega = \sum_{i=1}^n w_i$ is the total weight, as before.*

We leave the proof to the reader (Exercise 7.2).

### 7.1.2 Analysis of the Estiamtes

A multilinear regression model has $p+1$ unknown coefficients, $\beta_0, \ldots, \beta_p$, which are estimated by Equation (7.15), and an additional parameter $\sigma^2$ that represents the unknown variance. In this section, we discuss basic results regarding estimates for these parameters.

### Standard Errors

We begin with a basic result about our estimates Equation (7.15) for the unknown coefficients $\beta_0, \ldots, \beta_p$.

**Theorem 7.1.** *Let $\hat{\beta} \in \mathbb{R}^{p+1}$ be the least-squares solution of a linear regression problem. Then*

$$\mathbb{E}[\hat{\beta}] = \beta \tag{7.16}$$

$$\mathbb{V}(\hat{\beta}) = \sigma^2 (\mathbf{X}^T \mathbf{W} \mathbf{X})^{-1}. \tag{7.17}$$

*Proof.* We know $\mathbb{E}[\mathbf{y}] = \mathbf{X}\beta$; hence Equation (7.16) follows immediately from Equation (7.15).

To derive Equation (7.17), first recall that $\mathbb{V}(\mathbf{y}) = \mathbf{W}^{-1}\sigma^2$. In addition, it is a general property of the variance–covariance matrix that $\mathbb{V}(\mathbf{Ay}) = \mathbf{A}\mathbb{V}(\mathbf{y})\mathbf{A}^T$. The result follows by setting $\mathbf{A} = (\mathbf{X}^T\mathbf{W}\mathbf{X})^{-1}\mathbf{X}^T\mathbf{W}$. □

Our next result is to show that, for linear regression, the total sum of squares equals the sum of the sum of squared errors and the regression sum of squares. This result, however, will rely on the following lemma.

**Lemma 7.1.** *Let $\mathbf{H} = \mathbf{X}(\mathbf{X}^T\mathbf{W}\mathbf{X})^{-1}\mathbf{X}^T \in \mathbb{R}^{n \times n}$, so that the least squares estimates are given by*

$$\hat{\mathbf{y}} = \mathbf{H}\mathbf{W}\mathbf{y}.$$

*Then the following results hold:*

*(a)* $\mathbf{H}$ *is symmetric and* $\mathbf{HWH} = \mathbf{H}$,
*(b)* $\mathbf{X}^T\mathbf{W}(\mathbf{y} - \hat{\mathbf{y}}) = \mathbf{0}$,
*(c)* $\sum_{i=1}^{n} w_i(y_i - \hat{y}_i) = 0$.
*(d) Assuming* $n > p + 1$, *the sum* $\text{trace}(\mathbf{HW}) = p + 1$.

*Proof.* Symmetry of $\mathbf{H}$ follows from its definition. Similarly, the result $\mathbf{HWH} = \mathbf{H}$ is straightforward to verify.

For (b), we use $\hat{\mathbf{y}} = \mathbf{X}\hat{\boldsymbol{\beta}}$ with Equation (7.15) to obtain

$$\mathbf{X}^T\mathbf{W}(\mathbf{y} - \hat{\mathbf{y}}) = \mathbf{X}^T\mathbf{W}\left(\mathbf{y} - \mathbf{X}(\mathbf{X}^T\mathbf{W}\mathbf{X})^{-1}\mathbf{X}^T\mathbf{W}\mathbf{y}\right)$$
$$= \mathbf{X}^T\mathbf{W}\mathbf{y} - \mathbf{X}^T\mathbf{W}\mathbf{y} = \mathbf{0}.$$

Part (c) follows directly from part(b). (This is the first component of the vector equation given in part (b), since the first row of $\mathbf{X}^T$ is a row of 1s.)

For part (d), let $\lambda$ be an eigenvalue of $\mathbf{HW}$, so that there exists a nonzero $\mathbf{v}$ with $\mathbf{HWv} = \lambda\mathbf{v}$. Next, consider the quantity

$$\lambda\mathbf{v}^T\mathbf{v} = \mathbf{v}^T\mathbf{WHv}$$
$$= \mathbf{v}^T\mathbf{WHWHv}$$
$$= (\mathbf{HWv})^T\mathbf{WHv}$$
$$= \lambda\mathbf{v}^T\mathbf{WHv}$$
$$= \lambda^2\mathbf{v}^T\mathbf{v}.$$

It follows that $\lambda(1 - \lambda) = 0$. Since $\text{rank}(\mathbf{HW}) = p + 1$, it follows that there are exactly $p + 1$ unity eigenvalues and $n - p - 1$ zero eigenvalues of the matrix $\mathbf{HW}$. Since the trace is the sum of the eigenvalues, the result follows. $\square$

**Theorem 7.2.** *In a linear regression problem, the total sum of squares is the sum of the residual and regression sums of squares; i.e.,*

$$\text{SST} = \text{SSE} + \text{SSR}.$$

*Proof.* This follows as long as the cross terms sum to zero; i.e., as long as

$$\sum_{i=1}^{n} w_i(y_i - \hat{y}_i)(\hat{y}_i - \overline{y}) = 0.$$

Separating this into two sums, we have

$$\sum_{i=1}^{n} w_i(y_i - \hat{y}_i)(\hat{y}_i - \overline{y}) = \sum_{i=1}^{n} w_i(y_i - \hat{y}_i)\hat{y}_i - \overline{y}\sum_{i=1}^{n} w_i(y_i - \hat{y}_i).$$

The second term on the right-hand side vanishes due to part (c) of Lemma 7.1. The first term on the right-hand side vanishes as well: since $\hat{\mathbf{y}} = \mathbf{HWy}$, it follows that

$$\sum_{i=1}^{n} w_i(y_i - \hat{y}_i)\hat{y}_i = (\mathbf{HWy})^T \mathbf{W}(\mathbf{I} - \mathbf{HW})\mathbf{y}$$

$$= \mathbf{y}^T \mathbf{WHW}(\mathbf{I} - \mathbf{HW})\mathbf{y}$$
$$= \mathbf{y}^T \mathbf{W}(\mathbf{HW} - \mathbf{HWHW})\mathbf{y}$$
$$= \mathbf{y}^T \mathbf{W}(\mathbf{HW} - \mathbf{HW})\mathbf{y},$$

where we used part (a) of Lemma 7.1. The result follows.    □

**Theorem 7.3.** *The quantity*

$$\hat{\sigma}^2 = \text{MSE} = \frac{\text{SSE}}{n - p - 1} \tag{7.18}$$

*is an unbiased estimator for* $\sigma^2$.

*Proof.* This result is equivalent to the statement

$$\mathbb{E}[\text{SSE}] = \sigma^2(n - p - 1).$$

We start by writing the sum of squared errors as

$$\text{SSE} = (\mathbf{y} - \mathbf{X}\hat{\beta})^T \mathbf{W}(\mathbf{y} - \mathbf{X}\hat{\beta})$$
$$= \left[(\mathbf{y} - \mathbf{X}\beta) + (\mathbf{X}\beta - \mathbf{X}\hat{\beta})\right]^T \mathbf{W} \left[(\mathbf{y} - \mathbf{X}\beta) + (\mathbf{X}\beta - \mathbf{X}\hat{\beta})\right]$$
$$= (\mathbf{y} - \mathbf{X}\beta)^T \mathbf{W}(\mathbf{y} - \mathbf{X}\beta) + (\mathbf{X}\beta - \mathbf{X}\hat{\beta})^T \mathbf{W}(\mathbf{X}\beta - \mathbf{X}\hat{\beta})$$
$$+ 2(\mathbf{y} - \mathbf{X}\beta)^T \mathbf{W}(\mathbf{X}\beta - \mathbf{X}\hat{\beta}).$$

Now, $\mathbb{E}[\mathbf{y}] = \mathbf{X}\beta$ and $\mathbb{E}[\hat{\beta}] = \beta$, so that

$$\mathbb{E}[\text{SSE}] = \mathbb{E}[\mathbf{y}^T \mathbf{W}(\mathbf{y} - \mathbf{X}\beta)] - \mathbb{E}[\hat{\beta}^T \mathbf{X}^T \mathbf{W}(\mathbf{X}\beta - \mathbf{X}\hat{\beta})]$$
$$+ 2\mathbb{E}[\mathbf{y}^T \mathbf{W}(\mathbf{X}\beta - \mathbf{X}\hat{\beta})]$$
$$= \mathbb{E}[\mathbf{y}^T \mathbf{Wy}] - \beta^T \mathbf{X}^T \mathbf{WX}\beta$$
$$- \beta^T \mathbf{X}^T \mathbf{WX}\beta + \mathbb{E}\left[\hat{\beta}^T \mathbf{X}^T \mathbf{WX}\hat{\beta}\right]$$
$$+ 2\beta^T \mathbf{X}^T \mathbf{WX}\beta - 2\mathbb{E}\left[\mathbf{y}^T \mathbf{WX}\hat{\beta}\right].$$

However, recalling the fact that $\mathbf{X}\hat{\beta} = \mathbf{HWy}$, we find that both

$$\hat{\beta}^T \mathbf{X}^T \mathbf{WX}\hat{\beta} = \mathbf{y}^T \mathbf{WHWy} \text{ and } \mathbf{y}^T \mathbf{WX}\hat{\beta} = \mathbf{y}^T \mathbf{WHWy},$$

which therefore reduces our expression to

$$\mathbb{E}[\text{SSE}] = \mathbb{E}[\mathbf{y}^T \mathbf{Wy}] - \mathbb{E}\left[\mathbf{y}^T \mathbf{WHWy}\right]. \tag{7.19}$$

To complete the proof, we rely on a result from linear algebra:

$$\mathbb{E}[\mathbf{y}^T \mathbf{A} \mathbf{y}] = \text{trace}(\mathbf{A}\mathbb{V}(\mathbf{y})) + \mathbb{E}[\mathbf{y}]^T \mathbf{A} \mathbb{E}[\mathbf{y}]. \tag{7.20}$$

(See Seber and Lee [2003] for a proof.) Now, $\mathbb{E}[\mathbf{y}] = \mathbf{X}\boldsymbol{\beta}$ and $\mathbb{V}(\mathbf{y}) = \mathbf{W}^{-1}\sigma^2$. We therefore have

$$\mathbb{E}[\mathbf{y}^T \mathbf{W} \mathbf{y}] = \text{trace}(\mathbf{W}\mathbf{W}^{-1}\sigma^2) + \boldsymbol{\beta}^T \mathbf{X}^T \mathbf{W} \mathbf{X} \boldsymbol{\beta}$$

and

$$\mathbb{E}\left[\mathbf{y}^T \mathbf{W} \mathbf{H} \mathbf{W} \mathbf{y}\right] = \text{trace}(\mathbf{W}\mathbf{H}\mathbf{W}\mathbf{W}^{-1}\sigma^2) + \boldsymbol{\beta}^T \mathbf{X}^T \mathbf{W} \mathbf{H} \mathbf{W} \mathbf{X} \boldsymbol{\beta}.$$

But

$$\boldsymbol{\beta}^T \mathbf{X}^T \mathbf{W} \mathbf{H} \mathbf{W} \mathbf{X} \boldsymbol{\beta} = \boldsymbol{\beta}^T \mathbf{X}^T \mathbf{W} \mathbf{X} (\mathbf{X}^T \mathbf{W} \mathbf{X})^{-1} \mathbf{X}^T \mathbf{W} \mathbf{X} \boldsymbol{\beta} = \boldsymbol{\beta}^T \mathbf{X}^T \mathbf{W} \mathbf{X} \boldsymbol{\beta},$$

so that Equation (7.19) is equivalent to

$$\mathbb{E}[\text{SSE}] = \text{trace}(\mathbf{I}\sigma^2) - \text{trace}(\mathbf{W}\mathbf{H}\sigma^2) = \sigma^2(n - (p+1)),$$

following part (d) of Lemma 7.1. This completes the result.    $\square$

By replacing $\sigma^2$ with $\hat{\sigma}^2$ in Equation (7.17), we can estimate the variance of our least-squares solution $\hat{\boldsymbol{\beta}}$. The square roots of the diagonal elements are therefore estimates of our standard errors:

$$\hat{\text{se}}(\hat{\beta}_j) = \sqrt{\hat{\sigma}^2 (\mathbf{X}^T \mathbf{W} \mathbf{X})_{jj}^{-1}}. \tag{7.21}$$

### Distributional Results

So far, the only assumptions we've made on the random variables $Y_i$ are independence and that $\mathbb{E}[\mathbf{y}|\mathbf{X}] = \mathbf{X}\boldsymbol{\beta}$ and $\mathbb{V}(\mathbf{y}) = \mathbf{W}^{-1}\sigma^2$. In order to derive results regarding the distribution of the least-squares estimates $\hat{\boldsymbol{\beta}}$, we must further make an assumption regarding the distribution of the error; i.e., we will next consider the case of *normal* linear regression problems. For normal error, we have

$$Y_i = \beta_0 + \sum_{j=1}^{p} X_{ij}\beta_j + \epsilon_i,$$

$$\epsilon_i \sim \text{N}(0, \sigma^2/w_i).$$

Alternatively, this is sometimes expressed as

$$Y_i \sim \text{N}\left(\beta_0 + \sum_{j=1}^{p} X_{ij}\beta_j, \frac{\sigma^2}{w_i}\right).$$

**Theorem 7.4.** *The least squares estimates $\hat{\beta}$ in a normal linear regression model are normally distributed as*

$$\hat{\beta} \sim \mathrm{N}\left(\beta, \sigma^2 (\mathbf{X}^T \mathbf{W} \mathbf{X})^{-1}\right). \tag{7.22}$$

*Proof.* From Equation (7.15), we know that the coefficients $\hat{\beta}$ are linear combinations of the training target variables $Y_1, \ldots, Y_n$, which are normally distributed. Therefore the coefficient vector $\hat{\beta}$ must have a multivariate normal distribution.

From Equations (7.16) and (7.17) we further know the expected value and variance matrix for $\hat{\beta}$. The form given in Equation (7.22) follows.  □

Theorem 7.2 tells us that the total sum of squares in a linear regression model is equal to the sum of the residual sum of squares (sum of squared errors) and the regression sum of squares. Under the normality conditions, each of these components is further distributed as a chi-squared distribution, as shown in our next theorem.

**Theorem 7.5.** *In a normal linear regression model,*

$$\frac{\mathrm{SSR}}{\sigma^2} \sim \chi_p^2 \tag{7.23}$$

$$\frac{\mathrm{SSE}}{\sigma^2} \sim \chi_{n-p-1}^2 \tag{7.24}$$

$$\frac{\mathrm{SST}}{\sigma^2} \sim \chi_{n-1}^2. \tag{7.25}$$

For a proof, we direct the reader to Seber and Lee [2003]. Given the number of degrees of freedom of each of the sum of squares, we define their respective mean-squared errors as

$$\mathrm{MSR} = \frac{\mathrm{SSR}}{p}, \qquad \mathrm{MSE} = \frac{\mathrm{SSE}}{n-p-1}, \qquad \text{and} \qquad \mathrm{MST} = \frac{\mathrm{SST}}{n-1}.$$

*Note 7.3.* The adjusted $R^2$ metric given by Definition 6.8 for linear regression should be

$$\bar{R}^2 = 1 - \frac{\mathrm{MSE}}{\mathrm{MST}} = 1 - \frac{\mathrm{SSE}/(n-p-1)}{\mathrm{SST}/(n-1)},$$

as there are a total of $p+1$ (not $p$) independent parameters.  ▷

It can further be shown that $\hat{\sigma}^2$ and $\hat{\beta}$ are statistically independent.

### Hypothesis Tests and Confidence Intervals

The sums of squares and mean squares are summarized nicely in an analysis of variance (ANOVA) table, as shown in Table 7.1. Moreover, due to Theorem 7.5, we know that under our normality assumptions, the $F$-statistic is distributed as $F \sim F_{p,n-p-1}$. We can therefore use this statistic to test whether or not the result of our regression should be accepted, as opposed to modeling the output with the average value.

| Source | Degrees of freedom | Sum of squares | Mean square |
|---|---|---|---|
| Regression | $p$ | $\text{SSR} = \sum_{i=1}^{n} w_i (\hat{y}_i - \bar{y})^2$ | $\text{MSR} = \dfrac{SSB}{p}$ |
| Error | $n - p - 1$ | $\text{SSE} = \sum_{i=1}^{n} w_i (y_i - \hat{y}_i)^2$ | $\text{MSE} = \dfrac{SSW}{n - p - 1}$ |
| Total | $n - 1$ | $\text{SST} = \sum_{i=1}^{n} w_i (y_i - \bar{y})^2$ | $F = \text{MSR}/\text{MSE}$ |

Table 7.1: ANOVA table.

**Theorem 7.6.** *For a normal linear regression problem, the $F$-statistic,*

$$F = \frac{\text{MSR}}{\text{MSE}} = \frac{\text{SSR}/p}{\text{SSE}/(n-p-1)} \tag{7.26}$$

*is distributed as $F \sim \text{F}_{p,n-p-1}$.*

Intuitively, a large value of $F$ means that our predictions are farther off from the overall mean than they are to their true values. The $F$ statistic can therefore be used to test whether or not our model performs better than had we simply used a constant prediction $\hat{y} = \bar{y}$; i.e., had we chosen $\hat{\beta}_1 = \cdots = \hat{\beta}_p = 0$. For a significance level $\alpha$, we should therefore reject this null hypothesis whenever $F > \text{F}_{p,n-p-1}^{\alpha}$.

To test a null hypothesis about a particular coefficient, $H_0 : \beta_j = 0$, the distribution given by Equation (7.22) implies that, under the null hypothesis, the statistic

$$Z_j = \frac{\hat{\beta}_j}{\text{se}(\hat{\beta}_j)}$$

has a standard normal distribution. The standard error, however, depends on the variance $\sigma^2$, which is typically unknown. When we replace our estimate Equation (7.18) for the variance, we obtain the statistic

$$T_j = \frac{\hat{\beta}_j}{\hat{\text{se}}(\hat{\beta}_j)}, \tag{7.27}$$

which uses the estimated standard error given in Equation (7.21). This statistic has a Student's T distribution with $(n-p-1)$ degrees of freedom: $T_j \sim \text{t}_{n-p-1}$. Naturally the null hypothesis is to be rejected whenever $|T_j| > \text{t}_{n-p-1}^{\alpha/2}$. A $(1-\alpha)$ confidence interval is therefore given by

$$\hat{\beta}_j \pm \hat{\text{se}}(\hat{\beta}_j)\text{t}_{n-p-1}^{\alpha/2}.$$

Often, we may be interested in testing not only if a single parameter is significant, but if a subset of parameters are significant. For example, suppose a categorical feature is one-hot encoded into a set of $k$ binary features. We may want to test whether that set of $k$ features, as a group, is beneficial to the model. To proceed, we consider a *reduced model* with those features eliminated.

In general, consider a reduced model consisting of the constant term and the first $q$ coefficients: $\{\beta_0, \ldots, \beta_q\}$, with $q < p$. Our null hypothesis is

$$H_0 : \beta_{q+1} = \cdots = \beta_p = 0,$$

against the alternative hypothesis $H_A$ that at least one $\beta_j \neq 0$, for some $j > q$. In short, our null hypothesis is that the reduced model is better than the full model. Rejecting the null hypothesis implies that it is beneficial to include (at least some of) the additional features $\beta_{q+1}, \ldots, \beta_p$.

To test this null hypothesis, we train both models, and compute the sum of squared errors for both cases: $\text{SSE}_0$ is the sum of squared errors when using the reduced model and $\text{SSE}_1$ is the sum of squared errors when using the full model. We then express our $F$-statistic as

$$F = \frac{(\text{SSE}_0 - \text{SSE}_1)/(p - q)}{\text{SSE}_1/(n - p - 1)}. \tag{7.28}$$

Under the null hypothesis, $F \sim F_{p-q,n-p-1}$. We call this the *partial F test*. For the special case $q = 0$, $\text{SSE}_0 = \text{SST}$ and Equation (7.28) reverts back to Equation (7.26).

The the selection of a submodel for the partial F test must be carried out *before* the regression fit. If one performs a regression and then formulates a test with the worst performers removed, the resulting $F$ statistic will be too high and the submodel will be prone to selection bias.

*Example 7.1.* Load the Boston housing price dataset from `sklearn.datasets` and fit a linear regression model using a variable number of features. Plot the test error and the generalization error (expected test error) as a function of the number of features. Use `print(data.DESCR)` for a description of the dataset. Use both the given features as well as their squared values.

We can accomplish this in Code Block 7.1. We build a pipeline (lines 11–20) that imputes missing values with the feature's median value, passes through the `X[:, 0]` column (of all 1s), and filters out the columns in excess of `n_features`. We then conduct `n_sims` simulations, where we use a test size of 80%, so that each model is trained on approximately 100 randomly selected instances. Note that fitting the intercept $\beta_0$ is done automatically by the `LinearRegression` class, so we do not need to append a column of 1s in front of X.

The code to produce the plots is given in Code Block 7.2. The results are shown in Figure 7.1.

In Fig3702, we plot the test (blue) and training (red) error for our first fifty simulations. We then plot the average test error (dark blue) and average

```
1   data = load_boston()
2   X, y = data.data, data.target
3   X = np.concatenate([X, X**2], axis=1)
4   n_sims = 500
5   n, p = X.shape
6   err_test = np.zeros((p+1, n_sims))
7   err_train = np.zeros((p+1, n_sims))
8
9   for n_features in range(1, p+1):
10
11      select_columns = ColumnTransformer(
12              [('standardize', RobustScaler(quantile_range=(0.2,0.8)) ,
                    np.arange(n_features))],
13              remainder='drop')
14
15      transformers = [
16              ('impute', SimpleImputer(strategy='median')),
17              ('scale_select', select_columns),
18              ('model', LinearRegression(normalize=False))]
19
20      pipe = Pipeline(transformers)
21
22      for i in range(n_sims):
23          X_train, X_test, y_train, y_test = train_test_split(X, y,
                test_size=0.80)
24          pipe.fit(X_train, y_train)
25          y_hat = pipe.predict(X_test)
26          y_hat_train = pipe.predict(X_train)
27          err_test[n_features, i] = mean_squared_error(y_test, y_hat)
28          err_train[n_features, i] = mean_squared_error(y_train,
                y_hat_train)
```

Code Block 7.1: Linear Regression and Bias–Variance Tradeoff

training error (dark red) over all 500 of our simulations. These latter two plots are unbiased estimates for the expected test error and the expected training error.

Note that the training error continues to decrease as we add additional dimensions to the feature set. However, after about 14 dimensions, the variance of our model increases dramatically, resulting in an increased generalization error. This is why it is important to select a model that minimizes its generalization error.                                                                                    ▷

```
1  for i in range(50):
2      plt.plot(range(1,p+1), err_test[1:, i], color='#1f77b4',
           alpha=0.3)
3      plt.plot(range(1,p+1), err_train[1:, i], color='#d62728',
           alpha=0.3)
4
5  plt.plot(range(1,p+1), err_test.mean(axis=1)[1:], color='b',
       linewidth=2)
6  plt.plot(range(1,p+1), err_train.mean(axis=1)[1:], color='r',
       linewidth=2)
7  plt.axis([0, 26, 0, 100])
```

Code Block 7.2: Code to produce Figure 7.1.

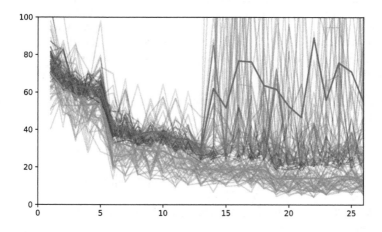

Fig. 7.1: Bias–variance tradeoff. Test error (blue) and training error (red) as a function of number of features. On the left: high bias, low variance models. On the right: low bias, high variance models.

### 7.1.3 Regularization

Linear regression suffers from two primary afflictions: it is a high-variance model and suffers from interpretability when many features are used. The presence of correlated explanatory variables leads to a high model variance: it is all too-often the case that two correlated variables receive large coefficients with opposite sign, the effect being one of cancellation. Slight perturbations in the training data set are therefore likely to throw the estimates off. It is therefore often the case that we are interested in finding a *subset* of features for our model that lead to a better generalization error than our full feature set.

**Lasso and Ridge Regression**

**Definition 7.4.** *Given a loss function $L$ for a parametric predictive model $f$ and the projection $\pi : \mathbb{R}^{k+1} \to \mathbb{R}^k$ defined by $\pi(\beta_0, \beta_1, \ldots, \beta_k) = (\beta_1, \ldots, \beta_k)$, the $\ell_p$-regularized loss for a given fixed $\lambda > 0$ is the augmented loss*

$$L_p(Y, f(X; \beta)) = L(Y, f(X; \beta)) + \lambda ||\pi(\beta)||_p, \tag{7.29}$$

*where $|| \cdot ||_p$ is the $\ell_p$ norm, defined, for $x \in \mathbb{R}^k$, by*

$$||x||_p = \left( \sum_{i=1}^{k} |x_i|^p \right)^{1/p}.$$

*The $\ell_1$ regularization of a linear regression problem is referred to as* lasso regression, *whereas the $\ell_2$ regularization is referred to as* ridge regression.

Conceptually, it is useful to keep the following equivalency in mind: minimization of the regularized loss is equivalent to the minimization of the training loss subject to an inequality constraint, as laid out in the following proposition.

**Proposition 7.4.** *There is a one-to-one correspondence between the optimization problems*

$$\min_{\beta \in B} L_p(Y, f(X; \beta))$$

*and*

$$\min_{\beta \in B} L(Y, f(X; \beta)) \text{ subject to } ||\pi(\beta)||_p \leq t,$$

*such that every $\lambda > 0$ of the first corresponds to a $t > 0$ of the second, and such that*

$$\lim_{t \to \infty} \lambda(t) = 0 \qquad and \qquad \lim_{\lambda \to \infty} t(\lambda) = 0.$$

*Each of these problems is known as the other problem's dual.*

Outside of the strict confines of Definition 7.4 and Proposition 7.4, we will refer to only the $\ell_1$ and $\ell_2$ regularized loss, and resume our use of the variable $p$ to represent the number of explanatory variables or features. For the squared-error loss used in linear regression models, we will consider the regularized training error functions

$$L_1(\beta; \mathcal{D}) = \sum_{i=1}^{n} w_i (y_i - f(x_i; \beta))^2 + \lambda \sum_{j=1}^{p} |\beta_j|, \tag{7.30}$$

$$L_2(\beta; \mathcal{D}) = \sum_{i=1}^{n} w_i (y_i - f(x_i; \beta))^2 + \lambda \sum_{j=1}^{p} \beta_j^2. \tag{7.31}$$

*Note 7.4.* The intercept $\beta_0$ is *not* penalized in the regularized loss function. The intercept helps control the overall balance and does not correspond to any of the explanatory variables. In fact, under $\lambda = \infty$ (or $t = 0$), all of the slopes $\beta_1 = \cdots = \beta_p = 0$ would be turned off, and $\beta_0 = \bar{y}$ would represent the average value of the target variable over the training set.    ▷

*Note 7.5.* The standard least squares solution to an ordinary linear regression problem is found by minimizing the training error Equation (7.12) over the parameter set $\beta$. Regularization can be therefore viewed as a method that penalizes large coefficients. Both ridge and lasso regression are tunable through the parameter $\lambda$, which controls the overall complexity of the model. The case $\lambda = 0$ corresponds to the ordinary least squares solution, whereas increased values of $\lambda$ result in diminished coefficient scales and thus reduced levels of complexity.    ▷

The lasso method has the tendency of pushing coefficients to zero, whereas ridge regression tends to shrink the coefficients more uniformly. This is visualized in Figure 7.2 for the case $p = 2$. The darkly shaded diamond corresponds to the region $||\pi(\beta)||_1 \leq 1$, and the lightly shaded circle corresponds to the region $||\pi(\beta)||_2 \leq 2$. The ordinary least squares loss is marked with an **x** at $(3, 1)$, whereas the solutions to the constrained optimization dual problems (Proposition 7.4) are represented by dots. For

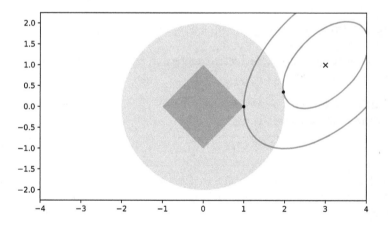

Fig. 7.2: Least squares solution with lasso and ridge regression.

$p > 2$, the region $||\pi(\beta)||_1 \leq t$ represents a family of *rhomboids* in $\mathbb{R}^p$. The solution of the lasso regression occurs when the contours of the average training loss function first touch this rhomboidal surface, providing ample opportunity for a selection of the $\beta$s to be set to zero. It is for this reason, lasso regression serves as a form of continuous subset selection, with the benefit of automatically selecting the most prevalent subset of features.

**Effective Degrees of Freedom**

Now, as anyone can see, the solution to the ridge regression problem is

$$\hat{\beta}^{\text{ridge}} = \left(\mathbf{X}^T \mathbf{W} \mathbf{X} + \lambda \boldsymbol{\Pi}\right)^{-1} \mathbf{X}^T \mathbf{W} \mathbf{y}, \qquad (7.32)$$

where

$$\boldsymbol{\Pi} = \frac{\partial \pi(\boldsymbol{\beta})}{\partial \boldsymbol{\beta}} = \mathbf{I}_{p+1} - \mathbf{e}_0 \mathbf{e}_0^T$$

is the $(p+1) \times (p+1)$ identity with the $(00)$ component removed[1]. Inspired by Lemma 7.1, we therefore define the *effective degrees of freedom* of the model as

$$df(\lambda) = \text{trace}(\mathbf{X} \left(\mathbf{X}^T \mathbf{W} \mathbf{X} + \lambda \boldsymbol{\Pi}\right)^{-1} \mathbf{X}^T \mathbf{W}). \qquad (7.33)$$

From the lemma, we have $df(0) = p + 1$. However, this number shrinks as we increase $\lambda$.

The lasso regression problem, on the other hand, has no closed-form solution. Instead, we use a numeric solver, typically involving the gradient

$$\frac{\partial L_1}{\partial \boldsymbol{\beta}} = -2\mathbf{X}^T \mathbf{W}(\mathbf{y} - \mathbf{X}\boldsymbol{\beta}) + \lambda \text{sign}(\pi(\boldsymbol{\beta})),$$

where $\text{sign}(\pi(\boldsymbol{\beta}))_j = \mathbb{I}[\beta_j > 0] - \mathbb{I}[\beta_j < 0]$, for $j = 1, \ldots, p$, and $\text{sign}(\pi(\boldsymbol{\beta}))_0 = 0$.

In order to quantify the effective degrees of freedom, we will rely on the following definitions.

**Definition 7.5.** *For a given coefficient vector $\beta$ in a linear regression problem, define the* active set $\mathcal{B}$ *as*

$$\mathcal{B} = \{j : \text{sign}(\beta_j) \neq 0\}. \qquad (7.34)$$

*If $\hat{\beta}(\lambda)$ is the solution to a regularized linear regression problem, as it depends on the parameter $\lambda$, we similarly define the active set in the obvious way: $\mathcal{B}_\lambda = \{j : \text{sign}(\beta_j(\lambda)) \neq 0\}$.*

*For a lasso regression problem, there is a discrete set of points*

$$\lambda_0 > \lambda_1 > \cdots > \lambda_k = 0,$$

*called the* transition points, *such that*

---

[1] The quantity $\mathbf{e}_0 \mathbf{e}_0^T$ is dyadic notation, where $\mathbf{e}_0 = \langle 1, 0, \ldots, 0 \rangle^T \in \mathbb{R}^{p+1}$ is a standard (column) unit vector.

- *For $\lambda > \lambda_0$, $\hat{\beta}_j(\lambda) = 0$ for $j > 0$,*
- *The active sets $\mathcal{B}_\lambda$ are constant on the interior intervals $(\lambda_m, \lambda_{m+1})$,*
- *As $\lambda$ decreases through $\lambda_m$, new features are added to the active set.*

*The transition points are ordered in terms of model complexity.*

These definition were proposed in Zou *et al.* [2007], along with the following proposition regarding the degrees of freedom for the lasso problem.

**Proposition 7.5.** *For a lasso regression problem, the number of effective degrees of freedom is*

$$df(\lambda) = \mathbb{E}\left[|\mathcal{B}_\lambda|\right],$$

*where the expectation is relative to $\mathbf{y} \sim \mathrm{N}(\beta, \mathbf{W}^{-1}\sigma^2)$, which determines the $\hat{\beta}(\lambda)$, and hence $\mathcal{B}_\lambda$.*

**Corollary 7.2.** *For a lasso regression problem, the estimate*

$$\hat{df}(\lambda) = |\mathcal{B}_\lambda| \tag{7.35}$$

*is an unbiased estimate of the effective degrees of freedom.*

## 7.2 Logistic Regression

We next consider a model for classification. We will primarily consider the case of binary classification, but will show how the model can be extended to a general classification problem at the end of the section.

### 7.2.1 The Model

Whereas linear regression is the cornerstone of regression problems, logistic regression is the cornerstone of classification problems. Logistic regression is also considered a *linear model*, as it is linear in the model parameters. It often results in a set of linear *decision boundaries* (more on this later), though this is not the case when applying the model on nonlinear features.

**Definition 7.6.** *A (binary) logistic regression model is a predictive model, for a binary target variable $Y \in \mathbb{B}$ over a p-dimensional feature vector $X$, of the form*

$$Y \sim \mathrm{Bern}(\mu) \tag{7.36}$$

$$\mathrm{logit}(\mu) = \beta_0 + \sum_{j=1}^{p} X_j \beta_j, \tag{7.37}$$

*where $\beta_0, \ldots, \beta_p$ are unknown coefficients and $\mathrm{logit}(\mu)$ is the logit function, defined as the* log-odds

$$\text{logit}(\mu) = \ln\left(\frac{\mu}{1-\mu}\right).\tag{7.38}$$

*(Recall that the ratio $\mu/(1-\mu)$ is referred to as the* odds ratio.*)*

Note that the inverse of the logit function is the *logistic function*; i.e., if $\text{logit}(\mu) = \eta$, then

$$\mu = \text{logit}^{-1}(\eta) = \frac{1}{1+e^{-\eta}} = \frac{e^{\eta}}{1+e^{\eta}}.\tag{7.39}$$

Note that the logit function is an invertible mapping $\text{logit} : (0,1) \to \mathbb{R}$.

*Note 7.6.* The form of the logistic regression problem, as defined in Definition 7.6, is the Bernoulli form. For the general case in dealing with proportions, please see Section 7.5, where we will discuss the binomial formulation of the problem.                                                                                  ▷

It can be shown that, under the modeling assumptions of Definition 7.1, the least-squares solution (Definition 7.3) for the linear regression model is equivalent to the maximum-likelihood solution (see Exercise 7.4). It is therefore natural to proceed by determining the value of the coefficients $\beta$ that maximizes the likelihood function for a given set of data. We do, however, require a slightly updated concept of likelihood for the given context.

**Definition 7.7.** *Let $(X_1, Y_1), \ldots, (X_n, Y_n)$ constitute a set of pairs of random variables, such that the random variables $Y_1, \ldots, Y_n$ are conditionally independent given the $X$s and $Y_i \sim F_{Y|X}(Y|X_i; \theta)$, for some parameter $\theta$. Then the* conditional likelihood *is defined as*

$$\mathcal{L}(\theta) = \prod_{i=1}^{n} f_{Y|X}(Y_i|X_i; \theta).$$

In the context of logistic regression problems, we will exclusively consider the conditional likelihood and, therefore, typically refer to it simply as the *likelihood*. Note that we do not require any independence assumptions for the explanatory (feature) variables.

**Proposition 7.6.** *Given a weighted data set $\{(x_{i1}, \ldots, x_{ip}, y_i, w_i)\}_{i=1}^{n}$, with binary target variable $Y$, the conditional likelihood function for the coefficients $\beta = \langle \beta_0, \ldots, \beta_p \rangle$ of a logistic regression model is given by*

$$\mathcal{L}(\beta) = \prod_{i=1}^{n} \mu^{w_i y_i} (1-\mu)^{w_i(1-y_i)}.\tag{7.40}$$

*The log-likelihood may therefore be expressed as*

$$\ell(\beta) = \sum_{i=1}^{n} w_i y_i \ln\left(\mu(\mathbf{x}_i; \beta)\right) + \sum_{i=1}^{n} w_i(1-y_i) \ln\left(1 - \mu(\mathbf{x}_i; \beta)\right),\tag{7.41}$$

*or, upon further simplification, as*

$$\ell(\boldsymbol{\beta}) = \sum_{i=1}^{n} w_i \left[ y_i \eta_i - \ln(1 + e^{\eta_i}) \right], \qquad (7.42)$$

*where $\eta_i = \beta_0 + \sum_{j=1}^{p} x_{ij}\beta_j = \mathbf{x}_i^T \boldsymbol{\beta} = \hat{\mathbf{e}}_{\mathbf{i}}^{\mathbf{T}} \cdot \mathbf{X} \cdot \boldsymbol{\beta}$, with matrix $\mathbf{X}$ and vector $\boldsymbol{\beta}$ defined as before (see text around Equation (7.13)).*

We will leave the proof as an exercise for the reader (see Exercise 7.5). Note, however, that for a positive-integer weight $w \in \mathbb{Z}_+$, the likelihood function is equivalent to simply having a repeated training instance $(x, y)$ with multiplicity $w$. Moreover, Equation (7.41) is equivalent to the negative log-loss.

## 7.2.2 Solution

As we saw in Chapter 5, the Newton–Raphson method is typically employed to solve maximum-likelihood problems. However, this method requires the score statistic and information matrix for a given set of data. Luckily, these are easily derived from Equation (7.42).

**Proposition 7.7.** *Given a weighted data set, as in Proposition 7.6, the score statistic and observed information are*

$$U_n(\boldsymbol{\beta}) = \sum_{i=1}^{n} w_i \left[ y_i - \mu(\mathbf{x}_i; \boldsymbol{\beta}) \right] \mathbf{x}_i^T, \qquad (7.43)$$

$$\mathcal{J}_n(\boldsymbol{\beta}) = \sum_{i=1}^{n} w_i \mu(\mathbf{x}_i; \boldsymbol{\beta}) \left[ 1 - \mu(\mathbf{x}_i; \boldsymbol{\beta}) \right] \mathbf{x}_i \mathbf{x}_i^T, \qquad (7.44)$$

*where $\mu(\mathbf{x}_i; \boldsymbol{\beta}) = \text{logit}^{-1}(\mathbf{x}_i^T \boldsymbol{\beta})$. Moreover, the Fisher information is equivalent to the observed information: $\mathcal{I}_n(\boldsymbol{\beta}) = \mathcal{J}_n(\boldsymbol{\beta})$.*

*Proof.* If we let $\eta_i = \mathbf{x}_i^T \boldsymbol{\beta}$, first observe that

$$\frac{\partial \eta_i}{\partial \boldsymbol{\beta}} = \mathbf{x}_i^T.$$

The results given in Equations (7.43) and (7.44) follow from simple calculus and the definitions

$$U_n(\boldsymbol{\beta}) = \frac{\partial \ell(\boldsymbol{\beta})}{\partial \boldsymbol{\beta}} \quad \text{and} \quad \mathcal{J}_n(\boldsymbol{\beta}) = -\frac{\partial^2 \ell(\boldsymbol{\beta})}{\partial \boldsymbol{\beta}^T \partial \boldsymbol{\beta}}.$$

Since the observed information does not depend on $Y_i$, it is its own expected value, so that $\mathcal{I}_n(\boldsymbol{\beta}) = \mathcal{J}_n(\boldsymbol{\beta})$. $\qquad \square$

The logistic regression model is, however, built into the scikit-learn package. Our next example shows how to train a logistic regression model using $k$-fold cross validation for a sample data set.

*Example 7.2.* Import the data set `sklearn.datasets.load_breast_cancer` that predicts a binary target variable (diagnosis for breast cancer) based on a 30-dimensional numeric feature set.

We can use the scikit-learn library to train this model, using 5-fold cross validation, as shown in Code Block 7.3. The average accuracy of the five test sets is 95.96%, with a standard deviation of 1.62%. The ROC curve is shown in Figure 7.3, with AUC of 98.68%.

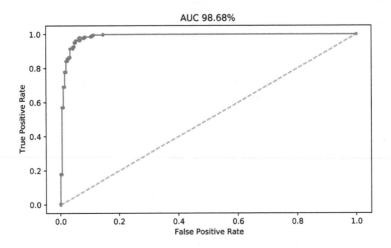

Fig. 7.3: ROC curve for logistic regression model for breast-cancer diagnostic. Red dot corresponds to the final cutoff threshold used in the `sklearn` model.

The calibration curve is plotted in Figure 7.4. Since the probability model is directly incorporated into the logistic regression model, the final probability predictions are well calibrated. (This is not necessarily the case with other classification algorithms, for which the calibration curve will diverge more dramatically from the diagonal.)

In addition, the precision and recall for the model can easily be computed to be 96.13% and 97.48%, as shown in lines 32 and 33.      ▷

Now that we've seen the overarching picture of training a logistic regression model–from model pipelining to AUC to cross validation—let's take a look under the hood and try to build our own logistic regression model from scratch, using the techniques of maximum likelihood.

```
 1  from sklearn.datasets import load_breast_cancer
 2  from sklearn.linear_model import LogisticRegression
 3  from sklearn.model_selection import KFold
 4  from sklearn.calibration import calibration_curve
 5  from sklearn.metrics import roc_curve, auc
 6
 7  data = load_breast_cancer()
 8  X, y = data.data, data.target
 9  transformers = [
10      ('standardize', RobustScaler(quantile_range=(0.2,0.8))),
11      ('model', LogisticRegression(max_iter=500))]
12
13  pipe = Pipeline(transformers)
14  kf = KFold(n_splits=5)
15  results, accuracy = [], []
16
17  for train_index, test_index in kf.split(X):
18      X_train, X_test = X[train_index], X[test_index]
19      y_train, y_test = y[train_index], y[test_index]
20      pipe.fit(X_train, y_train)
21      y_hat = pipe.predict(X_test)
22      p_hat = pipe.predict_proba(X_test)
23      results.append({'y_test': y_test, 'y_hat': y_hat, 'p_hat':
               p_hat[:, 1]})
24      accuracy.append(pipe.score(X_test, y_test))
25
26  y_total = np.concatenate([result['y_test'] for result in results])
27  y_hat   = np.concatenate([result['y_hat'] for result in results])
28  p_hat   = np.concatenate([result['p_hat'] for result in results])
29  p_true, p_pred = calibration_curve(y_total, p_hat, n_bins=5)
30  fpr, tpr, thresholds = roc_curve(y_total, p_hat)
31  print(auc(fpr, tpr), np.mean(accuracy)) # 98.68%, 95.96%
32  precis=np.sum((y_total==1)&(y_hat==1))/np.sum((y_hat==1)) # 96.13%
33  recall=np.sum((y_total==1)&(y_hat==1))/np.sum((y_total==1)) # 97.48%
```

Code Block 7.3: Logistic Regression Problem

*Example 7.3.* Solve the logistic regression problem from Example 7.2 directly using Newton's method and Equations (7.43) and (7.44).

We can create our own class to handle logistic regression problems using the method of scoring; see Code Block 7.4.

With the `fit`, `predict`, and `predict_proba` methods, this class is designed to fit into the same scikit-learn pipeline shown in Code Block 7.3. Precision and recall were 95.2% and 95.0%, respectively, with an overall accuracy of 93.8%. This is slightly worse than scikit-learn's built-in methods, though our method converges in fewer iterations.                    ▷

```
1   class NewtonLogisticRegression:
2       def __init__(self, max_iter=20, tol=1e-6, intercept=True):
3           self.max_iter = max_iter
4           self.tol = tol
5           self.intercept = intercept
6
7       def logistic(self, eta):
8           return 1 / (1 + np.exp(-eta))
9
10      def train(self, X, y):
11          n, m = X.shape
12          if self.intercept:
13              X = np.concatenate([np.ones(n).reshape((n,1)), X], axis=1)
14              m += 1
15          y = y.reshape((n, 1))
16          beta = np.zeros(m).reshape((m, 1))
17          beta[0] = 1
18          i = 0
19          err = 1
20          while (i < self.max_iter) and (err > self.tol):
21              UT = X.T @ (y - self.logistic(X@beta))
22              p_hat = self.logistic(X@beta)
23              FI = X.T @ np.diag( (p_hat * (1 - p_hat)).reshape(n)) @ X
24              err = np.linalg.norm(UT)
25              i += 1
26              beta += np.linalg.inv(FI) @ UT
27
28          self.converged = err < self.tol
29          self.beta = beta
30
31      def predict(self, X, threshold=0.5):
32          return (self.predict_proba(X) > threshold).astype(int)
33
34      def predict_proba(self, X):
35          if self.intercept:
36              X = np.concatenate([np.ones(len(X)).reshape((len(X),1)),
37                  X], axis=1)
37          return self.logistic(X @ self.beta)
```

Code Block 7.4: Newton's Method Solution to Logistic Regression Problem

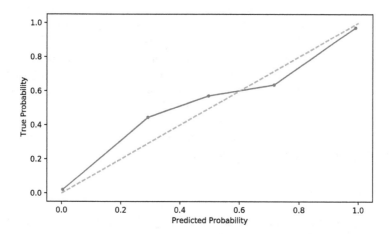

Fig. 7.4: Calibration plot for logistic regression model for breast-cancer diagnostic; 5 calibration bins and $n = 569$ data points. Orange dashed line represents perfect calibration.

### 7.2.3 Imbalanced Data Sets

The relative frequency of classes in the training instances of a classification problem can have a significant impact on the efficacy of the model, especially if one of the target classes is under or overrepresented. In the case of binary classification, we will, without loss of generality, think of the minority class as being the positive instances, as is common in most applications. (One typically tries to detect when an instance is going to be a *rare* event, not when it is going to be a typical event.) We will present some basic definitions and approaches for handling class imbalance here; for additional detail, see Kuhn and Johnson [2013].

**Definition 7.8.** *In a classification model, a* class imbalance *is said to exist whenever there is an unequal distribution among the various classes comprising the target variable. The severity of such an imbalance can vary from mild to extreme.*

*In a binary classification problem, we define the* class ratio *as the ratio of majority to minority classes. A* moderate class imbalance *is said to exist if the class ratio exceeds 3:1 (25% positive instances). A* severe class imbalance *is said to exist if the class ratio exceeds 19:1 (5% positive instances). An* extreme class imbalance *is said to exist if the class ratio exceeds 99:1 (1% positive instances).*

It is not uncommon in practice to encounter cases where it is necessary to model extreme class imbalances, exceeding 100:1. In the era of big data, one can even encounter extreme class imbalances on the order of 1,000:1 or even 10,000:1. For example, in online advertising, a classification problem could be constructed to predict who might click on an ad, where the overall probability of a click (i.e., the *click-through rate*) might be only 1%. Similarly, the probability of installing a mobile app (that is being advertised), given that a user clicks on the ad, could likewise be on the order of 100:1. Thus, the overall probability of installing an app, given that a user is served an advertising impression, is on the order of 10,000:1. Severe class imbalances can occur in a variety of fields, not just advertising, such as outlier detection (where it occurs almost by definition), spam detection, insurance claims, or even predicting when an insurance claim is fraudulent.

**Accuracy Paradox**

To see why class imbalance in classification problems can be so delicate, consider the case of a severe class imbalance of 200:1. Your current production model provides 99.5% accuracy, for which your business stakeholders are immeasurably pleased. Is it worth precious bandwidth to look into and improve this model? It depends. Quite possibly it is an absolute imperative.

If the model was selected for its accuracy, one needn't have looked far to find a highly accurate model: the model

$$\hat{Y} = 0,$$

which predicts *every* instance to be in the negative class, has an accuracy of 99.5%. The model is going to be right 99.5% of the time, because it is trying to detect something that is so rare that it only occurs in about 0.5% of all cases. Note, the precision of this model is undefined, as there are no positive predictions, and the recall is 0%; i.e., 0% of the actual positives were correctly classified. Similarly, the model has 0% sensitivity (which is the true-positive rate, or recall) and 100% specificity (which is the true-negative rate, or one minus the false positive rate). On an ROC curve, this model would be at the origin $(0, 0)$, quite far from the ideal model at $(0, 1)$.

A simple approach to this problem might be to tune the model to optimize for recall; i.e., tune the model to maximize the accuracy of the minority class. This approach, however, leads to the opposite problem: always predicting $\hat{Y} = 1$ leads to 100% recall, because it correctly predicts *all* positive instances. (The problem is that it all negative instances are also classified as positives.) In the remainder of this section, we will explore some other (better) approaches for handling class imbalance.

**ROC Curve and Threshold Tuning**

As discussed in the preceding paragraph, problems can easily arise when tuning classification models with imbalanced data sets to optimize for a single metric, such as accuracy or recall. This is why classification metrics often come in pairs: *precision and recall* or *sensitivity and specificity*. One common way to optimize for both simultaneously is to optimize for the area under the ROC curve; i.e., the AUC. Not only does optimizing for AUC incorporate both sensitivity and specificity into the optimization problem, but also it eliminates the particular classification threshold from model comparison. Recall from Equation (6.14) that a binary classification model outputs a probability-like function $f(X; \mathcal{D})$, that is then used in conjunction with a classification threshold $t$, so that an instance is classified according to

$$\hat{Y}_t(X; \mathcal{D}) = \mathbb{I}\left[f(X; \mathcal{D}) > t\right].$$

A single predictive model might perform dramatically different based on the classification threshold that is ultimately used in classification. The AUC method therefore compares different models *without regard for* their particular classification thresholds. A model with a superior AUC will be able to generate better precision and recall with an appropriate choice of threshold cutoff. A higher AUC means there is more area under the curve, which means the curve must pass more closely to the ideal model $(0, 1)$. This approach therefore entails two steps in model selection:

1. Optimize for AUC.
2. Once the optimal model is chosen, choose the classification threshold corresponding to the point on the ROC curve that minimizes the distance to the ideal model $(0, 1)$.

## Sampling Methods

Another approach is simply to adjust for the class imbalance in the training dataset. This typically entails either oversampling the underrepresented class or undersampling the overrepresented class[2]. Oversampling can also be achieved by adjusting the training weights in the algorithm, which is typically equivalent to duplicating training instances.

A more recent approach introduced by Chawla *et al.* [2002] is known as SMOTE, or *synthetic minority over-sampling technique*, which incorporates both upsampling and downsampling. Essentially, overrepresented classes are downsampled, and underrepresented classes are upsampled *synthetically*. The new synthetic instances are generated by using a random combination of features from a randomly selected data point from the minority class and its "nearest neighbors."

Regardless of the particular sampling method deployed, adjusting for class imbalance must be limited to the training set. It is crucial that the imbalance is preserved in the test and validation sets in order to accurately estimate the generalization error. Furthermore, the functional com-

---

[2] Dunn was under Oveur and I was under Dunn.

ponent $f(X; \mathcal{D})$ of the model should no longer be interpreted as a probability, as the training set was artificially altered to control for the relative frequency between the classes.

### Cost-sensitive Training

Instead of optimizing for a performance metric, such as AUC, we could instead optimize for a given loss function. *Cost-sensitive training* is a method in which different types of errors are given different weights in the loss function used for training a classification model. The decision on how to weight various types of errors depends on the particular business use case. For example, for a model used to classify whether images of a tumor are cancerous, the cost of a false negative is much greater than that of a false positive; whereas a false positive could lead to some undue emotional distress for the patient, a subsequent test can easily catch the error, whereas missing the detection of an actual positive could lead to dire consequences.

### 7.2.4 Multinomial Logistic Regression

Next, we show how the multinomial Bernoulli distribution can be used to extend logistic regression to the case of multiclass classification.

**Definition 7.9.** *A multiclass (or multinomial) logistic regression model is a predictive model, for a categorical target variable $Y \in \{0, \ldots, k-1\}$ over a $p$-dimensional feature vector $\mathbf{x}$, of the form*

$$Y \sim \text{MultiBern}(\boldsymbol{\pi}) \tag{7.45}$$
$$\text{alr}(\boldsymbol{\pi}) = \boldsymbol{\eta} = \mathbf{x}^T \mathbf{B}, \tag{7.46}$$

*where $\mathbf{B} \in \mathbb{R}^{(p+1) \times (k-1)}$ is a matrix of unknown coefficients, $\boldsymbol{\eta} \in \mathbb{R}^{k-1}$, and $\text{alr}(\boldsymbol{\pi})$ is the* additive logratio transform $\text{alr} : \Delta^{k-1} \to \mathbb{R}^{k-1}$ *defined by*

$$\text{alr}(\boldsymbol{\pi}) = \left\langle \ln\left(\frac{\pi_1}{\pi_0}\right), \ldots, \ln\left(\frac{\pi_{k-1}}{\pi_0}\right) \right\rangle. \tag{7.47}$$

*As usual, $\Delta^{k-1} = \{\boldsymbol{\pi} \in \mathbb{R}_*^k : \sum_{i=0}^{k-1} \pi_i = 1\}$ is the probability simplex in $\mathbb{R}^k$.*

*Note 7.7.* The choice of feature index used in the denominators for Equation (7.47) is an arbitrary choice, typically chosen as some reference point. For this reason, this transform is referred to as the *baseline-category logits* by Agresti [2019]. ▷

**Proposition 7.8.** *The inverse of the alr transform is given by*

$$\text{alr}^{-1}(\boldsymbol{\eta}) = \left(1 + \sum_{l=1}^{k-1} e^{\eta_l}\right)^{-1} \langle 1, e^{\eta_1}, \ldots, e^{\eta_{k-1}} \rangle. \tag{7.48}$$

*Proof.* Let $\text{alr}(\boldsymbol{\pi}) = \boldsymbol{\eta}$, for $\boldsymbol{\pi} \in \Delta^{k-1}$ and $\boldsymbol{\eta} \in \mathbb{R}^{k-1}$. From Equation (7.47), this implies that

$$\ln\left(\frac{\pi_i}{\pi_0}\right) = \eta_i,$$

for $i = 1, \ldots, k - 1$. It follows that $\pi_i = \pi_0 e^{\eta_i}$, again for $i = 1, \ldots, k - 1$. Moreover, from our normalization condition, we have

$$\sum_{i=0}^{k-1} \pi_i = \pi_0 + \sum_{i=1}^{k-1} \pi_i = \pi_0 \left(1 + \sum_{i=1}^{k-1} e^{\eta_i}\right) = 1,$$

which yields $\pi_0 = \left(1 + \sum_{i=1}^{k-1} e^{\eta_i}\right)^{-1}$. The result follows. $\qquad\square$

*Note 7.8.* The *central logratio transform* is defined as the mapping clr : $\Delta^{k-1} \to U \subset \mathbb{R}^k$ defined by

$$\text{clr}(\boldsymbol{\pi}) = \left\langle \ln\left(\frac{\pi_0}{g(\boldsymbol{\pi})}\right), \ldots, \ln\left(\frac{\pi_{k-1}}{g(\boldsymbol{\pi})}\right) \right\rangle, \qquad (7.49)$$

where $g(\boldsymbol{\pi})$ is the geometric mean:

$$g(\boldsymbol{\pi}) = \left(\prod_{i=0}^{k-1} \pi_i\right)^{1/k}.$$

It can be shown (see Exercise 7.6) that the inverse transform is the softmax:

$$\text{clr}^{-1}(\boldsymbol{\eta}) = \left\langle \frac{e^{\eta_0}}{\sum_{i=0}^{k-1} e^{\eta_i}}, \ldots, \frac{e^{\eta_{k-1}}}{\sum_{i=0}^{k-1} e^{\eta_i}} \right\rangle.$$

The reason the additive logratio transform is preferred, is that the model resulting from the central logratio transform is not identifiable, meaning the resulting parameter matrix $\mathbf{B} \in \mathbb{R}^{(p+1) \times k}$ would have a linearly dependent column. (Notice it would require $k$ columns, not $k-1$, as we had before.) To see this, note that the softmax transform is invariant under scalar addition; i.e., $\text{clr}^{-1}(\boldsymbol{\eta} + \alpha\mathbf{1}) = \text{clr}^{-1}(\boldsymbol{\eta})$. Therefore the probability vector corresponding to the coefficient matrix $\mathbf{B} = [\boldsymbol{\beta}_0, \ldots, \boldsymbol{\beta}_{k-1}]$ is invariant relative to vector translation of each of its columns; i.e., $\mathbf{B} = [\boldsymbol{\beta}_0 + \mathbf{c}, \ldots, \boldsymbol{\beta}_{k-1} + \mathbf{c}]$. By selecting $\mathbf{c} = -\boldsymbol{\beta}_0$, we recover the same inverse transform given by Equation (7.48). (Of course with a redefined set of $\beta$s.) The additive logratio transform is therefore a special case of the central logratio transform, and its inverse softmax transform, with the redundancy removed.

This has the effect of shifting $\eta_i \to \eta_i - \eta_0$, which further reduces Equation (7.49) to Equation (7.47) (with an extra zero component). $\qquad\triangleright$

**Proposition 7.9.** *Given a weighted data set $\{(\mathbf{x}_i, y_i, w_i)\}_{i=1}^{n}$, where $\mathbf{x}_i \in \mathbb{R}^p$, with categorical target variable $Y$, the conditional likelihood function for the coefficients $\mathbf{B} \in \mathbb{R}^{(p+1)\times(k-1)}$ of a logistic regression model is given by*

$$\mathcal{L}(\mathbf{B}) = \prod_{i=1}^{n} \prod_{j=0}^{k-1} \pi_j(\mathbf{x}_i; \mathbf{B})^{w_i y_{ij}}, \tag{7.50}$$

*where $y_{ij} = \mathbb{I}[y_i = j]$, for $i = 1, \ldots, n$ and $j = 0, \ldots, k-1$, and where*

$$\boldsymbol{\pi}(\mathbf{x}_i, \mathbf{B}) = \mathrm{alr}^{-1}(\mathbf{x}_i^T \mathbf{B}).$$

*The log-likelihood may therefore be expressed as*

$$\ell(\boldsymbol{\beta}) = \sum_{i=1}^{n} \sum_{j=0}^{k-1} w_i y_{ij} \ln \pi_j(\mathbf{x}_i; \mathbf{B}). \tag{7.51}$$

*or, upon further simplification, as*

$$\ell(\boldsymbol{\beta}) = \sum_{i=1}^{n} w_i \left[ y_{ij} \eta_{ij} - \ln\left(1 + \sum_{j=1}^{k-1} e^{\eta_{ij}}\right)\right], \tag{7.52}$$

*where $\eta_{ij} = \mathbf{x}_i^T \boldsymbol{\beta}_j$, such that $\boldsymbol{\beta}_j$ is the $j$th column of $\mathbf{B}$, for $j = 1, \ldots, k-1$, and where we define $\boldsymbol{\beta}_0 = \mathbf{0}$ for convenience.*

Compare Equation (7.52) with Equation (7.42). Note again that Equation (7.51) is just the negative of the log-loss, defined in Equation (6.20); minimizing log-loss is equivalent to maximizing the likelihood of the multinomial distribution.

As it turns out, scikit-learn's `LogisticRegression` class also handles multinomial logistic regression, following the same interface we saw in Example 7.4.

*Example 7.4.* The NIST handwritten digit data set consists of a set of 1,797 instances of handwritten digits (i.e., 0, 1, 2, 3, 4, 5, 6, 7, 8, 9) along with labels representing their true values. Each handwritten digit is represented as a 64-dimensional feature vector, with each component being an integer between 0 and 16. If the 64-dimensional feature vector is rearranged into an $8 \times 8$ square, the value represents the shading of the handwritten sample for that pixel. Some sample feature vectors, represented visually, from the data set are shown in Figure 7.5.

We see that the multinomial logistic regression problem can be trained with minor modification as compared with how we trained a binary logistic regression problem; see Code Block 7.5. Scikit-learn's logistic regression class achieves a 91.6% accuracy on the holdout sets from 5-fold cross validation.

```
1   from sklearn.datasets import load_digits
2   from sklearn.metrics import confusion_matrix
3   data = load_digits()
4   X, y = data.data, data.target
5
6   plt.figure(figsize=(8, 9/2)) # Plot sample feature vectors.
7   for i in range(30):
8       ax = plt.subplot(5,6,i+1, xticklabels=[], yticklabels=[],
            xticks=[], yticks=[])
9       ax.matshow(X[i,:].reshape(8,8))
10
11  pipe = LogisticRegression(multi_class='multinomial', max_iter=500)
12  kf = KFold(n_splits=5)
13  accuracy = []
14  results = []
15
16  for train_index, test_index in kf.split(X):
17      X_train, X_test = X[train_index], X[test_index]
18      y_train, y_test = y[train_index], y[test_index]
19      pipe.fit(X_train, y_train)
20      y_hat = pipe.predict(X_test)
21      results.append({'y_test': y_test, 'y_hat': y_hat})
22      accuracy.append(pipe.score(X_test, y_test))
23  np.mean(accuracy) # 91.598
24  y_total = np.concatenate([result['y_test'] for result in results])
25  y_hat  = np.concatenate([result['y_hat'] for result in results])
26  C = confusion_matrix(y_hat, y_total) # Predictions rows, actual
        columns
```

Code Block 7.5: Multinomial Logistic Regression Problem

|  |  | Actual Digit | | | | | | | | | |
|---|---|---|---|---|---|---|---|---|---|---|---|
|  |  | 0 | 1 | 2 | 3 | 4 | 5 | 6 | 7 | 8 | 9 |
|  | 0 | 174 | 0 | 0 | 0 | 0 | 0 | 0 | 3 | 0 | 1 |
|  | 1 | 0 | 160 | 7 | 0 | 2 | 2 | 2 | 1 | 11 | 4 |
| Predicted Digit | 2 | 1 | 0 | 170 | 1 | 0 | 1 | 0 | 0 | 2 | 0 |
|  | 3 | 0 | 1 | 0 | 160 | 0 | 2 | 0 | 1 | 1 | 3 |
|  | 4 | 1 | 1 | 0 | 0 | 169 | 1 | 1 | 3 | 0 | 0 |
|  | 5 | 1 | 0 | 0 | 4 | 0 | 165 | 3 | 0 | 5 | 1 |
|  | 6 | 1 | 5 | 0 | 0 | 6 | 1 | 173 | 0 | 0 | 0 |
|  | 7 | 0 | 0 | 0 | 2 | 1 | 1 | 0 | 161 | 0 | 2 |
|  | 8 | 0 | 7 | 0 | 12 | 0 | 1 | 2 | 1 | 153 | 8 |
|  | 9 | 0 | 8 | 0 | 4 | 3 | 8 | 0 | 9 | 2 | 161 |

Table 7.2: Confusion matrix for Example 7.4.

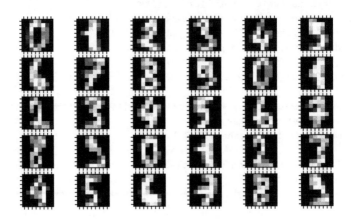

Fig. 7.5: Examples of NIST handwritten digit data set.

Our code also combines the result from each holdout set and computes the resulting confusion matrix, as shown in Table 7.2. Note that in order to get the predicted values on the rows, in agreement with Definition 6.11, one reverses the order of the inputs as per scikit-learn's documentation (which suggests entering `confusion_matrix(y_true, y_pred)`.)    ▷

Previously, we saw how the linear regression model can be broken up into two primary components:

$$Y_i \sim \mathrm{N}(\mu_i, \sigma^2/w_i) \qquad \text{random component}$$
$$g(\mu_i) = \mu_i = \mathbf{x}_i^T \boldsymbol{\beta} \qquad \text{systematic component.}$$

The logistic regression model falls neatly into a similar structure:

$$Y_i \sim \mathrm{Bern}(p_i) \qquad \text{random component}$$
$$g(p_i) = \mathrm{logit}(p_i) = \mathbf{x}_i^T \boldsymbol{\beta} \qquad \text{systematic component.}$$

It is natural to ask whether or not this structure can be generalized to a broader class of problems. And, if so, what types of distributions are conducive to such a generalization? The answer, of course, is that the structure is indeed generalizable for a specific class of distributions. We will begin by discussion the exponential family of distributions. We will go on to lay out the general theory of *generalized linear models*. Finally, we will discuss specific examples of our general theory that are suited to specific types of data: count data, proportions, positive-continuous data, and so forth. For additional references on generalized linear models, we recommend Agresti [2015], Dobson and Barnett [2018], Dunn and Smyth [2018], or Gill and Torres [2020], with Dunn and Smyth [2018] being our favorite.

# 7.3 Exponential Dispersion Models

Exponential dispersion models are simply a generalization of the exponential family of distributions, in which a parameter for the dispersion is called out explicitly, rather than being treated implicitly as a nuisance parameter.

## 7.3.1 Canonical Form

All distributions within the family of exponential dispersion models can be converted into a certain form known as the *canonical form*. This conversion entails a reparameterization of the model using what is known as the *canonical parameter*. We will see examples of this conversion later on; to begin, however, let us examine what these models look like in their canonical form.

**Definition 7.10.** *A distribution is said to belong to the* exponential dispersion model (EDM) *family of distributions if its* PDF *or* PMF *can be written (via a reparameterization) in the form*

$$f(y; \theta, \phi) = a(y, \phi) \exp \left[ \frac{y\theta - \kappa(\theta)}{\phi} \right], \qquad (7.53)$$

*for a random variable* $Y \sim$ EDM, *where*

- *the parameter $\theta$ is called the* canonical parameter,
- *the parameter $\phi$ is called the* dispersion parameter, *and*
- *the function $\kappa(\theta)$ is called the* cumulant function.

*When an* EDM *is written in the form of Equation (7.53), it is said to be in* canonical form.

### Cumulants of an EDM

Given an EDM in canonical form, it is easy to derive its cumulants; i.e., its mean, variance, and so forth. The result is given in the following.

**Theorem 7.7.** *The $r$th cumulant $\kappa_r$ of a random variable in the* EDM *family is determined by*

$$\kappa_r = \phi^{r-1} \frac{d^r \kappa(\theta)}{d\theta^r}. \qquad (7.54)$$

*In particular, the mean and variance (i.e., the first and second cumulants) are given by*

$$\mathbb{E}[Y] = \mu = \kappa'(\theta), \qquad (7.55)$$
$$\mathbb{V}(Y) = \sigma^2 = \phi\kappa''(\theta), \qquad (7.56)$$

*respectively. The function $\theta = (\kappa')^{-1}(\mu)$ is referred to as the* canonical link.

*Proof.* From the definition of the moment generating function(MGF), we have

$$M(t) = \mathbb{E}[e^{tY}] = \int a(y, \phi) \exp\left[\frac{y\theta - \kappa(\theta)}{\phi} + ty\right] dy.$$

Next, define $\psi = \theta + t\phi$, so that

$$M(t) = \int a(y, \phi) \exp\left[\frac{y\psi - \kappa(\theta) + \kappa(\psi) - \kappa(\psi)}{\phi}\right] dy$$

$$= \exp\left[\frac{\kappa(\psi) - \kappa(\theta)}{\phi}\right] \int a(y, \phi) \exp\left[\frac{y\psi - \kappa(\psi)}{\phi}\right] dy$$

$$= \exp\left[\frac{\kappa(\psi) - \kappa(\theta)}{\phi}\right]. \tag{7.57}$$

The last equality follows because the integral in an EDM in its own right (and therefore normalized).

Now, the *cumulant-generating function* $K(t)$ is defined by the relation

$$K(t) = \ln M(t),$$

and has the property that the $r$th cumulant is determined by its $r$th derivative:

$$\kappa_r = K^{(r)}(0).$$

It follows that, for an EDM, the cumulant-generating function is given by

$$K(t) = \frac{\kappa(\theta + t\phi) - \kappa(\theta)}{\phi}.$$

The result given in Equation (7.54) follows.                                    □

**Corollary 7.3.** *The mean $\mu$ of an EDM is a monotonically increasing (and therefore one-to-one) function of the canonical parameter $\theta$.*

*Proof.* Due to Equation (7.56), it follows that $\kappa''(\theta) > 0$, which means that $\mu = \kappa'(\theta)$ is an increasing function of $\theta$. The result follows.        □

**Corollary 7.4.** *A GLM is uniquely specified by its dispersion parameter $\phi$ and cumulant function $\kappa(\theta)$.*

*Proof.* A distribution is uniquely determined by knowing its cumulants, which are given in Equation (7.54), expressed in terms of $\phi$ and $\kappa(\theta)$. Therefore, $\phi$ and $\kappa(\theta)$ uniquely determine each cumulant of the distribution, hence the distribution itself. This proves the result.                               □

*Note 7.9.* As a result of Corollary 7.4, the coefficient function $a(y, \phi)$ must be uniquely specified based on the cumulant function and dispersion. For this reason, it is sometimes referred to as a *normalizing function*. This term is not consistent in the literature, however, as some authors refer to the cumulant function as the normalizing function.                            ▷

Our next result provides an interpretation for the *weight* of an instance from a GLM, as being synonymous of the average value of an IID sample from a fixed GLM.

**Theorem 7.8.** *Let $Y_1, \ldots, Y_w \sim \text{EDM}(\mu, \phi)$ represent an IID sample from a single EDM. Then the sample mean*

$$\overline{Y} = \frac{1}{w} \sum_{i=1}^{w} Y_i$$

*is also distributed as an EDM from the same family. Moreover,*

$$\overline{Y} \sim \text{EDM}(\mu, \phi/w). \tag{7.58}$$

*Proof.* Since the data $Y_1, \ldots, Y_w$ are IID, the MGF of their sample mean is given by

$$M_{\overline{Y}}(t) = \mathbb{E}\left[e^{t\overline{Y}}\right] = \mathbb{E}\left[e^{t(Y_1 + \cdots + Y_w)/w}\right] = [M_Y(t/w)]^w.$$

Combining this expression with Equation (7.57), we find that

$$M_{\overline{Y}}(t) = \exp\left[\frac{\kappa(\theta + t\phi/w) - \kappa(\theta)}{\phi/w}\right].$$

However, this is identical to the MGF for $Y$, except with $\phi$ replaced with $\phi/w$. We conclude that $\overline{Y} \sim \text{GLM}(\mu, \phi/w)$, which shows our result. □

**Definition 7.11.** *Given a cumulant function $\kappa(\theta)$ for an EDM, the* variance function *is defined as*

$$V(\mu) = \kappa''\left((\kappa')^{-1}(\mu)\right). \tag{7.59}$$

Equation (7.59) is simply the derivative $d\mu/d\theta$ expressed as a function of $\mu$. To see this, recall that $\mu = \kappa'(\theta)$, so that $\theta = (\kappa')^{-1}(\mu)$ (the canonical link). Moreover, $d\mu/d\theta = \kappa''(\theta)$. Combining these results yields Equation (7.59). The variance of a random variable $Y \sim \text{EDM}(\mu, \phi)$ may therefore be expressed in terms of its mean by the relation

$$\mathbb{V}(Y) = \phi V(\mu), \tag{7.60}$$

which is an equivalent expression of Equation (7.56). Moreover, it follows that when $\theta$ is expressed as a function of $\mu$, we have

$$\frac{d\theta}{d\mu} = \frac{1}{V(\mu)}.$$

This follows since $V(\mu)$ is simply $d\mu/d\theta$ expressed as a function of $\mu$.

## 7.3.2 Dispersion Form

Corollary 7.3 guarantees that the mean $\mu$ of an EDM has a one-to-one relationship to its canonical parameter $\theta$. We will often be interested in parameterizing an EDM with respect to its mean, as opposed to its canonical parameter. In order to facilitate this formalism, we first introduce an important function.

**Definition 7.12.** *The* unit deviance *of an EDM is the function*

$$d(y, \mu) = 2\left[t(y, y) - t(y, \mu)\right], \tag{7.61}$$

*where* $t(y, \mu)$ *is defined by the relation*

$$t(y, \mu) = y\theta(\mu) - \kappa(\theta(\mu)), \tag{7.62}$$

*where* $\theta(\mu) = (\kappa')^{-1}(\mu)$, *as before.*

Our next proposition reveals that the unit deviance can be thought of as a measure of distance between $y$ and $\mu$.

**Proposition 7.10.** *The unit deviance satisfies the inequality*

$$d(y, \mu) \geq 0,$$

*where there equality holds if and only if* $\mu = y$.

*Proof.* From the definition of $t(y, \mu)$, observe that, for fixed $y$,

$$\frac{\partial t(y, \mu)}{\partial \mu} = [y - \kappa'(\theta(\mu))]\frac{d\theta}{d\mu}.$$

However, since $\theta = (\kappa')^{-1}(\mu)$, and since $d\theta/d\mu = 1/V(\mu)$, we have

$$\frac{\partial t(y, \mu)}{\partial \mu} = \frac{y - \mu}{V(\mu)},$$

which vanishes only if $\mu = y$. Moreover, the second derivative is given by

$$\frac{\partial^2 t(y, \mu)}{\partial \mu^2} = -\frac{V(\mu) + (y - \mu)V'(\mu)}{V(\mu)^2}.$$

Now, since

$$\frac{\partial^2 t(y, y)}{\partial \mu^2} = -V(\mu)^{-1},$$

it follows that $t(y, \mu)$ must, for fixed $y$, have its global maximum at the point $\mu = y$. It therefore follows that the unit deviance, defined in Equation (7.61), is positive for $\mu \neq y$ and zero for $\mu = y$.    □

**Proposition 7.11.** *An EDM may be expressed in terms of* $\mu$ *as*

$$f(y; \mu, \phi) = b(y, \phi) \exp\left[\frac{-d(y, \mu)}{2\phi}\right], \tag{7.63}$$

*where* $b(y, \phi) = a(y, \phi) \exp[t(y, y)/\phi]$. *This is called the* dispersion form *of the EDM.*

### 7.3.3 A Catalogue of EDMs

We next consider a number of common EDMs. A summary is provided in Table 7.3.

### Normal Distribution

From the definition of the normal distribution, we have

$$
\begin{aligned}
f(y; \mu, \sigma^2) &= \frac{1}{\sqrt{2\pi\sigma^2}} \exp\left[\frac{-(y-\mu)^2}{2\sigma^2}\right] \\
&= \frac{\exp[-y^2/(2\sigma^2)]}{\sqrt{2\pi\sigma^2}} \exp\left[\frac{y\mu - \mu^2/2}{\sigma^2}\right].
\end{aligned}
$$

We thus identify the canonical parameter $\theta = \mu$, the cumulant function $\kappa(\theta) = \theta^2/2$, and the dispersion parameter $\phi = \sigma^2$. Since $\kappa''(\theta) = 1$, it follows that $V(\mu) = 1$. Moreover, the function $a(y, \theta)$ is given by the coefficient

$$
a(y, \phi) = \frac{\exp[-y^2/(2\phi)]}{\sqrt{2\pi\phi}}.
$$

Finally, the unit deviance is obtained by computing

$$
d(y, \mu) = 2\left[(y^2 - y^2/2) - (y\mu - \mu^2/2)\right] = (y-\mu)^2.
$$

### Bernoulli Distribution

From the definition of the Bernoulli distribution, we have

$$
f(y; \mu) = \mu^y (1-\mu)^{1-y} = \exp\left[y \ln\left(\frac{\mu}{1-\mu}\right) + \ln(1-\mu)\right].
$$

From this, it is apparent that $a(y, \phi) = 1$ and $\phi = 1$. Furthermore, the canonical parameter is given by

$$
\theta = \ln\left(\frac{\mu}{1-\mu}\right),
$$

which we recognize as the logit function, which we may invert to obtain

$$
\mu = \frac{e^\theta}{1 + e^\theta}.
$$

The cumulant function is therefore obtained by replacing this expression for $\mu$ in $-\ln(1-\mu)$:

$$
\kappa(\theta) = \ln(1 + e^\theta).
$$

Since $d\mu/d\theta = \mu(1-\mu)$ (since $\mu(\theta)$ is the logistic function), it follows that the variance function is given by

| EDM | $d(y,\mu)$ | $\kappa(\theta)$ | $\phi$ | $\theta = \theta(\mu)$ | $V(\mu)$ | $\mathcal{Y}$ | $\mathrm{dom}(\mu)$ | $\mathrm{dom}(\theta)$ |
|---|---|---|---|---|---|---|---|---|
| Normal | $(y-\mu)^2$ | $\theta^2/2$ | $\sigma^2$ | $\mu$ | $1$ | $\mathbb{R}$ | $\mathbb{R}$ | $\mathbb{R}$ |
| Bernoulli | $2\left[y\ln\left(\dfrac{y}{\mu}\right)+(1-y)\ln\left(\dfrac{1-y}{1-\mu}\right)\right]$ | $\ln(1+e^\theta)$ | $1$ | $\ln\left(\dfrac{\mu}{1-\mu}\right)$ | $\mu(1-\mu)$ | $\mathbb{B}$ | $(0,1)$ | $\mathbb{R}$ |
| Poisson | $2\left[y\ln\left(\dfrac{y}{\mu}\right)-(y-\mu)\right]$ | $\exp(\theta)$ | $1$ | $\ln(\mu)$ | $\mu$ | $\mathbb{N}$ | $\mathbb{R}_+$ | $\mathbb{R}$ |
| Negative Binomial | $2\left[y\ln\left(\dfrac{y}{\mu}\right)-(y+k)\ln\left(\dfrac{y+k}{\mu+k}\right)\right]$ | $-k\ln(1-e^\theta)$ | $1$ | $\ln\left(\dfrac{\mu}{\mu+k}\right)$ | $\mu+\mu^2/k$ | $\mathbb{N}$ | $\mathbb{R}_+$ | $\mathbb{R}_-$ |
| Gamma | $2\left[-\ln\left(\dfrac{y}{\mu}\right)+\dfrac{y-\mu}{\mu}\right]$ | $-\ln(-\theta)$ | $\phi$ | $-1/\mu$ | $\mu^2$ | $\mathbb{R}_+$ | $\mathbb{R}_+$ | $\mathbb{R}_-$ |
| Inverse Gaussian | $\dfrac{(y-\mu)^2}{\mu^2 y}$ | $-\sqrt{-2\theta}$ | $\phi$ | $-1/(2\mu^2)$ | $\mu^3$ | $\mathbb{R}_+$ | $\mathbb{R}_+$ | $\mathbb{R}_-$ |

Table 7.3: Common EDMs.

$$V(\mu) = \mu(1 - \mu).$$

Finally, the unit deviance works out to be

$$d(y, \mu) = 2\left[y \ln \frac{y}{\mu} + (1 - y) \ln \frac{1 - y}{1 - \mu}\right].$$

## Poisson Distribution

We may express the Poisson distribution as

$$f(y; \mu) = \frac{e^{-\mu}\mu^y}{y!} = \frac{1}{y!} \exp\left[y \ln \mu - \mu\right].$$

We thus identify

$$\theta = \ln \mu$$
$$\phi = 1$$
$$a(y, \phi) = \frac{1}{y!}$$
$$\kappa(\theta) = e^{\theta}$$
$$V(\mu) = \mu.$$

Finally, the unit deviance is easily shown to yield

$$d(y, \mu) = 2\left[y \ln \frac{y}{\mu} + (\mu - y)\right].$$

## Gamma Distribution

The gamma distribution is given by

$$f(y; \alpha, \beta) = \frac{1}{\Gamma(\alpha)\beta^{\alpha}} y^{\alpha-1} e^{-y/\beta},$$

with $\mathbb{E}[Y] = \alpha\beta$ and $\mathbb{V}(Y) = \alpha\beta^2$. By substituting $\alpha = 1/\phi$ and $\beta = \mu\phi$, we can reparametrize this distribution as

$$f(y; \mu, \phi) = \frac{\phi^{-1/\phi} y^{1/\phi-1}}{\Gamma(1/\phi)} \exp\left[\frac{-y/\mu - \ln(\mu)}{\phi}\right].$$

From this, we find

$$\theta = -1/\mu$$
$$\kappa(\theta) = \ln(-1/\theta)$$
$$V(\mu) = \mu^2$$
$$d(y, \mu) = 2\left[\frac{y}{\mu} - 1 + \ln \frac{\mu}{y}\right].$$

## 7.4 Generalized Linear Models

Now that we have defined and explored the class of distributions known as exponential dispersion models, we are ready for our main course. In this section, we introduce the main definition of generalized linear models and then proceed to lay the theory on how such models can be analyzed and applied.

### 7.4.1 Generalized Linear Models

We would like to develop a model in which the target variable $Y$ is distributed relative to a member of the EDM family, such that each observation is allowed to be taken from a distribution with a different mean and dispersion:

$$Y_i \sim \text{EDM}(\mu_i, \phi/w_i).$$

As stated, if there are $n$ observations, there would be $n$ unknown parameters $\mu_i$ to solve for, which would make the inference task ill posed and prone to overfitting. To remedy this, we require some additional structure on *how* the parameter $\mu$ is allowed to vary, based on a lower-dimensional feature vector $\mathbf{x} \in \mathbb{R}^p$, in analogy to the structure we established for linear and logistic regression models. This is achieved with the following.

**Definition 7.13.** *A* generalized linear model (GLM) *is a predictive model for a random variable $Y$ consisting of three components:*

1. *A* random component*: each observed target variable is conditionally distributed relative to the same member of the EDM family:*

$$Y_i|\mathbf{x}_i \sim \text{EDM}(\mu_i, \phi/w_i), \qquad (7.64)$$

   *where $w_i$ is the (known) weight of the ith observation, and $\mu_i$ varies per observation, based on the feature vector $\mathbf{x}_i$;*
2. *A* linear predictor*: a scalar parameter $\eta_i$ is constructed using the ith feature vector $\mathbf{x}_i$ and an (unknown) vector of coefficients $\boldsymbol{\beta}$:*

$$\eta_i = \mathbf{x}_i^T \boldsymbol{\beta}; \qquad (7.65)$$

   *moreover, when $x_{i0} = 1$, then $\beta_0$ is called the* intercept *of the model; and*
3. *A* link function*: a function $g$ that links the expected mean of the EDM to the linear predictor:*

$$g(\mu_i) = \eta_i. \qquad (7.66)$$

*The particular link function $g = (\kappa')^{-1}$, where $\kappa$ is the cumulant function of the EDM, is called the* canonical link function.

*Note 7.10.* When using the canonical link function, the linear predictor coincides $\eta$ with the canonical parameter $\theta$:

$$\eta = g(\mu) = (\kappa')^{-1}(\mu) = \theta,$$

which follows from $\mu = \kappa'(\theta)$, as stated in Equation (7.55).     ▷

*Example 7.5.* The linear regression model is a GLM using the normal distribution and the identity link $\mu = \eta$; i.e.,

$$Y_i \sim \mathrm{N}(\mu_i, \sigma^2/w_i)$$
$$\mu_i = \mathbf{x}_i^T \boldsymbol{\beta}.$$

Note that the link function is the identity: $g(\mu) = \mu$. This is also the canonical link function for the normal distribution, so that $\mu$ and $\eta$ each coincide with the canonical parameter: $\mu = \eta = \theta$.     ▷

*Example 7.6.* The logistic regression model is a GLM using the Bernoulli distribution and the logit link function; i.e.,

$$Y_i \sim \mathrm{Bern}(\mu_i)$$
$$\mathrm{logit}(\mu_i) = \mathbf{x}_i^T \boldsymbol{\beta}.$$

Note that the link function is the logit function: $g(\mu) = \mathrm{logit}(\mu)$. Moreover, this represents the canonical link function, so that $\eta$ is the canonical parameter of the Bernoulli GLM.     ▷

## Saturated Models and Total Deviance

A useful benchmark for comparison is obtained by breaking the link function and overfitting a model to the data.

**Definition 7.14.** *Given a set of data* $(\{\mathbf{x}_i, y_i, w_i\})_{i=1}^n$ *and a dispersion parameter* $\phi$, *we define the* saturated model, *relative to a particular member of the* EDM *family, as the model*

$$\hat{Y}_i \sim \mathrm{EDM}(y_i, \phi/w_i);$$

*i.e., the saturated model is the model obtained by overfitting the parameters* $\mu_i$ *with* $\mu_i = y_i$.

Now, for a given set of data $(\{\mathbf{x}_i, y_i, w_i\})_{i=1}^n$, the likehood of a GLM relative to the canonical parameter $\theta$ is given by

$$L_\theta(\boldsymbol{\theta}; \mathbf{y}) = \prod_{i=1}^n f(y_i; \theta_i, \phi/w_i)$$

$$= \exp\left[\frac{1}{\phi}\left(\sum_{i=1}^n w_i y_i \theta_i - \sum_{i=1}^n w_i \kappa(\theta_i)\right)\right] \prod_{i=1}^n a(y_i, \phi/w_i).$$

The subscript $L_\theta$ reminds us that we are expressing the likelihood relative to the canonical parameters $\theta$. The log likelihood is therefore given by

$$\ell_\theta(\boldsymbol{\theta}; \mathbf{y}) = \frac{1}{\phi} \sum_{i=1}^{n} w_i \left[ y_i \theta_i - \kappa(\theta_i) \right] + \sum_{i=1}^{n} \ln a(y_i, \phi/w_i). \tag{7.67}$$

In order to write the log likehood in terms of the parameters $\mu_i$, we can use Equation (7.62), thereby obtaining

$$\ell_\mu(\boldsymbol{\mu}; \mathbf{y}) = \frac{1}{\phi} \sum_{i=1}^{n} w_i t(y_i, \mu_i) + \sum_{i=1}^{n} \ln a(y_i, \phi/w_i). \tag{7.68}$$

This motivates the following.

**Definition 7.15.** *Given a set of data* $(\{\mathbf{x}_i, y_i, w_i\})_{i=1}^{n}$ *and a member of the* EDM *family, the* total deviance *is the function*

$$D(\mathbf{y}, \boldsymbol{\mu}) = \sum_{i=1}^{n} w_i d(y_i, \mu_i), \tag{7.69}$$

*where the function d is the unit deviance, defined in Equation (7.61).*

**Proposition 7.12.** *The total deviance is equivalent to the log likelihood ratio between a given model and its saturated counterpart:*

$$\frac{1}{\phi} D(\mathbf{y}, \boldsymbol{\mu}) = -2 \ln \frac{L_\mu(\boldsymbol{\mu}, \mathbf{y})}{L_\mu(\mathbf{y}, \mathbf{y})} = -2 \left[ \ell_\mu(\boldsymbol{\mu}, \mathbf{y}) - \ell_\mu(\mathbf{y}, \mathbf{y}) \right].$$

*The term on the left is called the* scaled (total) deviance.

   This result follows directly from Equations (7.68) and (7.69). As a result, the total scaled deviance is equivalent to the likelihood ratio test statistic of a given model as compared with its saturated counterpart. As we discussed earlier, $D(\mathbf{y}, \boldsymbol{\mu}) \geq 0$, with larger values representing poorer fits. In some cases, the scaled deviance has an approximate chi-squared distribution, with degrees of freedom equal to the difference of the number of parameters used to fit the two models; i.e., $df = n - p - 1$.

### 7.4.2 Maximum Likelihood Redux

We next consider the special form of some key results from maximum-likelihood theory as applied to generalized linear models. We begin with the score statistics.

**Theorem 7.9.** *Given a set of data $(\{\mathbf{x}_i, y_i, w_i\})_{i=1}^n$ and a* GLM, *the score statistics (Definition 5.3) are given by*

$$U_j = \frac{\partial \ell}{\partial \beta_j} = \frac{1}{\phi} \sum_{i=1}^n \frac{w_i(y_i - \mu_i)x_{ij}}{V(\mu_i)g'(\mu_i)}, \tag{7.70}$$

*for $j = 0, \ldots, p$. Moreover, when the canonical link function is used, these equations reduce to*

$$U_j = \frac{\partial \ell}{\partial \beta_j} = \frac{1}{\phi} \sum_{i=1}^n w_i(y_i - \mu_i)x_{ij}. \tag{7.71}$$

*Here, $\ell = \ell(\boldsymbol{\beta})$ is the log-likelihood function with respect to the unknown modeling parameters $\boldsymbol{\beta}$.*

*Proof.* The log-likelihood function is given by expressing Equation (7.67) in terms of the unknown modeling parameters $\boldsymbol{\beta}$ via the chain

$$\ell(\boldsymbol{\beta}) = \ell_\theta(\theta(\mu(\eta(\boldsymbol{\beta})))).$$

Note that the second term in Equation (7.67) is a function of $\mathbf{y}$ and $\phi$ only, and therefore vanishes when differentiating with respect to the parameters $\boldsymbol{\beta}$.

To proceed, we invoke the chain rule:

$$\frac{\partial \ell}{\partial \beta_j} = \sum_{i=1}^n \frac{\partial \ell}{\partial \theta_i} \frac{\partial \theta_i}{\partial \mu_i} \frac{\partial \mu_i}{\partial \eta_i} \frac{\partial \eta_i}{\partial \beta_j}.$$

Now, from Equation (7.67), the first factor is given by

$$\frac{\partial \ell}{\partial \theta_i} = \frac{w_i(y_i - \kappa'(\theta_i))}{\phi} = \frac{w_i(y_i - \mu_i)}{\phi}.$$

For the second factor, recall from the definition of the variance function that

$$\frac{d\theta}{d\mu} = \frac{1}{V(\mu)}.$$

For the penultimate factor, recall from the definition of the link function that $\mu = g^{-1}(\eta)$, so that

$$\frac{d\mu}{d\eta} = (g^{-1})'(\eta) = \frac{1}{g'(g^{-1}(\eta))} = \frac{1}{g'(\mu)},$$

with the second equality following from elementary calculus. Finally, due to the definition of the linear predictor (Equation (7.65)), it follows that

$$\frac{\partial \eta_i}{\partial \beta_j} = x_{ij},$$

for $i = 1, \ldots, n$ and $j = 0, \ldots, p$. Combining the preceding equations yields the first result.

For the second result, note that when the canonical link function is used to define the GLM, we have $\theta = g(\mu)$, which implies that

$$\frac{d\theta}{d\mu} = g'(\mu).$$

It therefore follows that, for the case of the canonical link function, $g'(\mu)V(\mu) = 1$, thereby yielding the simplification of Equation (7.71).    □

**Theorem 7.10.** *Given a set of data $(\{x_i, y_i, w_i\})_{i=1}^n$ and a GLM, the Fisher information (Definition 5.4) is given by*

$$\mathcal{I}_{jk} = \mathbb{E}\left[-\frac{\partial^2 \ell}{\partial \beta_k \partial \beta_j}\right] = \frac{1}{\phi} \sum_{i=1}^n \frac{w_i x_{ij} x_{ik}}{V(\mu_i) g'(\mu_i)^2}. \tag{7.72}$$

*When $g(\mu)$ is the canonical link, this expression reduces to*

$$\mathcal{I}_{jk} = \frac{1}{\phi} \sum_{i=1}^n \frac{w_i x_{ij} x_{ik}}{g'(\mu_i)}. \tag{7.73}$$

*Proof.* From Equation (7.70), we have

$$\frac{\partial^2 \ell}{\partial \beta_k \partial \beta_j} = \frac{1}{\phi} \sum_{i=1}^n (y_i - \mu_i) \frac{\partial}{\partial \beta_k} \left[\frac{w_i x_{ij}}{V(\mu_i) g'(\mu_i)}\right] - \frac{1}{\phi} \sum_{i=1}^n \frac{w_i x_{ij}}{V(\mu_i) g'(\mu_i)} \frac{\partial \mu_i}{\partial \eta_i} \frac{\partial \eta_i}{\partial \beta_k}.$$

The expectation in Equation (7.72) is the conditional expectation of $Y_i$ given $x_i$, which is simply $\mathbb{E}[Y_i | x_i] = \mu_i$. We therefore find that

$$\mathbb{E}\left[-\frac{\partial^2 \ell}{\partial \beta_k \partial \beta_j}\right] = \frac{1}{\phi} \sum_{i=1}^n \frac{w_i x_{ij}}{V(\mu_i) g'(\mu_i)} \frac{\partial \mu_i}{\partial \eta_i} \frac{\partial \eta_i}{\partial \beta_k},$$

which simplifies to Equation (7.72) following the same steps from Theorem 7.9. Equation (7.73) again follows from the relation $V(\mu_i)g'(\mu_i) = 1$, when $g(\mu)$ is the canonical link function. This completes the result.    □

## Matrix Formalism

We may economically represent the score statistic, which is really a gradient vector, and Fisher information using matrix notation as follows.

**Proposition 7.13.** *Let $\mathbf{V}$ and $\mathbf{L}$ represent the $n \times n$ diagonal matrices with diagonal elements*

$$V_{ii} = \frac{V(\mu_i)}{w_i} \qquad and \qquad L_{ii} = g'(\mu_i),$$

*respectively. Then the score statistic and Fisher information are equivalent to the following expressions*

$$\mathbf{u} = \frac{1}{\phi}\mathbf{X}^T\mathbf{V}^{-1}\mathbf{L}^{-1}(\mathbf{y} - \boldsymbol{\mu}) \tag{7.74}$$

$$\boldsymbol{\mathcal{I}} = \frac{1}{\phi}\mathbf{X}^T\mathbf{V}^{-1}\mathbf{L}^{-2}\mathbf{X}, \tag{7.75}$$

*respectively, where* $\mathbf{X}$ *is the design matrix. For the canonical link, we further obtain the simplification* $\mathbf{VL} = \mathbf{I}$.

*The diagonal matrix* $\mathbf{W} = \mathbf{V}^{-1}\mathbf{L}^{-2}$ *is sometimes referred to as the matrix of working weights.*

*Proof.* It's obvious.    □

## Method of Scoring

As we saw in Section 5.1, the maximum likelihood solution is the solution to the score equation

$$U_j = 0.$$

Moreover, the Newton–Raphson method, which requires a calculation of the *observed information*, is typically replaced with the method of scoring, which uses the Fisher information. As we saw in Equation (5.15), the method of scoring is an iterative algorithm that updates the current "guess" for the parameters $\boldsymbol{\beta}$ with

$$\boldsymbol{\beta}^{k+1} = \boldsymbol{\beta}^k + \boldsymbol{\mathcal{I}}(\boldsymbol{\beta}^k)^{-1} \cdot \mathbf{u}(\boldsymbol{\beta}^k). \tag{7.76}$$

Of course, the score statistic and Fisher information depend on the unknown parameters explicitly through the parameters $\boldsymbol{\mu}$ and the link function.

## Estimating Dispersion

Oftentimes the dispersion parameter $\phi$ is not known, but must be estimated. As we saw in the case of linear regression, the MLE for $\phi$ tends to be a biased estimate, hence it is seldom used in practice. Another approach is to estimate $\phi$ using the *mean deviance estimator*

$$\hat{\phi} = \frac{D(\mathbf{y}, \hat{\boldsymbol{\mu}})}{n - (p+1)},$$

where $p+1$ is the total number of parameters. This estimate, however, only behaves well under restricted circumstances: the GLM must be normal or inverse Gaussian, or the dispersion must be small. Details of the pros and cons of these estimates are discussed in Dunn and Smyth [2018].

In practice, the go-to estimate for $\phi$, due to its ease-of-use and robustness across use cases, is based on the following.

**Definition 7.16.** *Given a set of data* $(\{\mathbf{x}_i, y_i, w_i\})_{i=1}^n$, *let* $\hat{\boldsymbol{\beta}}$ *represent the* MLE *for a* GLM, *and let* $\hat{\eta}_i = \mathbf{x}_i^T \hat{\boldsymbol{\beta}}$ *and* $\hat{\mu}_i = g(\hat{\eta}_i)$. *Then the* Pearson *statistic is defined as the quantity*

$$X^2 = \sum_{i=1}^n \frac{w_i(y_i - \hat{\mu}_i)^2}{V(\hat{\mu}_i)}. \tag{7.77}$$

*Moreover, the* Pearson *estimator for* $\phi$ *is defined by*

$$\hat{\phi} = \frac{X^2}{n - (p+1)}, \tag{7.78}$$

*where* $p + 1$ *is the total number of unknown parameters; i.e.,* $\boldsymbol{\beta} \in \mathbb{R}^{p+1}$.

*Note 7.11.* The idea behind the Pearson statistic is to construct a quantity similar to the SSE, defined in Equation (6.2), using the working weights from the matrix $\mathbf{W}$ and the differential response in the $\eta$-space, to obtain

$$\text{SSE} = \sum_{i=1}^n W_i(z_i - \hat{\eta}_i)^2,$$

where

$$W_i = \frac{w_i}{V(\mu_i)g'(\mu_i)^2}$$

is the working weight and $z_i = g(y_i)$. Next, when $z_i \approx \hat{\eta}_i$, we can use the approximation

$$g'(\mu_i) \approx \frac{\Delta \eta_i}{\Delta \mu_i} = \frac{z_i - \hat{\eta}_i}{y_i - \hat{\mu}_i}$$

to obtain the result.                                                                          ▷

### 7.4.3 Inference

We next consider the sampling distribution for our estimates. From this we will construct approximate confidence intervals.

### Asymptotic Distribution of Maximum-likelihood Estimates

As we saw in Chapter 5, the maximum-likelihood estimates of our parameter set are approximately normally distributed for large $n$. In particular, we have the following.

**Proposition 7.14.** *Given a set of data* $(\{\mathbf{x}_i, y_i, w_i\})_{i=1}^n$, *let* $\hat{\boldsymbol{\beta}}$ *represent the* MLE *for a* GLM. *For large* $n$, *the* MLE *is approximately normally distributed*

$$\hat{\boldsymbol{\beta}} \approx \mathrm{N}(\boldsymbol{\beta}_0, \boldsymbol{\mathcal{I}}(\boldsymbol{\beta}_0)^{-1}), \tag{7.79}$$

*where* $\boldsymbol{\beta}_0$ *is the true value of the parameter* $\boldsymbol{\beta}$.

*Proof.* This result is a generalization of Corollary 5.2.    □

**Corollary 7.5.** *Given a set of data* $(\{\mathbf{x}_i, y_i, w_i\})_{i=1}^n$, *let* $\hat{\boldsymbol{\beta}}$ *represent the* MLE *for a* GLM. *For large* $n$, *the variance of this estimate is approximately given by*

$$\mathbb{V}(\hat{\boldsymbol{\beta}}) = \boldsymbol{\mathcal{I}}^{-1}. \tag{7.80}$$

*Moreover, the standard error of any individual* $\hat{\beta}_j$ *is given by*

$$\mathrm{se}(\hat{\beta}_j) = \sqrt{\phi\,(\mathbf{X}^T\mathbf{W}\mathbf{X})_{jj}^{-1}}, \tag{7.81}$$

*where* $\mathbf{W}$ *is the working weight matrix defined in Proposition 7.13.*

*Proof.* This follows directly form Proposition 7.14.    □

**Corollary 7.6.** *Given* $\mathbf{x} \in \mathbb{R}^{p+1}$ *and the* MLE $\hat{\boldsymbol{\beta}}$, *construct* $\eta = \mathbf{x}^T\hat{\boldsymbol{\beta}}$, *as usual. The variance of this estimate is given by*

$$\mathbb{V}(\hat{\eta}) = \mathbf{x}^T\boldsymbol{\mathcal{I}}^{-1}\mathbf{x}. \tag{7.82}$$

*Proof.* This follows from Equation (7.80) and properties of variance:

$$\mathbb{V}(\mathbf{x}^T\hat{\boldsymbol{\beta}}) = \mathbf{x}^T\mathbb{V}(\boldsymbol{\beta})\mathbf{x}.$$

The result follows.    □

### Tests for Individual Estimates

Given that the MLE $\hat{\boldsymbol{\beta}}$ is an asymptotically normal and unbiased estimate of the true parameter values, we may use the Wald test (Definition 3.11) to check the null hypothesis $H_0 : \beta_j = \beta_j^0$. Often, one selects $\beta_j^0$, in order to test whether or not the $j$th coefficient is nonzero, in which case the model can be simplified. When the dispersion parameter $\phi$ is known, the Wald statistic is given by Equation (3.8) as

$$Z = \frac{\hat{\beta}_j - \beta_j^0}{\mathrm{se}(\hat{\beta}_j)}. \tag{7.83}$$

Theorem 3.1 guarantees that the Wald statistic $Z$ is approximately a standard normal distribution for large $n$. We may therefore construct a $(1 - \alpha)$ confidence interval for $\hat{\beta}_j$, or for $\hat{\eta}$, by the relations

$$\hat{\beta}_j \pm z_{\alpha/2}\mathrm{se}(\hat{\beta}_j)$$
$$\hat{\eta} \pm z_{\alpha/2}\mathrm{se}(\hat{\eta}), \tag{7.84}$$

respectively, where the standard errors are given in Equations (7.81) and (7.82). A confidence interval for $\hat{\mu}$ is then obtained by applying the mapping

$g^{-1}$ to the upper and lower bounds of the confidence interval for $\hat{\eta}$. Note that the confidence interval for $\hat{\eta}$ is symmetric, whereas the confidence interval for $\hat{\mu}$ is typically asymmetric, due to the nonlinearity of the link function.

When the dispersion parameter $\phi$ is not known, we must estimate the standard error using a suitable estimate $\hat{\phi}$ of the dispersion; typically the Pearson estimator. For a consistent estimator and a very large sample size, $\hat{\phi} \approx \phi$, and the normal approximation is still often suitable. For small or moderate sample sizes, however, we instead formulate the $T$-statistic (Proposition 3.3) as

$$T = \frac{\hat{\beta}_j - \beta_j^0}{\hat{se}(\hat{\beta}_j)}, \tag{7.85}$$

which is distributed as a $t_{n-p-1}$ distribution. The corresponding $(1 - \alpha)$ confidence intervals are therefore obtained from the relations

$$\hat{\beta}_j \pm t_{n-p-1,\alpha/2}\hat{se}(\hat{\beta}_j)$$
$$\hat{\eta} \pm t_{n-p-1,\alpha/2}\hat{se}(\hat{\eta}). \tag{7.86}$$

The corresponding confidence region for $\hat{\mu}$ can again be constructed by applying the function $g^{-1}$ to the upper and lower bounds of the confidence region for $\hat{\eta}$.

### 7.4.4 Implementation

We can easily write an abstract class for a generic GLM, with the score function, Fisher information, and method of scoring built in. This is achieved in Code Blocks 7.6 and 7.7. Note that GLM is a subclass of our abstract Model class, defined in Code Block 6.22. Compare with the implementation of logistic regression in Code Block 7.4.

## 7.5 Models for Proportions

In Section 7.2, we formulated the logistic regression problem in terms of a Bernoulli random variable:

$$Y_i \sim \text{Bern}(\mu_i)$$
$$\text{logit}(\mu_i) = \mathbf{x}_i^T \boldsymbol{\beta}.$$

In Equation (7.40), we formulated the likelihood and log-likelihood functions for a weighted set of data. In this formulation, the target variable $Y_i$ still takes on a binary value $Y_i \in \mathbb{B}$, and a specific weight $w_i$ is applied to each observation. This formulation, however, is more commonly used for purely binary data, for which the weights are all $w_i = 1$. Purely binary data are likely to turn up when the feature vector has at least one continuous

```python
class GLM(Model):
    def __init__(self, max_iter=20, tol=1e-6, intercept=True,
            canonical_link=False, classifier=False):
        self.max_iter = max_iter
        self.tol = tol
        self.intercept = intercept
        self.beta = []
        self.canonical_link = canonical_link # Override default in
            subclass
        self.classifier = classifier # Override default in subclass

    @abstractmethod
    def variance(self, mu):
        pass

    @abstractmethod
    def link(self, mu):
        pass

    @abstractmethod
    def linkinv(self, eta):
        pass

    @abstractmethod
    def dlink(self, mu):

        if self.canonical_link:
            return 1 / self.variance(mu)

    def score(self, X, y, mu, w=None):
        if not w:
            w = np.ones(len(y))

        return X.T @ (w*(y - mu) / self.variance(mu) /
            self.dlink(mu))

    def information(self, X, y, mu, w=None):
        if not w:
            w = np.ones(len(y))

        D = np.diag(w / self.variance(mu) / self.dlink(mu)**2 )
        return X.T @ D @ X
## Continued...
```

Code Block 7.6: Abstract parent class for GLM implementation. (Continued in Code Block 7.7.)

```
1   def train(self, X, y, weights=None):
2       n, p = X.shape
3       w = weights if weights else np.ones(n)
4       y = y.reshape((n, 1))
5       w = w.reshape((n, 1))
6
7       if self.intercept:
8           X = np.concatenate([np.ones(n).reshape((n,1)), X], axis=1)
9           p += 1
10      beta = np.zeros(p).reshape((p, 1))
11      beta[0] = 1
12      i = 0
13      err = 1
14      while (i < self.max_iter) and (err > self.tol):
15          i += 1
16          mu = self.linkinv(X@beta)
17          score = self.score(X, y, mu, w=w)
18          info = self.information(X, y, mu, w=w)
19          err = np.linalg.norm(score)
20          beta += np.linalg.inv(info) @ score
21
22      self.converged = err < self.tol
23      self.beta = beta
24
25  def predict(self, X, threshold):
26      if self.intercept:
27          X = np.concatenate([np.ones(len(X)).reshape((len(X),1)),
                   X], axis=1)
28      assert self.converged, "Method of Scoring failed to converge"
29
30      mu = self.linkinv(X@self.beta)
31      if self.classifier:
32          return (mu > threshold).astype(int)
33      return mu
34
35  def predict_proba(self, X):
36      if self.intercept:
37          X = np.concatenate([np.ones(len(X)).reshape((len(X),1)),
                   X], axis=1)
38      assert self.converged, "Method of Scoring failed to converge"
39      return self.linkinv(X@self.beta)
```

Code Block 7.7: Abstract parent class for GLM implementation. (Continued from Code Block 7.6.)

variable. When the feature vector consists purely of categorical features, it is more likely to observe data with weights and proportions. When a weight is ascribed to each datum, we prefer a slight reformulation of the problem as a binomial random variable. As it turns out, the expressions derived for a Bernoulli random variable are still valid, except now the target variable is allowed to vary on the unit interval $[0, 1]$.

### 7.5.1 Binomial GLM

We begin by showing how a binomial random variable can be cast as a GLM. The result is closely related to the Bernoulli random variable formulation.

**Proposition 7.15.** *Let* $wY \sim \mathrm{Binom}(w, \mu)$ *represent a binomial random variable, for* $w \in \mathbb{Z}_+$, *so that the random variable* $Y$ *represents the proportion of successes. Then* $Y$ *is distributed as a weighted Bernoulli* GLM, *with weight* $w$; *i.e., the distribution of* $Y$ *is given by*

$$f_Y(y) = a(y, 1/w) \exp\left[yw\theta + w \ln(1 + e^\theta)\right],$$

*where* $\theta = \mathrm{logit}(\mu)$ *is the canonical parameter for the Bernoulli* GLM.

*Note 7.12.* Thus, when $Y$ represents the proportion of successes out of an IID sequence of $w$ Bernoulli trials, the Bernoulli GLM, as represented in Table 7.3, can still be used by applying weight $w$ and allowing the target variable $Y$ to range over the unit interval $Y \in [0, 1]$, as opposed to over the set $Y \in \mathbb{B}$.

Technically, for a fixed observation (i.e., fixed $\mu$) with weight $w$, the target variable $Y$ must take on a discrete set of values $Y \in \{0, 1/w, 2/w, \ldots, 1\}$. However, as $w$ typically varies for each observation, it is simpler to represent the acceptable values of $Y$ as the unit interval $[0, 1]$, with it being understood that the value of $Y$, for each observation, may only take on an appropriate discrete subset of values.    ▷

*Proof.* This result follows directly from Theorem 7.8, as the target variable $Y$ represents the sample mean of a sequence of $w$ IID Bernoulli trials. □

Though the preceding result follows from Theorem 7.8, it is nonetheless to see how it may be obtained directly. Since $wY \sim \mathrm{Binom}(w, \mu)$, it follows that the PMF for $Y$ is given by

$$f_Y(y) = \binom{w}{wy} \mu^{wy}(1 - \mu)^{w(1-y)}$$

$$= \binom{w}{wy} \exp\left[wy \ln(\mu) + w(1 - y)\ln(1 - \mu)\right]$$

$$= \binom{w}{wy} \exp\left[wy \ln\left(\frac{\mu}{1 - \mu}\right) + w \ln(1 - \mu)\right].$$

Here, we identify $a(y, 1/w) = \binom{w}{wy}$, along with the canonical variable $\theta = \text{logit}(\mu)$ and cumulant function $\kappa(\theta) = \ln(1 + e^\theta)$. The only difference between this GLM and the Bernoulli GLM is the weight $w$ and the fact that the variable $y$ is now allowed to take on the set of values $Y \in \{0, 1/w, 2/w, \ldots, 1\}$.

The other simplification gained from reformulating this as a binomial GLM is found in the representation of the data. In the weighted Bernoulli formulation of Equation (7.40), the results for a given feature vector $\mathbf{x}_0$ would be broken into two individual records: one for successes and one for failures. For example, we might have:

$$w_0 = 70, \qquad y_0 = 0, \qquad \mathbf{x}_0 = \mathbf{x}^*$$
$$w_1 = 30, \qquad y_1 = 1, \qquad \mathbf{x}_1 = \mathbf{x}^*$$

in the formulation of Equation (7.40), for which $Y$ is only allowed to take on the values $Y \in \mathbb{B} = \{0, 1\}$. In this new formulation, we would represent the same datum as

$$w_0 = 100, \qquad y_0 = 0.3, \qquad \mathbf{x}_0 = \mathbf{x}^*.$$

Thus, instead of requiring two records for each unique feature vector $\mathbf{x}$, one representing the positive outcomes $Y = 1$ and one representing the negative outcomes $Y = 0$, we now only require a single record, in which $Y$ represents the proportion of positive outcomes.

### 7.5.2 Link Functions for Logistic Regression

In Section 7.2, we used the logit function as our link function. We now recognize that the logit link function for binary or binomial (proportion) data represents the canonical link function, so that the linear predictor coincides with the natural parameter of the Bernoulli or binomial GLM. However, there are several alternatives that are worth being aware of. We enumerate a few main link functions that are sometimes used with binary data as follows.

1. The *logit link function*, as we saw previously, represents the canonical link function and is defined by the log-odds

$$g(\mu) = \text{logit}(\mu) = \ln \left( \frac{\mu}{1 - \mu} \right).$$

2. The *probit link function* is defined by

$$g(\mu) = \Phi^{-1}(\mu),$$

where $\Phi(\cdot)$ is the CDF of the standard normal distribution.

3. The *complementary log-log link function* is defined by

$$g(\mu) = \log\left[-\log(1 - \mu)\right],$$

with inverse function $g^{-1}(\eta) = 1 - \exp[-\exp(\eta)]$.
4. The *log-log link function* is defined by

$$g(\mu) = -\log\left[-\log(\mu)\right],$$

with inverse function $g^{-1}(\eta) = \exp[-\exp(-\eta)]$.

Each of the preceding link functions have an important quality in common, and that is they each represent a mapping $g : (0,1) \to \mathbb{R}$, so that the expected probability may be expressed without constraint in terms of the linear predictor. Each of these link functions is plotted concurrently in Figure 7.6. Aside from this important quality, there are some similarities and some differences among the various link functions.

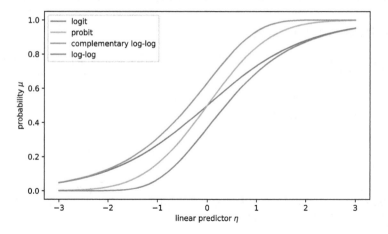

Fig. 7.6: Different (inverse) link functions for logistic regression; $\mu = g^{-1}(\eta)$.

To begin, both the probit and logit link functions are antisymmetric about the probability $\mu = 0.50$; i.e.,

$$\text{logit}(\mu) = -\text{logit}(1 - \mu) \qquad \text{and} \qquad \text{probit}(\mu) = -\text{probit}(1 - \mu).$$

In particular, this implies that $\text{logit}(1/2) = \text{probit}(1/2) = 0$. While both the probit and logit functions follow the same similar shape, the probit function is slightly more sensitive to changes in the underlying linear predictor.

This symmetry is broken for the log-log and complementary log-log link functions. These final two link functions are closely related, as the complementary log-log link function is the log-log link function applied to the complementary probability $1 - \mu$. Note in Figure 7.6 that the complementary log-log link function closely approximates the logit function for small values of $\mu \approx 0$. Similarly, the log-log link function closely approximates the logit function for large values of $\mu \approx 1$.

## 7.6 Models for Counts

In this section, we consider generalized linear models for which the target variable represents *count* data. People in erudite circles count things all the time; we love to count! And when we're not counting, we're modeling what our counts might total. Such is the topic of the present section.

Example of count data include: the number of potholes per mile of street or highway; the number of typos in a book; the number of signups for a website; the number of text messages; the number of sales in a particular region. Count data also frequently appear in contingency tables, for which observations are cross-classified relative to a number of factors. This arises, for instance, in analyzing the results of a multi-level experiment (Section 4.4).

### 7.6.1 Poisson Regression

The simplest model for count data is obtained using the Poisson distribution, which is parameterized by a single parameter $\mu$, equivalent to the distribution's mean.

**Definition 7.17.** *The* Poisson GLM *is the* GLM *defined by*

$$Y_i = \text{Poiss}(\mu_i), \tag{7.87}$$

$$g(\mu_i) = \mathbf{x}_i^T \boldsymbol{\beta}, \tag{7.88}$$

*for $Y_i \in \mathbb{N}$, $\mu_i \in \mathbb{R}_+$, and for a given link function $g : \mathbb{R}_+ \to \mathbb{R}$. When the link function is the canonical link $g(\mu_i) = \ln(\mu_i)$, the* GLM *is referred to as the* (Poisson) loglinear model.

Since the canonical link function, which connects the mean to the linear predictor, is the log function, the model, under the canonical link, is sometimes referred to as a *loglinear model*. As this is the only link function we will consider for the Poisson regression problem, we will often simply refer to the model as Poisson, with *loglinear* being understood.

**Proposition 7.16.** *The score function and Fisher information for the Poisson loglinear model are given by*

$$U_j = \sum_{i=1}^{n} (y_i - \mu_i) x_{ij}, \tag{7.89}$$

$$\mathcal{I}_{jk} = \sum_{i=1}^{n} \mu_i x_{ij} x_{ik}. \tag{7.90}$$

*respectively.*

*Proof.* Since the canonical link function is used, these relations follows from Equations (7.71) and (7.73), since $\phi = 1$ and $V(\mu) = 1/g'(\mu) = \mu$.    □

Note that for a Poisson GLM, the Pearson statistic, defined in Equation (7.77), simplifies to

$$X^2 = \sum_{i=1}^{n} \frac{(y_i - \hat{\mu}_i)^2}{\hat{\mu}_i},$$

which is equivalent to the Pearson chi-squared test statistic, defined in Definition 3.18.

The relationship between Poisson regression and logistic and multinomial models is discussed in Agresti [2015].

## 7.6.2 Contingency Tables

Count data often arise in the context of contingency tables. Interestingly, the method of data collection affects the choice of model, depending on the constraints used during the data collection process. We will focus on two-factor contingency tables, though the methods presented herein can easily be generalized to higher-order tables. A typical two-factor contingency table is shown in Table 7.4.

|  | $j = 1$ | $j = 2$ | $\cdots$ | $j = s$ | row totals |
|---|---|---|---|---|---|
| $i = 1$ | $Y_{11}$ | $Y_{12}$ | $\cdots$ | $Y_{1s}$ | $Y_{1\cdot}$ |
| $i = 2$ | $Y_{21}$ | $Y_{22}$ |  | $Y_{2s}$ | $Y_{2\cdot}$ |
| $\vdots$ | $\vdots$ | $\vdots$ | $\ddots$ | $\vdots$ | $\vdots$ |
| $i = m$ | $Y_{m1}$ | $Y_{m2}$ | $\cdots$ | $Y_{ms}$ | $Y_{m\cdot}$ |
| column totals | $Y_{\cdot 1}$ | $Y_{\cdot 2}$ | $\cdots$ | $Y_{\cdot s}$ | $Y_{\cdot\cdot}$ |

Table 7.4: Layout of a two-factor contingency table.

The random variables $Y_{ij}$, for $i = 1, \ldots, m$ and $j = 1, \ldots, s$ represent the total observed counts for the $i$th level of the first factor and the $j$th level of the second factor. As usual, the quantities

$$Y_{i\cdot} = \sum_{j=1}^{s} Y_{ij} \quad \text{and} \quad Y_{\cdot j} = \sum_{i=1}^{m} Y_{ij}$$

represent the marginal counts for row $i$ and column $j$, respectively, and the quantity $Y_{\cdot\cdot} = \sum_{i=1}^{m} \sum_{j=1}^{s} Y_{ij}$ represents the total count of all the observations.

**Contingency Tables with No Constraints**

We begin by analyzing an *unconstrained* contingency table. This is common, for example, when data are collected for a specified period of time, and the results are random and tabulated at the end of the data. (We will discuss constrained variations, e.g., when the total number of counts is specified a priori, momentarily.)

In modeling a two-factor contingency table, the responses are modeled as independent Poisson random variables by the relation

$$Y_{ij} = \text{Poiss}(\mu_{ij}). \tag{7.91}$$

Further, we assume independence in the two factors, such that the Poisson parameters $\mu_{ij}$ satisfy the relation

$$\mu_{ij} = \mu \phi_i \psi_j, \tag{7.92}$$

where the parameters $\phi_i$ and $\psi_j$ are normalized by the conditions

$$\sum_{i=1}^{m} \phi_i = 1 \quad \text{and} \quad \sum_{j=1}^{s} \psi_j = 1, \tag{7.93}$$

so that $\mu = \sum_{i=1}^{m} \sum_{j=1}^{s} \mu_{ij}$. In the unconstrained case, note that the total count $Y_{..}$ is distributed as a Poisson random varaible, according to

$$Y_{..} \sim \text{Poiss}(\mu). \tag{7.94}$$

When the log link function is used, the independence assumption in Equation (7.92) results in

$$\ln \mu_{ij} = \beta_0 + \beta_i^A + \beta_j^B, \tag{7.95}$$

where we use the superscript $A$ to denote the first factor (the rows) and $B$ to denote the second factor (the columns). Moreover, the constraints defined in Equation (7.93) place a constraint on $\beta_i^A$ and $\beta_j^B$. Without loss of generality, we may take $\beta_0^A = \beta_0^B = 0$ to eliminate the dependency among the parameters. Note that the feature vector consists of the multi-index iteself; i.e., it is the rows and columns of the contingency table that constitute an encoding of the two categorical features. To see this, we may write the following one-hot encoding for the features as

$$\mathbf{x}_{ij} = \begin{bmatrix} 1 \\ \mathbb{I}[i=2] \\ \vdots \\ \mathbb{I}[i=m] \\ \mathbb{I}[j=2] \\ \vdots \\ \mathbb{I}[j=s] \end{bmatrix} \quad \text{and} \quad \boldsymbol{\beta} = \begin{bmatrix} \beta_0 \\ \beta_2^A \\ \vdots \\ \beta_m^A \\ \beta_2^B \\ \vdots \\ \beta_s^B \end{bmatrix},$$

so that Equation (7.95) is equivalent to the more familiar

$$\ln \mu_{ij} = \mathbf{x}_{ij}^T \boldsymbol{\beta}.$$

Note that $\boldsymbol{\beta} \in \mathbb{R}^{m+s-1}$, as there is one (1) parameter due to the intercept, $m-1$ parameters due to the first factor, and $s-1$ parameters due to the second factor.

The likelihood function for the data is given by

$$L = \prod_{i=1}^{m} \prod_{j=1}^{s} \frac{e^{-\mu_{ij}} \mu_{ij}^{y_{ij}}}{y_{ij}!} = \exp\left[\sum_{i=1}^{m} \sum_{j=1}^{s} (y_{ij} \ln \mu_{ij} - \mu_{ij})\right] \prod_{i=1}^{m} \prod_{j=1}^{s} \frac{1}{y_{ij}!}.$$

Next, we can substitute in the link function Equation (7.95) to express the log-likelihood in terms of $\boldsymbol{\beta}$ as

$$\ell(\boldsymbol{\beta}) = y_{..}\beta_0 + \sum_{i=1}^{m} y_{i.}\beta_i^A + \sum_{j=1}^{s} y_{.j}\beta_j^B - \sum_{i=1}^{m} \sum_{j=1}^{s} \exp(\beta_0 + \beta_i^A + \beta_j^B) + \ln a(y_{ij}),$$

where $a(y_{ij}) = \prod_{i=1}^{m} \prod_{j=1}^{s} (y_{ij}!)^{-1}$. The score equations follow:

$$\frac{\partial \ell}{\partial \beta_0} = y_{..} - \sum_{i=1}^{m} \sum_{j=1}^{s} \exp(\beta_0 + \beta_i^A + \beta_j^B)$$

$$\frac{\partial \ell}{\partial \beta_i^A} = y_{i.} - \sum_{j=1}^{s} \exp(\beta_0 + \beta_i^A + \beta_j^B)$$

$$\frac{\partial \ell}{\partial \beta_j^B} = y_{.j} - \sum_{i=1}^{m} \exp(\beta_0 + \beta_i^A + \beta_j^B).$$

Recognizing that $\exp(\beta_0 + \beta_i^A + \beta_j^B) = \mu_{ij} = \mu\phi_i\psi_j$, and upon setting the preceding equations to zero, we have the following MLE estimates:

$$\hat{\mu} = y_{..}, \qquad \hat{\phi}_i = \frac{y_{i.}}{y_{..}}, \qquad \hat{\psi}_j = \frac{y_{.j}}{y_{..}}.$$

From this, it follows that

$$\hat{\mu}_{ij} = \frac{y_{i.} y_{.j}}{y_{..}}, \tag{7.96}$$

as one might well expect.

**Contingency Tables with Constrained Overall Total**

Next, let us reformulate the problem supposing that the total count $Y_{..}$ is not random, but fixed $y_{..} = n$ in advanced. For example, it is determined *in advanced* that the data shall stop being collected once the first $n = 1000$

samples have arrived. How those samples sort among the rows and columns is random, but the overall total count is held fixed.

The conditional probability of $Y_{ij}$ given the total $Y_{..} = n$ is given by

$$f(y_{ij}|y_{..} = n) = \frac{f(y_{ij})}{\mathbb{P}(y_{..} = n)}.$$

However, recalling Equation (7.94), we have

$$f(y_{ij}|y_{..} = n) = \frac{n!}{e^{-\mu}\mu^n} \prod_{i=1}^{m} \prod_{j=1}^{s} \frac{e^{-\mu_{ij}}\mu_{ij}^{y_{ij}}}{y_{ij}!}$$

$$= \frac{n!}{\prod_{i=1}^{m}\prod_{j=1}^{s} y_{ij}!} \frac{1}{e^{-\mu}\mu^n} \prod_{i=1}^{m}\prod_{j=1}^{s} e^{-\mu_{ij}}\mu_{ij}^{y_{ij}}$$

$$= \frac{n!}{\prod_{i=1}^{m}\prod_{j=1}^{s} y_{ij}!} \prod_{i=1}^{m}\prod_{j=1}^{s} \pi_{ij}^{y_{ij}},$$

$$= \frac{n!}{\prod_{i=1}^{m}\prod_{j=1}^{s} y_{ij}!} \exp\left[\sum_{i=1}^{m}\sum_{j=1}^{s} y_{ij}\ln \pi_{ij}\right],$$

where $\pi_{ij} = \mu_{ij}/\mu$ represents the probability of the $ij$th cell. We immediately recognize this as the multinomial distribution:

$$f(Y_{ij}|Y_{..} = n) = \mathrm{Multi}(\pi_{ij}).$$

We may therefore express the log-likelihood kernel as

$$\ell(\phi, \psi) = \sum_{i=1}^{m} y_{i\cdot}\ln(\phi_i) + \sum_{j=1}^{s} y_{\cdot j}\ln(\psi_j),$$

which is subject to the constraints $\sum_{i=1}^{m} y_{i\cdot} = \sum_{j=1}^{s} y_{\cdot j} = n$. Solving the constrained MLE problem yields the estimates

$$\hat{\phi}_i = \frac{y_{i\cdot}}{n} \qquad \text{and} \qquad \hat{\psi}_j = \frac{y_{\cdot j}}{n},$$

which results in the final probability estimate $\pi_{ij} = y_{i\cdot} y_{\cdot j}/n^2$.

### 7.6.3 Negative-binomial GLMs

In the Poisson model, the variance of the response is equal to the mean; i.e., $\mathbb{V}(Y) = \mu$. In practice, however, the observed variance often exceeds this amount, resulting in a phenomenon known as *overdispersion*. A common approach to dealing with overdispersion is by using a *mixture model*. The idea is that the Poisson parameter $\mu_i$, even for a fixed feature vector, is not homogeneous across the group, but rather exhibits a certain level of heterogeneity, which can be modeled by a secondary distribution. Mixing the Poisson distribution with a gamma distribution, in order to model the population heterogeneity, yields the negative binomial distribution. For details of this derivation, see Agresti [2015] or Dunn and Smyth [2018].

**Definition 7.18.** *The negative-binomial distribution (NBD) is the distribution for a discrete random variable $Y \in \mathbb{N}$ with probability mass function*

$$f(y; \mu, k) = \frac{\Gamma(y+k)}{\Gamma(k)\Gamma(y+1)} \left(\frac{\mu}{\mu+k}\right)^y \left(\frac{k}{\mu+k}\right)^k, \qquad (7.97)$$

*for parameters $\mu > 0$ and $k > 0$.*

**Proposition 7.17.** *For fixed $k > 0$, the negative-binomial distribution is an* EDM *with canonical parameter, cumulant function, and variance function given by*

$$\theta = \ln\left(\frac{\mu}{\mu+k}\right) \qquad (7.98)$$

$$\kappa(\theta) = -k \ln\left(1 - e^\theta\right) \qquad (7.99)$$

$$V(\mu) = \mu + \mu^2/k, \qquad (7.100)$$

*respectively.*

*Proof.* The proof is straight-forward. We begin by rewriting Equation (7.97) as

$$f(y; \mu, k) = \frac{\Gamma(y+k)}{\Gamma(k)\Gamma(y+1)} \exp\left[y \ln\left(\frac{\mu}{\mu+k}\right) + k \ln\left(\frac{k}{\mu+k}\right)\right].$$

Immediately, we recognize the canonical link, as given in Equation (7.98). Moreover, since

$$\frac{k}{\mu+k} = 1 - \frac{\mu}{\mu+k} = 1 - e^\theta,$$

the cumulant function is clearly given by Equation (7.99). Finally, the variance function given in Equation (7.100) follows from the relation $V(\mu) = 1/g'(\mu)$, which holds for the canonical link $g(\mu) = \ln[\mu/(\mu+k)]$. This completes the result. □

*Note 7.13.* Notice, by the nature of Equation (7.98), that the canonical parameter takes values in $\mathbb{R}_-$. For this reason, the canonical link is seldom, if ever, used as the link function when the negative-binomial distribution is used as part of a GLM. Instead, the log link function $g(\mu) = \ln(\mu)$ is typically used, ensuring that the parameter $\eta$, representing the linear predictor, can take on all values in $\mathbb{R}$.    ▷

**Proposition 7.18.** *For the* GLM *consisting of the negative-binomial distribution and the log link function* $g(\mu) = \ln(\mu)$, *the score and Fisher information are given by*

$$U_j = \sum_{i=1}^{n} \frac{(y_i - \mu_i)x_{ij}}{1 + \mu_i/k}, \tag{7.101}$$

$$\mathcal{I}_{jk} = \sum_{i=1}^{n} \frac{\mu x_{ij} x_{ik}}{1 + \mu_i/k}, \tag{7.102}$$

*respectively.*

*Proof.* This follows from Equations (7.70) and (7.72) and the fact that $V(\mu) = \mu + \mu^2/k$ and $g'(\mu) = 1/\mu$. $\qquad\square$

## 7.7 Models for Positive Continuous Data

In this section, we discuss models for positive continuous data. Such models are useful when the target variable measures a physical quantity.

### 7.7.1 Gamma GLM

We have already seen how the gamma distribution can be represented as an EDM in Section 7.3. However, since the canonical parameter takes on strictly negative values, another link function is in order. Typically, one uses the log link function.

**Proposition 7.19.** *For the* GLM *consisting of the gamma distribution and log link function* $g(\mu) = \ln(\mu)$, *the score and information are given by*

$$U_j = \frac{1}{\phi} \sum_{i=1}^{n} \frac{w_i(y_i - \mu_i)x_{ij}}{\mu} \tag{7.103}$$

$$\mathcal{I}_{jk} = \frac{1}{\phi} \sum_{i=1}^{n} w_i(y_i - \mu_i)x_{ij}, \tag{7.104}$$

*respectively.*

*Proof.* This result follows from Equations (7.70) and (7.72) and the fact that $V(\mu) = \mu^2$ and $g'(\mu) = 1/\mu$. $\qquad\square$

### 7.7.2 Inverse Gaussian GLM

When the target variable is even more right-skewed than the gamma distribution, a popular alternative is the inverse Gaussian distribution, which is defined as follows.

**Definition 7.19.** *The* inverse Gaussian distribution *is the probability distribution defined over* $(0, \infty)$ *by the* PDF

$$f(y; \mu, \phi) = \frac{1}{\sqrt{2\pi y^3 \phi}} \exp\left[\frac{-(y - \mu)^2}{2\phi y \mu^2}\right], \tag{7.105}$$

*for* $y > 0$ *and positive parameters* $\mu, \phi > 0$.

We next verify the result that is tabulated in Table 7.3.

**Proposition 7.20.** *The inverse Gaussian distribution is an* EDM *with dispersion parameter* $\phi$ *and canonical parameter, cumulant function, and variance function given by*

$$\theta = \frac{-1}{2\mu^2}, \tag{7.106}$$

$$\kappa(\theta) = -\sqrt{-2\theta}, \tag{7.107}$$

$$V(\mu) = \mu^3, \tag{7.108}$$

*respectively, where* $\theta < 0$.

*Proof.* The PDF given in Equation (7.105) can be rewritten as

$$\frac{\exp[-1/(2\phi y)]}{\sqrt{2\pi y^3 \phi}} \exp\left[\frac{-y}{2\phi\mu^2} + \frac{1}{\phi\mu}\right].$$

Immediately, we recognize the canonical parameter $\theta$, as given in Equation (7.106). This relationship can be inverted to find $\mu = 1/\sqrt{-2\theta}$. The cumulant function Equation (7.107) follows. Finally, we note that the derivative of the canonical link function given by Equation (7.106) is $g'(\mu) = 1/\mu^3$, from which Equation (7.108) follows. This completes the result. □

Since the canonical parameter $\theta$ only takes negative values, the canonical link function is not preferred when constructing a GLM. Instead, we can use the log-link function, as we did with the gamma distribution.

**Proposition 7.21.** *For the* GLM *consisting of the inverse Gaussian distribution and log link function* $g(\mu) = \ln(\mu)$, *the score and information are given by*

$$U_j = \frac{1}{\phi} \sum_{i=1}^{n} \frac{w_i(y_i - \mu_i)x_{ij}}{\mu^2} \tag{7.109}$$

$$\mathcal{I}_{jk} = \frac{1}{\phi} \sum_{i=1}^{n} \frac{w_i(y_i - \mu_i)x_{ij}}{\mu}, \tag{7.110}$$

*respectively.*

*Proof.* This result follows from Equations (7.70) and (7.72) and the fact that $V(\mu) = \mu^3$ and $g'(\mu) = 1/\mu$. □

## Problems

**7.1.** Compute the gradient of Equation (7.13), set it equal to zero, and show that the only critical point is the one given by Equation (7.14). Why must this be the global minimum?

**7.2.** Prove Corollary 7.1.

**7.3.** Prove Equation (7.32).

**7.4.** Derive the likelihood function and maximum-likelihood solution for the linear regression problem, given by Definition 7.1. Show that your result is equivalent to the least-squares solution.

**7.5.** Show that the likelihood function for a logistic regression model is given by Equation (7.40).

**7.6.** Show that the inverse of the central logratio transform, defined in Equation (7.49), is the softmax transform.

**7.7.** Prove that both the probit and logit link functions are antisymmetric about $\mu = 0.50$; i.e., that

$$g(\mu) = -g(1 - \mu).$$

**7.8.** Prove that the MLE estimator for the natural parameter $\theta$ of an EDM (eq4001) is related to the sample mean by

$$\frac{1}{n} \sum_{i=1}^{n} y_i = \kappa'(\hat{\theta}).$$

# 8

## It's not my Fault

*Correlation does not imply causation* is the mantra heard 'round the world—at least in statistics courses. Two random variables can be correlated without one begin a *cause* of the other. Ice cream sales might be correlated with people going to the beach, but you can't boost beach attendance in the middle of winter by selling ice cream. In other words, it is not the ice cream vendors who *cause* beachgoers to don their trunks and lather up in SPF 42; instead, the correlation is explained by the fact that sunbathing and ice cream sales have a common cause: hot weather.

If correlation is not causation, then what is? As it turns out, the framework for causality lies outside the scope of traditional statistics, instead relying on graph theory and graphical models. Moreover, causality can be used to answer important questions related to organic data sets, i.e., data sets that are not the result of a properly controlled experiment. We begin by introducing these *observational studies* and the problem they present to classical statistics. We then go on to define graphs, introduce causal models, and show how causal implications can be inferred from graph structures and data.

Our go-to reference is Pearl, *et al.* [2016], is it serves as a concise introduction to the field. From their, we recommend both Rosenbaum [2002] for a light introduction to observational studies and Pearl [2009] for a more indepth test on causality. For introductory texts into causality that feature social science and biomedical applications, see Imbens and Rubin [2015] and Morgan and Winship [2015].

## 8.1 Simpson's Paradox

We begin with a brief introduction to the definition of an observational study. We then discuss the shortcoming of the ability for traditional statistical techniques to draw inferences from such data. This will establish the need for a proper framework around the notion of causality, the topic of this chapter.

### 8.1.1 Observational Studies

We discussed experimentation in depth in Chapter 4. In particular, we discussed the importance of randomization in constructing certain *controlled* experiments. But what if it is not possible to control an experiment, for either ethical or logistical reasons? For instance, how would you prove that smoking causes lung cancer? It is not ethical to take a large sample of a population and randomly select half of them to smoke three packs of cigarettes a day for the next twenty years. It is readily apparent, for such cases, that a controlled experiment is not possible. We instead have entered the realm of observational studies.

**Definition 8.1.** *An* observational study *is a statistical analysis which draws a conclusion from a set of data about the effect of a treatment, for which the assignment of treatment to individuals is not under control of the statistician.*

Instead of choosing our subject's fate by toss of a coin, an observational study would seek to infer the effect of smoking by looking at large amounts of data, consisting of smokers and nonsmokers alike, and seeking to draw an appropriate conclusion about the effect of smoking. In this context, the idea of *causality*—i.e., the notion that one event causes another—becomes critically important. For example, let us suppose that the data show that smokers have a dramatically increased rate of lung cancer as compared with their nonsmoking piers. Are we able to draw a conclusion from this fact alone? *Correlation does not imply causation.* What if the tendency to smoke is itself genetic? What if we have a gene that determines whether or not we ultimately become smokers later in life? And what if this gene is also responsible for lung cancer? We would then be back in the realm of our failed business enterprise to drive sunbathers to the beach by opening ice cream stands during the winter, except, this time, ice cream is smoking and the beach is lung cancer. In short, how can we draw conclusions from data when there is potentially an unknown common cause that explains the data? How do we know that there is not an undiscovered gene that both *causes* smoking and *causes* lung cancer? We will answer that question momentarily. But first, we will explore another substantial obstacle to drawing conclusions from data, and that is the certain reversal of trends at different levels of aggregation known as Simpson's paradox.

### 8.1.2 Simpson's Paradox

Simpson's paradox represents a conundrum for the field of traditional statistics, that must be accounted for if we are to formulate a proper theory of causality. In short, we may state it as follows.

**Definition 8.2.** Simpson's paradox *is the phenomenon that occurs when the aggregation of data at one level results in a different apparent conclusion than would be obtained by an aggregation at a different level.*

That is, Simpson's paradox occurs whenever different slices of the data seem to support different, often opposing, conclusions. Simpson's paradox is one danger of observational studies, and is best understood through an illustration.

*Example 8.1.* Consider the data shown in Table 8.1. We seek to draw a conclusion about the effect of a treatment for a given medical condition. The data are aggregated by gender, and we observe that treatment A has a higher recovery rate than treatment B for both men and women. Women recover 70% of the time with treatment A, as opposed to 65% of the time with treatment B, whereas men recover 50% of the the time with treatment A, as opposed to only 45% with treatment B. Thus, it seems that regardless of whether a patient is male or female, the optimal prescription for that patient should be treatment A.

| Treatment | Women | Men |
|:---------:|:-----:|:---:|
| A | 70/100 (70%) | 200/400 (50%) |
| B | 260/400 (65%) | 45/100 (45%) |

Table 8.1: Recovery rates under two treatment options, broken down by gender.

Seems straightforward enough, but let's aggregate all of our data together, as shown in Table 8.1. Here, we are surprised to observe the exact *opposite* conclusion: the overall recover rate of treatment B is 61%, whereas the overall recovery rate of treatment A is only 54%.

| Treatment | Overall |
|:---------:|:-------:|
| A | 270/500 (54%) |
| B | 305/500 (61%) |

Table 8.2: Overall recovery rates.

How can this be? Our overall recovery rate of treatment B is clearly better than our overall recovery rate of treatment A. But yet, when we divide the data by gender, we observe that treatment A in fact does better, for *both* men *and women*!                                                    ▷

The conundrum presented by Example 8.1 is an instance of Simpson's paradox: the apparent reversal of a conclusion by aggregating at different levels.

Now, it is tempting to say: *It's obvious! We just go with the conclusion of Table 8.1. Treatment A is better.* It is, after all, clear what is happening. Women have a higher recovery rate than men, but women are more likely to have been prescribed treatment B than men. This leads to the fact that treatment B in the overall aggregate has a better outcome: because it oversamples women, with their higher recovery rate, than men.

The reason, however, that this is a paradox is that such a conclusion is not supported by traditional statistics, but by our intuition. And not only is our intuition responsible for this conclusion, it is, in particular, our intuition on causality. That is, we have intuitively constructed a *causal* story about the data.

*Nonsense!* you say. *Let's just go to the most granular slice that has statistical significance and go with that.* Such an approach seems logical, except—

*Example 8.2.* Let us consider the data of Table 8.3, aggregated based on whether a patient exhibits side effects following the treatment.

| Treatment | No side effect | Side effect |
|:---:|:---:|:---:|
| A | 70/100 (70%) | 200/400 (50%) |
| B | 260/400 (65%) | 45/100 (45%) |

Table 8.3: Recovery rates under two treatment options, broken down by gender.

Note that the data are identical to Table 8.1, only we have a different interpretation of the aggregation groups. Can we draw the same conclusion? An important difference is that we are now aggregating by an effect of the treatment, the appearance of side effect. On one hand, any given patient will experience side effects or they will not, and in either case it looks like treatment A does better. But on the other hand, the side effects appear *after* the treatment is administered, and so they could be caused in part by the ultimate outcome. It therefore seems logical to go with the overall winner, treatment B.                                    ▷

So it seems like our intuition in Example 8.2 is the exact opposite of what it was in Example 8.1, despite the underlying data being identical. The difference is the causal story that underlies our intuition. In one case, our intuition tells us that gender is a causal factor in outcome. In the other case, our intuition leads us to at least suspect that the outcome might be a causal factor, along with the choice of treatment, in whether or not a patient experiences side effect.

Since different interpretations of the data lead to different conclusions, this truly is a paradox for classical statistics. Instead, our conclusion heavily relies on the causal story we prescribe for our data. We therefore need a new set of tools for weaving these causal stories, which we will develop shortly. But first, we conclude our introduction with a final twist on Example 8.1.

*Example 8.3.* Let us continue Example 8.1, with our original gender interpretation of Table 8.1. Our conclusion was that treatment A was the better option, as treatment A exhibited higher recovery rates for both men and women. However, Simpson's paradox can sometimes reoccur upon further subdivision of the data. To illustrate this, let us suppose that our data are further subdivided by the severity of illness, as shown in Table 8.4.

| Treatment | Women | | Men | |
|---|---|---|---|---|
| | Mild | Severe | Mild | Severe |
| A | 64/80 (80%) | 6/20 (30%) | 128/160 (80%) | 72/240 (30%) |
| B | 204/240 (85%) | 56/160 (35%) | 17/20 (85%) | 28/80 (35%) |

Table 8.4: Recovery rates under two treatment options, broken down by gender and severity of illness.

It appears that treatment B is the better choice after all. When we consider both gender and severity of illness, treatment B wins across the board. This effect was reversed at the gender level (Table 8.1), since men are more likely to present with a severe case of the ailment as compared to women. Moreover, if we look at the treatment–severity matrix, we notice that it is in fact the same for both men and women! Thus, gender actually has no effect on outcome. Patients with a mild case who take treatment A are 80% likely to recover, and so forth. The four recovery rates—80%, 30%, 85%, and 35%—do not depend on gender at all. Instead, the likelihood of a serious illness does depend on gender: 64% of men have a serious version of the illness, as opposed to only 36% of women. This imbalance is what leads to the faulty conclusion of Table 8.1. But when properly considering severity, it is clear, according to our causal intuition, that treatment B is the better choice.                                                                    ▷

## 8.2 Graphs and Causal Models

In this section, we lay out several key definitions of graph theory.

### 8.2.1 Graphs Defined

**Definition 8.3.** *A graph $G$ is a collection of* nodes $\mathcal{N}$ *and edges* $\mathcal{E}$. *A* node *is a point, and an edge $e \in \mathcal{E}$ connects two points; i.e., an edge $e$ is uniquely specified by two distinct points $a, b \in \mathcal{N}$, such that $e = (a, b)$.*

An edge is said to be directed *if the order matters, in which case we use the notation* $|a, b\rangle$, *so that* $|a, b\rangle$ *is distinct from* $|b, a\rangle$. *A graph with directed edges is called a* directed graph. *Two nodes are* adjacent *if they are connected by an edge.*

Given a directed edge $|a, b\rangle$, the node $a$ is referred to as the parent of node $b$, whereas node $b$ is referred to as the child of node $a$. We write $a = \mathrm{par}(b)$ and $b \in \mathrm{child}(a)$.

A directed edge $e = (a, b)$ is represented visually with an arrow traveling from $a$ to $b$, whereas an undirected edge is represented by a line segment. In studying graphs, one often wishes to traverse a graph from one point to another. The following definitions are therefore in order.

**Definition 8.4.** *Given a graph* $G = \{\mathcal{N}, \mathcal{E}\}$, *a* path *between two nodes* $a, b \in \mathcal{N}$ *is a sequence of nodes, beginning with* $a$ *and ending with* $b$, *such that each node is connected to the next by an edge; the* length *of a path is the number of edges comprising the path.*

A path is simple *if no node is repeated.*

A path in a directed graph is called a directed path *if, in addition, its connecting edges are directed edges of the graph; i.e., if* $|x_{i-1}, x_i\rangle \in \mathcal{E}$ *for* $i = 1, \ldots, l$.

A cycle *is a directed path that starts and returns to the same node:* $x_0 \to x_1 \to x_{l-1} \to x_0$. *A directed graph with no cycles is a* directed acyclic graph *or* DAG.

Given two points $a$ and $b$ in a directed graph, we say that $a$ is an ancestor *of* $b$, *denoted* $a \in \mathrm{anc}(b)$, *and that* $b$ *is a* descendant *of* $a$, *denoted* $b \in \mathrm{desc}(a)$, *if and only if there is a directed path starting at* $a$ *and ending at* $b$.

Examples of a directed cyclic and a directed acyclic graph are shown in Figure 8.1.

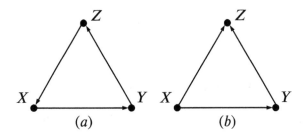

Fig. 8.1: Example of a directed cyclic graph (a) and a directed acyclic graph (b).

In other words, a path of length $l$ between nodes $a$ and $b$ is an ordered sequence $x_0, x_1, \ldots, x_l$, such that $x_0 = a$, $x_l = b$, and each pair $(x_{i-1}, x_i) \in$

$\mathcal{E}$, for $i = 1, \ldots, l$. With paths defined, we next define the important concept of connectedness.

**Definition 8.5.** *A graph is* connected *if there is a path between any two nodes; i.e., the graph does not have any isolated islands. A graph is* singly connected *if there is only one simple path between any two nodes.*

Singly connected graphs, commonly called *trees* are an important concept that are used extensively in statistical modeling and machine learning. They also constitute an important type of data structure in computer science. However, we typically require some additional structure.

**Definition 8.6.** *A* (rooted) tree *is a directed, singly connected graph that contains a single node without parents, known as the* root, *which determines the direction of the edges, such that all edges are directed away from the root.*

*A* branch *of a tree is a node that has children; a* leaf *of a tree is a node without children.*

*The* height *of a node is the length of the longest directed path from that node to a leaf. The* depth *of a node is the length of the path connecting the root node to that node. The* height *of the tree is the height of its root.*

*A* binary tree *is a tree in which each node has at most two children.*

An example of a tree is shown in Figure 8.2. Here, the root node is shaded red, and the leaves are shaded green. The height of the tree is three.

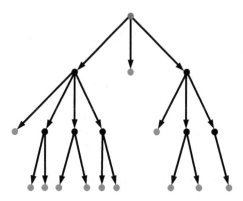

Fig. 8.2: Example of tree of height 3. Root node is red and the twelve leaves are green.

### 8.2.2 Causality, Structure, and Graphs

We next show how graphs can be applied to weave the *causal story* for a data set.

## Causal Models

We begin by defining a mathematical object used to model causal structure in data.

**Definition 8.7.** *A causal structure* over a set of variables $\mathcal{V}$ *is a directed graph* $G = \{\mathcal{V}, \mathcal{E}\}$, *in which each node corresponds to one of the variables and each directed edge represents a direct functional relationship among the corresponding variables.*

*A* causal feedback structure *is a causal structure that contains cycles. A* feedback loop *is any cycle in such a structure. A causal acyclic structure is a causal structure that does not contain cycles.*

Feedback loops occur frequently in engineering control systems. For example, consider a thermostat: the room temperature determines whether or not the thermostat will turn on or off a heater, but the heater, in turn, affects the room temperature. This is an example of a closed feedback loop.

In general, we will assume a causal structure to be acyclic unless otherwise specified.

Causal relationships can now be defined within the context of a given causal structure.

**Definition 8.8.** *Given a causal structure $G$ of a set of variables, variable $X$ is a* direct cause *of variable $Y$ if the graph $G$ contains a directed edge from $X$ to $Y$.*

*Two variables $X$ and $Y$ are said to be* causally connected *if there is a directed path connecting them. If $X$ and $Y$ are causally connected, we say that the ancestor is a* potential cause *of its descendant. We denote the condition that $X$ is an ancestor (hence a potential cause) of $Y$ with the notation $X \prec Y$, which defines a natural partial ordering on the causal structure[1].*

*We furthermore say that a potential cause $X$ of a variable $Y$ is, in fact, a* cause *of $Y$ if a variation in $X$ results in a variation in $Y$, such that variations are only allowed to propagate directionally through the graph; i.e., the variation in $X$ does not change the values for any of $X$'s parents. If a potential cause is not a cause, we say that the situation is* intransitive.

Thus, a causal structure is a mathematical object used to encode our causal picture of the world. Given a causal structure, we can now introduce the notion of a causal model.

---

[1] The ordering is only partial as two variables are comparable relative to the ordering if and only if they are causally connected. In general, a total ordering can be defined on any directed acyclic graph; such total ordering is, however, not uniquely defined. In particular, Kahn's algorithm can be used to define a total ordering on any DAG; see Kahn [1962].

**Definition 8.9.** *A structural causal model (SCM) is a triple* $(\mathcal{U}, \mathcal{V}, \mathcal{F})$ *that consists of*

1. *a set of $n$ variables $\mathcal{U}$, referred to as* exogenous variables, *that represent factors outside of the model;*
2. *a set of $n$ variables $\mathcal{V}$, referred to as* endogenous variables, *that represent variables internal to the model;*
3. *and a set of functions $\mathcal{F}$, in one-to-one correspondence with the variables in $\mathcal{V}$, that yield the* structural equations

$$V_i = f_{V_i}(P(V_i), U_i), \tag{8.1}$$

*for $i = 1, \ldots, n$; where $P(V_i) \subset \mathcal{V} \setminus \{V_i\}$ are the endogenous parents of the variable $V_i$;*

*such that any specification of the variables $\mathcal{U}$ uniquely determines the values of the variables $\mathcal{V}$, so that, as a whole, $V = V(U)$.*

Any causal model $M = (\mathcal{U}, \mathcal{V}, \mathcal{F})$ determines a unique causal structure $G(M) = (\mathcal{U} \cup \mathcal{V}, \mathcal{E})$, for which the edges $\mathcal{E}$ are comprised of the functional dependencies of $\mathcal{F}$; i.e.,

$$\mathcal{E} = \bigcup_{i=1}^{n} \bigcup_{P \in \mathrm{par}(V_i)} |P, V_i\rangle,$$

where $\mathrm{par}(V_i) = P(V_i) \cup \{U_i\}$ represents the complete set of dependent variables of the function $f_{V_i}$.

The condition in Definition 8.9 that the variables $\mathcal{U}$ uniquely determine the values of the variables $\mathcal{V}$ is equivalent to the requirement on the graph that each variable (node) in $\mathcal{V}$ have an ancestor in $\mathcal{U}$.

In general, however, the values of the exogenous variables are not known, but rather are characterized by a probability distribution. This is expressed as follows.

**Definition 8.10.** *A* probabilistic causal model *(or* causal model*) is a structural causal model $M = (\mathcal{U}, \mathcal{V}, \mathcal{F})$ and a joint probability function $f : \mathcal{U} \to \mathbb{R}_*$ defined over the sample space $\mathcal{U}$.*

Recalling that the endogenous variables in a causal model are uniquely determined by the exogenous variables, the probability distribution $f(u)$ over $\mathcal{U}$ determines a probability distribution over $\mathcal{V}$ by the relation

$$\mathbb{P}(V = v) = \int_{\{u : V(u) = v\}} f(u) \, du,$$

where the domain of integration consists of all possible combinations of $u$ that yield the value $V = v$, for any $V \subset \mathcal{V}$.

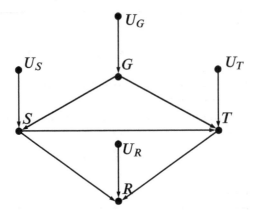

Fig. 8.3: The causal structure from Example 8.3. $G$ represents gender; $S$ represents severity; $T$ represents treatment; $R$ represents recovery.

*Example 8.4.* Let us again consider the Simpson's paradox example from Example 8.3. The causal model for the data is shown in Figure 8.3. The nodes are labeled $G$, $S$, $T$, and $R$ for gender, severity, treatment, and recovery, respectively.

Recall that gender influenced both treatment and severity, and that severity was another factor influencing treatment (52% of severe patients were assigned treatment A, whereas only 48% of mild patients were assigned the same treatment, reflecting a bias to assign severe cases to treatment A). Moreover, recovery rate is determined by severity and treatment, but not gender. Moreover, each of these factors could potentially be influenced by other (unknown) exogenous variables, represented by the $U$s. We leave the causal graphs for Examples 8.1 and 8.2 as an exercise for the reader.    ▷

### Spurious Association and Confounding Variables

We have previously examined the danger of making inferences from observational data during our discussion on Simpson's theory. Now that we have established a language for causality, we will add a few definitions to help describe what we observed.

**Definition 8.11.** *A spurious association is said to exist between two variables $X$ and $Y$ in a causal model if those variables are dependent, yet not causally connected.*

*Any common ancestor (whether or not observed) of two spuriously associated variables is said to be a confounding factor, or confounder.*

A classic example of a spurious association is ice cream sales and beach attendance. We discussed in the chapter introduction that both are caused by a confounding factor: the weather. In order to understand the causal effect that ice cream has on beach attendance, we therefore need to *control* for the sun. But, alas, we're getting ahead of ourselves. We will return to this issue shortly, but first we will examine several common structures in graphs and detail a few powerful results regarding their (conditional) independence.

### 8.2.3 Causal Building Blocks and Independence

Now that we have defined causal structures and models, we need a calculus for analyzing dependency relations in complex causal structures. There are three main classes of structures we will identify in our graphical models: chains, forks, colliders. Each of these has its own rules governing dependencies, which we will discuss in turn. We will conclude the section with a powerful result called *d*-separation. The goal is to be able to quickly identify likely dependencies within a given causal structure. Astonishingly, these building blocks will be valid regardless of the specific set of structural equations attributed to our model.

### Chains

We begin with a definition of the most basic structure one can spot in a graph: the chain.

**Definition 8.12.** *A* chain *of length* $n \geq 3$ *is a connected, directed acyclic graph comprised of* $n + 1$ *nodes and* $n$ *directed edges.*

Because the DAG is connected, and because there is one fewer edge than nodes, all chains are in the form of a *unidirectional path*. The simplest example consists of three nodes $(X, Y, Z)$ that are directional, in the sense that $X \to Y \to Z$. Such a chain is shown in Figure 8.4.

Fig. 8.4: A chain of length 3.

Of course, in the context of causal models, chains do not occur in a vacuum, but rather as part of a larger causal model. We can illustrate the dependency on the exogenous variables by adding them to our graph, as shown in Figure 8.5. As should be intuitively clear, the variables $X$ and $Y$ are likely dependent. Similarly, the variables $Y$ and $Z$ are likely dependent, as well as the variables $X$ and $Z$. We cannot conclude any of the variables are dependent, as there are certain intransitive cases. For example, suppose that the variable $X$ can only take on negative values, and the function $f_Y(X) = \text{sign}(X)$ is the sign function. In this case, the variable $Y$ will take the value $Y = -1$, regardless of the value of the variable $X$. We remind the reader of such degenerate cases by saying that $X$ and $Y$ are *likely* dependent.

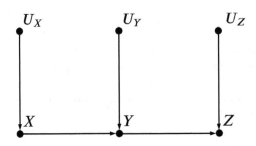

Fig. 8.5: A chain in a graphical causal model.

The other observation that should follow intuitively from Figure 8.5 is that the variables $X$ and $Z$ are *conditionally independent*, given the variable $Y$. In other words, variations of $X$ and $Z$ consistent with a single fixed $Y$ will be independent of each other. This is illustrated with the following example.

*Example 8.5.* Consider the causal model shown in Figure 8.5, and suppose the functional relationships are given by

$$f_X(U_X) = U_X,$$
$$F_Y(X, U_Y) = X + U_Y$$
$$F_Z(Y, U_Z) = Y + U_Z.$$

If we fix the value $Y = c$, then changes in $X$ will be commensurate with the opposite change in $U_Y$, but will not affect the variable $Y$. Since the variable $Y$ is held fixed, and hence not affected by changes in $X$, the variable $Z$ will also not be affected by changes in $X$, as $Z$ depends on $X$ only through its effect on $Y$. Thus, $X$ and $Z$ will vary independently of one another, when the value for $Y$ is fixed.                                         ▷

We can state this principle more formally as follows.

**Proposition 8.1 (Conditional Independence in Chains).** *Let $X$ and $Y$ represent two variables in a causal model that are connected by a single path, and let $\mathcal{Z}$ represent any set of variables that intercepts that path. If the unique path connecting $X$ and $Y$ is unidirectional (i.e., a chain), then $X$ and $Y$ are conditionally independent given $\mathcal{Z}$.*

## Forks

We now move on to another commonly found structure.

**Definition 8.13.** *A fork in a directed graph is a triple of nodes $(X, Y, Z)$ such that one of the nodes is a common parent to the other two. In a causal model, we call that common parent a* common cause *to the other two variables. If $X$ is the common cause of $Y$ and $Z$, we may represent this fork as $Y \leftarrow X \rightarrow Z$.*

The nodes $(X, Y, Z)$ in Figure 8.6 constitute a fork in a graphical causal model. As before, the variables $X$, $Y$, and $Z$ are all likely dependent, meaning the value of any one of the variables likely depends on the value of any other.

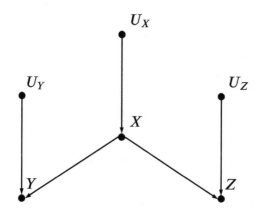

Fig. 8.6: A fork in a graphical causal model.

In addition, similar to the case of forks, the variables $Y$ and $Z$ are *conditionally independent*, given $X$. When $X$ is held fixed, variations in $Y$ and $Z$ must be due to $U_Y$ and $U_Z$, respectively, and, therefore, a variation in one will not influence a variation in the other. This is stated more formally as follows.

**Proposition 8.2 (Conditional Independence in Forks).** *If variable $X$ is a common cause to variables $Y$ and $Z$ in a causal model, and if there is only one path connecting the variables $Y$ and $Z$, then $Y$ and $Z$ are conditionally independent given $X$.*

## Colliders

Our third structure is slightly more sophisticated, though it is simply the causal reversal of a fork.

**Definition 8.14.** *A* collider *in a directed graph is a triple of nodes* $(X, Y, Z)$ *such that one node, referred to as the* collision node, *is a simultaneous child of the other two nodes. If $Z$ is a collision node, we may represent this collider as $X \rightarrow Z \leftarrow Y$.*

An example of a collider in a graphical model is shown in Figure 8.7. Note that by reversing the arrows $|X, Z\rangle$ and $|Y, Z\rangle$, we would instead have a fork. The analysis, however, is slightly more subtle.

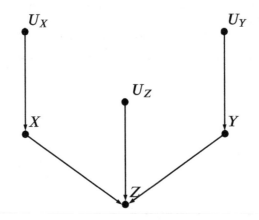

Fig. 8.7: A collider in a graphical causal model.

For the collider shown in Figure 8.7, the variables $X$ and $Z$ are likely dependent. Similarly, the variables $Y$ and $Z$ are likely dependent. (This follows because $X$ and $Y$ are both direct causes of the variable $Z$.) The variables $X$ and $Y$ themselves, however, are independent. The only factor that influences the variable $X$ is the variable $U_X$; therefore, a variation in $Y$ cannot manifest an influence in the variable $X$.

Now, in a fork, the two children are conditionally independent, given a fixed value for the common parent. Since a collider is a causally reversed fork, we might expect, therefore, that the two common parents in a collider are conditionally *dependent*, given a fixed value for their child. This turns out to be precisely the case: when we fix the value of the variable $Z$, we create a dependency among the variables $X$ and $Y$.

Despite following formally from reversing the logic of a fork, this result might seem at first counterintuitive. To understand it on a more intuitive level, consider the following example.

*Example 8.6.* Consider the collider shown in Figure 8.7. Suppose the variables have the following real-world meanings:

1. $X$ represents whether or not it is raining today;

2. $Y$ represents whether or not my sprinkler is on;
3. $Z$ represents whether or not my sidewalk is wet.

Clearly, $X$ and $Y$ are independent: the weather is not moved by my lawn-care habits. However, if we know that the sidewalk is wet, this creates a dependency in the random variables $X$ and $Y$! If my sidewalk is wet, but it is not raining, it is much more likely I am running my sprinkler, and vice versa!                                                                  ▷

**Proposition 8.3 (Independence and Conditional Dependence in Colliders).** *If a variable $Z$ in a causal model is the collision node between two variables $X$ and $Y$, and there is only one path between $X$ and $Y$, then $X$ and $Y$ are unconditionally independent, but dependent conditionally on $Z$ or any descendant of $Z$.*

### d-separation

So far, we have considered three basic causal structures: chains, forks, and colliders. Rarely, however, are causal models so simple. More commonly, one is likely to find multiple paths between nodes, each containing a combination of chains, forks, and colliders. Our goal then should be to determine a process or rule that can be applied to any graphical causal model, regardless its complexity, that reveals likely dependencies among the variables for any data set that is generated by the graph. The tool we will use to discover such dependencies is called d-separation, which we define as follows.

**Definition 8.15.** *A path $p$ in a directed graph is* blocked *by a set of nodes $\mathcal{Z}$ if and only if*

1. *the path $p$ contains a chain $A \to B \to C$ or a fork $A \leftarrow B \to C$, such that the middle node $B \in \mathcal{Z}$; or*
2. *the path $p$ contains a collider $A \to B \leftarrow C$ such that neither the collision node $B$, nor any of its descendants, are in $\mathcal{Z}$.*

*If the set $\mathcal{Z}$ blocks every path between $X$ and $Y$, we say that $X$ and $Y$ are* directionally separated, *or d-separated,* conditional on $\mathcal{Z}$.

*Two points $X$ and $Y$ in a directed graph that are not d-separated are said to be* d-connected.

The idea behind this definition is that if two points are d-separated, they must be independent. On the other hand, if they are d-connected, they are likely, but not necessarily, dependent.

**Theorem 8.1.** *If two nodes in a causal graph are d-separated by a set of nodes $\mathcal{Z}$, then those two nodes are independent conditionally on the set $\mathcal{Z}$.*

*Proof.* Consider a single path between two nodes $X$ and $Y$. Since $X$ and $Y$ are d-separated conditional on $\mathcal{Z}$, we know that $X$ and $Y$ must be conditionally independent, given $\mathcal{Z}$, as the conclusion of at least one of the results from Proposition 8.1–8.3 must hold. Therefore, $X$ and $Y$ would be conditionally independent, if the graph consisted of only this path. But since $\mathcal{Z}$ blocks *every* path between $X$ and $Y$, therefore $X$ and $Y$ must be conditionally independent, given $\mathcal{Z}$, considering the full set of paths between them. Since causal affect travels only through paths, the result holds.    □

*Example 8.7.* Consider the nodes $X$ and $Y$ in the causal graph shown in Figure 8.8. There are two paths (green and blue) connecting these two nodes. (Note that these paths need not be unidirectional.) In order to determine whether $X$ and $Y$ are d-separated, let us consider each path in turn.

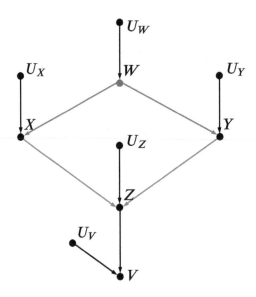

Fig. 8.8: d-separation in a causal graph; Example 8.7.

First, let us consider the *blue path*, consisting of the fork $X \leftarrow W \rightarrow Y$. Since this path contains (in fact, is) a fork, we know that it is blocked if and only if the branching point $W$ is in our set $\mathcal{Z}$. Therefore, from this path alone, we know that the condition $W \in \mathcal{Z}$ is a necessary condition for conditional independence.

Next, let us consider the *green path*, consisting of the collider $X \rightarrow Z \leftarrow Y$. Moreover, we note that $Z$ has a descendant, $V$. Our second necessary condition for conditional independence is, therefore, that neither $Z$ nor $V$ be included in our blocking set $\mathcal{Z}$.

Since these are our only two paths, we can conclude that $X$ and $Y$ are d-separated, conditional on a set $\mathcal{Z}$, if and only if $W \in \mathcal{Z}$ and $Z, V \notin \mathcal{Z}$. Now, since we have exhausted the entire graph, except for exogenous variables, we can state our conclusion, in its simplest form, that $X$ and $Y$ are conditionally independent given the set $\mathcal{Z} = \{W\}$, which is highlighted in red.                                                                        ▷

## 8.3 A Calculus of Interventions

So far, we have discussed observational studies and the danger of drawing conclusions from their data without a proper understanding of the underlying causal structure. We then went on to formalize how one can encode such a structure in mathematics, using graphical representations of the causal network. We then showed how one may make conclusions about conditional independence of various events in a causal model, using the concept of d-separation. But we still have not addressed the concerns of our main inquiry: how might we infer the causal effect between two events from observational data? This shall constitute our present goal, as we introduce a calculus for inferring such effects.

### 8.3.1 Interventions and Causal Effect

We begin by introducing the concept of interventions and the do-calculus. We then consider a simple example for which we are able to quantify the causal effect from observational data. We conclude the section with a generalization of the result. In subsequent sections of the chapter, we will further generalize the result to make it as broadly applicable as possible.

#### Graph Surgeries

Before formally defining interventions, we first require a few useful definitions regarding certain manipulations of graphs.

**Definition 8.16 (Backdoor/Frontdoor).** *Given a directed graph $G = \{\mathcal{V}, \mathcal{E}\}$, a* backdoor *of a node $X \in \mathcal{V}$ is any directed edge leading into $X$. Similarly, a* frontdoor *of a node $X$ is any directed edge emanating from $X$. Moreover, we define the sets*

$$\mathcal{E}^X = \{|P, X\rangle : P \in \mathrm{par}(X)\},$$
$$\mathcal{E}_X = \{|X, C\rangle : C \in \mathrm{child}(X)\}$$

*as the sets of backdoors and front doors to $X$, respectively.*

**Definition 8.17 (Graph Surgeries).** *Given a directed graph* $G = \{\mathcal{V}, \mathcal{E}\}$ *and a node* $X \in \mathcal{V}$, *we define the* manipulated graphs

$$G_{\overline{X}} = \{\mathcal{V}, \mathcal{E} \setminus \mathcal{E}^X\},$$
$$G_{\underline{X}} = \{\mathcal{V}, \mathcal{E} \setminus \mathcal{E}_X\}$$

*as the graphs obtained by (surgically) removing all backdoors or front doors of the node* $X$, *respectively.*

*Graph surgeries are additive; for example, given the nodes* $X$ *and* $Y$, *we can define the manipulated graph*

$$G_{\overline{X}\underline{Y}} = \{\mathcal{V}, \mathcal{E} \setminus \mathcal{E}^X \setminus \mathcal{E}_Y\}$$

*in the obvious way: by removing all backdoors of* $X$ *and front doors of* $Y$.

Each of the graph surgeries from this definition are illustrated in Figure 8.9.

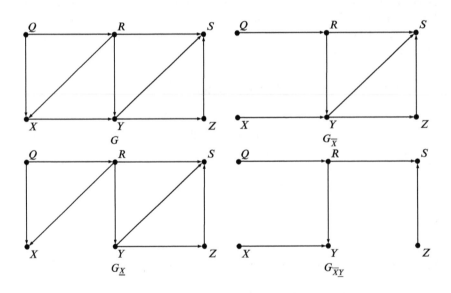

Fig. 8.9: Illustration of the graph surgeries defined in Definition 8.17.

**Interventions**

In observational studies, scientists are not able to properly randomize any aspect of the data. We therefore seek an alternative approach to properly *control* for potential spurious effects in order to draw the correct conclusion. In particular, we seek to draw conclusions about what effect a choice on the value of a variable would have made on a particular outcome, if we were able to make that choice. Such a choice is called en *intervention*.

In the context Example 8.1, we wish to know what the effect our choice of treatment has on recovery. The problem is that the data are purely observational: we had no *choice* on the treatment. Thus, an intervention is a form of counterfactual: what would the effect have been, had we been able to intervene on a variable $X$, removing all influences on it and forcing it to take a specific value? Alternatively: what would the effect have been, had we been able to go in and surgically remove all influences on a variable (i.e., the treatment) and instead *intervene* by prescribing a treatment?

There is an important distinction between intervention and conditioning:

1. When we intervene on a variable $X$, we change the system by fixing the value $X = x$ of a variable, thereby causing a change in the values of the other variables as a result;
2. When we condition on a variable $X$, we do not change the system, but instead narrow our focus on the subset of data for which $X = x$ organically.

To make this concept more formal, we have the following.

**Definition 8.18.** *Let $X$ be an endogenous variable in a causal model $M = (\mathcal{U}, \mathcal{V}, \mathcal{F}, \mathbb{P})$. Then an* intervention *on $X$ with value $x$, denoted* $\mathrm{do}(X = x)$, *represents a counterfactual reality described by a specific manipulated model $M_{X=x}$ obtained by replacing $G$ with $G_{\overline{X}}$ (see Definition 8.17) and the structural equation for $X$, i.e., $X = f_X(\mathrm{par}(X), U_X)$, with the constant equation $X = x$.*

*Given an intervention* $\mathrm{do}(X = x)$, *we define its* causal effect *(or,* causal influence*) on the variable $Y$ (alternatively, the* postintervention distribution *of outcome $Y$) as*

$$\mathbb{P}(Y = y | \mathrm{do}(X = x)) = \mathbb{P}_{M_{X=x}}(Y = y),$$

*where $\mathbb{P}_{M_{X=x}}$ represents the probability under the manipulated model $M_{X=x}$.*

Thus, an intervention on $X$ does two things: removes all parents (i.e., direct causes) of $X$ from the causal structure, and replaces the structural equation that determines $X$ with a constant, as prescribed by the intervener.

*Note 8.1.* Interventions are additive: the intervention $\mathrm{do}(X = x, Y = y)$ is obtained by the manipulated model $M_{X=x, Y=y}$, which is defined by replacing the causal structure

$$G \to G_{\overline{XY}}$$

and the *two* structural equations

$$\left\{ \begin{array}{l} X = f_X(\mathrm{par}(X), U_X) \\ Y = f_Y(\mathrm{par}(Y), U_Y) \end{array} \right\} \quad \rightarrow \quad \left\{ \begin{array}{l} X = x \\ Y = y \end{array} \right\}.$$

In other words, we remove all backdoors for $X$ *and* $Y$, and replace the corresponding structural equations with their respective constants.    ▷

*Note 8.2.* Conditional probabilities under interventions are defined in the obvious way; i.e.,

$$\mathbb{P}(Y = y | \mathrm{do}(X = x), Z = z) = \mathbb{P}_{M_{X=x}}(Y = y | Z = z);$$

i.e., by the conditional probability in the manipulated model.    ▷

*Note 8.3.* An alternate definition of an intervention is as follows: an intervention on $X$ is a node $I_X$ that can take on a null value $\emptyset$ or any value that the variable $X$ can take on. The structural equation for $X$ is then modified to account for the possibility of an intervention:

$$X = \hat{f}_X(\mathrm{par}(X), U_X, I_X) = \begin{cases} f_X(\mathrm{par}(X), U_X) & \text{if } I_X = \emptyset \\ I_X & \text{otherwise} \end{cases}.$$

In this case, the node $I_X$ acts as a switch that, when turned on, ignores all natural effects on the variable $X$, replacing them instead with a constant prescribed by the intervener.    ▷

*Example 8.8.* A classic example of an intervention involves three random variables: $R$ represents whether or not it rained today; $S$ represents whether or not the sprinklers rand; and $W$ represents whether or not the driveway pavement is wet. However, suppose that we have a smart sprinkler that is supposed to detect whether or not it has rained. It is not perfect, but the probability that the sprinkler runs does depend on whether or not it has rained. Naturally, whether or not the pavement is wet depends on both $R$ and $S$. A simple causal structure for this model is shown in Figure 8.10.

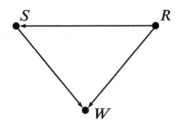

Fig. 8.10: DAG for Example 8.8: $R$ represents rain, $S$ represents sprinkler, and $W$ represents wet pavement.

Next, let us assign some specific probabilities. Let us suppose that there is a 30% chance of rain. If it does rain, the sprinkler will be activated with only a 10% probability. If it does not rain, the sprinkler will be activated with a 30% probability. (The sprinkler is programmed to run about twice a week.)

Finally, the probability that the pavement is wet depends on both the sprinkler and weather conditions. On days in which it does not rain, the pavement will be wet with 0% probability if the sprinkler is not on and a 10% probability if the sprinkler is on. On days in which it does rain, the pavement will be wet with 85% probability if the sprinkler is not on and a 90% probability if the sprinkler is on. These probabilities are illustrated graphically in the decision diagram of Figure 8.11.

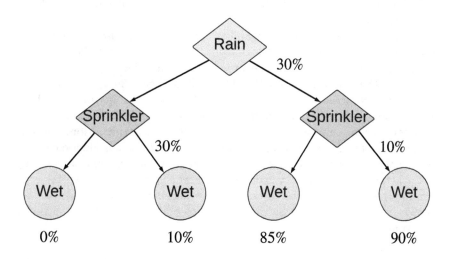

Fig. 8.11: Conditional probabilities in a decision diagram for Example 8.8. Right arrows coming out of decision nodes (diamonds) represent true; left arrows represent false.

In this model, we wish to illustrate the difference between the conditional probability $\mathbb{P}(W = 1|S = 1)$ and the causal effect $\mathbb{P}(W = 1|\mathrm{do}(S = 1))$. Since we haven't yet discussed the formulas to determine the causal effect, we will construct a data set that conforms to our prescribed probability structure. Such a data set, comprised of 1,000 total records, is given in Table 8.5.

First, let us consider the conditional probability $\mathbb{P}(W = 1|S = 1)$; i.e., what is the probability that the pavement is wet given that the sprinkler has run? To compute this, we must condition on the outcomes in which the sprinkler was running, which occurred 240 times. Out of those 240 times in which the sprinkler was running, the pavement was wet a total of 48 times. Hence, the conditional probability is given by

| | No Rain 700 (70%) | | Rain 300 (30%) | |
|---|---|---|---|---|
| | No Sprinkler 490 (70%) | Sprinkler 210 (30%) | No Sprinkler 270 (90%) | Sprinkler 30 (10%) |
| Wet Pavement | 0/490 (0%) | 21/210 (10%) | 229.5/270 (85%) | 27/30 (90%) |

Table 8.5: Data following our probability model for Example 8.8.

$$\mathbb{P}(W = 1 | S = 1) = \frac{21 + 27}{210 + 30} = 20\%.$$

In particular, notice that we by conditioning we have narrowed our focus on a subset of the overall data.

Next, let us compute the probability of wet pavement *given an intervention* do$(S = 1)$. In other words, what is the probability that the pavement is wet, had we stepped in and ensured that the sprinkler was running?

Under the intervention do$(S = 1)$, it will still rain 30% of the time, as the weather does not care whether or not we are running our sprinkler. What changes, however, is whether or not our sprinkler is running. Under an intervention, our sprinkler *is* running, 100% of the time. Moreover, our probability of wet pavement *given* both the weather and sprinkler condition is unchanged. (We are changing whether or not the sprinkler is running, not whether or not the pavement will get wet *given* that the sprinkler is running.) Thus, in our altered reality, we would have obtained the data set shown in Table 8.6, if we had intervened to make sure that the sprinkler was running every single day.

| | No Rain 700 (70%) | | Rain 300 (30%) | |
|---|---|---|---|---|
| | No Sprinkler 0 (0%) | Sprinkler 700 (100%) | No Sprinkler 0 (0%) | Sprinkler 300 (100%) |
| Wet Pavement | 0/0 (0%) | 70/700 (10%) | 0/0 (85%) | 270/300 (90%) |

Table 8.6: Data for Example 8.8 *under the intervention* do$(S = 1)$; i.e., what our data would have looked like in the altered world in which we had ensured our sprinkler had run every day.

In our altered world of Table 8.6, it is clear that we now have wet pavement on 340 days out of 1,000, or 34% of the time. Since this altered world is the world we obtained by conditioning on our intervention, the causal effect of running our sprinkler on the pavement is therefore

$$\mathbb{P}(W = 1 | \text{do}(S = 1)) = 34\%.$$

This simple example illustrates the divergence between the conditional probability $\mathbb{P}(W = 1|S = 1) = 20\%$ and the intervention effect $\mathbb{P}(W = 1|\mathrm{do}(S = 1))$.

$\triangleright$

### 8.3.2 The Adjustment Formula

Now that we have defined what an intervention is, we will develop a preliminary formula that can be used for computation. In order to account for spurious associations, we must devise a method to control for, or block, confounding factors. The most simplest approach is to control for a variable's direct causes: this basic result is known as the adjustment formula. We will first develop it for a simple case, and then more generally. Subsequent sections in the chapter will continue to build on this basic adjustment formula, extending it to more general contexts.

### Adjustment for Direct Causes

The adjustment formula will be our first, simple tool that will allow us to control for, or adjust for, confounding factors. With it, we will be able to describe the influence of an intervention based solely on the observed probabilities in our actual universe (as opposed to the probabilities within the manipulated model). This, however, is but a first step; we will extend this to more general settings in upcoming sections.

The adjustment formula, however, will rely on the following definition.

**Definition 8.19 (Direct Sum).** *Let* $\mathcal{X} = \{X_1, \ldots, X_n\}$ *be a finite set of* $n$ *random variables over a common probability space* $(\Omega, \mathcal{E}, \mathbb{P})$. *Then the direct sum over* $\mathcal{X}$, *denoted* $\oplus \mathcal{X}$, *is the random vector*

$$\oplus \mathcal{X} = \bigoplus_{i=1}^{n} X_i = \langle X_1, \ldots, X_n \rangle,$$

*with the natural range*

$$\mathrm{range}(\oplus \mathcal{X}) = \{\langle x_1, \ldots, x_n \rangle : x_i \in \mathrm{range}(X_i) \ for \ i = 1, \ldots, n\};$$

*i.e., the set of all possible combinations of the individual random variable.*

For example, if random variables $X$ and $Y$ are binary, the direct sum $X \oplus Y$ is a two-dimensional random binary vector that takes on four possible values in $\mathbb{B}^2$. More generally, if $\mathcal{X} = \{X_1, \ldots, X_n\}$ is a set of $n$ random binary variables, the direct sum $\oplus \mathcal{X}$ will be an $n$-dimensional random binary vector, that takes on $2^n$ possible values in $\mathbb{B}^n$.

We next devise a method method for determining causal effect of a variable $X$ on a variable $Y$ by explicitly controlling for each of the parents of variable $X$. We begin with two lemmas.

**Lemma 8.1.** *Let $X$ and $Z$ be two variables in an acyclic causal model $M$, such that $Z$ is a direct cause of $X$; i.e., $Z \in \text{par}(X)$. Let $m = M_{X=x}$ represent the manipulated model representing the intervention $\text{do}(X = x)$. Then*

$$\mathbb{P}_m(Z = z | X = x) = \mathbb{P}_m(Z = z) = \mathbb{P}(Z = z); \qquad (8.2)$$

*i.e., $X$ and $Z$ are unconditionally independent in the modified model and, moreover, the probability distribution of $Z$ is invariant under the intervention $\text{do}(X = x)$.*

*Proof.* First, let us consider the unconditional independence of $X$ and $Z$ in the modified model. All paths between $X$ and $Z$ have been removed, so the only possible way they are dependent is through a path connecting $X$ and $Z$ indirectly. However, since the model is acyclic, we know there is no directed path between $X$ and $Z$, as such a path, when added to the missing link $|X, Z\rangle$, would constitute a cycle. Any path connecting $X$ and $Z$ in the modified model must therefore necessarily contain a collider, making $X$ and $Z$ unconditionally independent. See Figure 8.12 for an illustration. Thus, in the modified model, $\mathbb{P}_m(Z = z | X = x) = \mathbb{P}_m(Z = z)$.

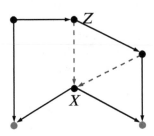

Fig. 8.12: Try as we might, with the backdoors of $X$ (shown in purple, dashed lines) removed, we cannot construct a path between $X$ and $Z$ that does not contain a collider (red nodes). Also, no directed path from $Z$ to $X$ exists, as all edges leading into $X$ have been removed.

Now, let us prove that the probability distribution of $Z$ is invariant under the distribution. The value of $Z$ is obtained from the structural equation

$$Z = f_Z(\text{par}(Z), U_Z).$$

Since each parent variable $Z$ is determined by a similar structural equation, we may more generally state that the value of $Z$ is uniquely determined given the values of all ancestors of $Z$ and their respective exogenous counterparts. Since the graph is acyclic, $\text{anc}(Z)$ does not contain any descendants of $X$, as $X$ is a child of $Z$. (If one of $Z$'s ancestors were a descendant of $X$, this would create a cycle.)

Now, the intervention removes the causal links between $X$ and $Z$, fixing the value of $X$. This potentially affects the probability distribution of the descendants of $X$, but cannot affect the value of any of the ancestors or parents of $Z$. Thus, the value of $Z$ is already determined, and remains invariant under the intervention. □

**Lemma 8.2.** *Let $X$ and $Y$ be two variables in an acyclic causal model $M$, such that $Y$ is not a direct cause of $X$; i.e., $Y \notin \mathrm{par}(X)$, and let $Z = \oplus\mathrm{par}(X)$ be the random vector consisting of all direct causes (parents) of $X$. Let $m = M_{X=x}$ represent the manipulated model representing the intervention $(X = x)$. Then*

$$\mathbb{P}_m(Y = y \mid X = x, Z = z) = \mathbb{P}(Y = y \mid X = x, Z = z); \qquad (8.3)$$

*i.e., the conditional probability of $Y$ given the value of $X$ and each of its parents in invariant under the intervention $\mathrm{do}(X = x)$.*

*Proof.* The manipulated model changes the functional relationship between $X$ and each of its parents. Such an intervention therefore cannot affect the conditional probability distribution over $Y$, as long as we condition on the values of both $X$ and its parents. Essentially, we have sealed off the effect of the intervention—the removal of the causal connections between $X$ and its parents—from the rest of the graph, by conditioning on the full set of variables that contain the intervention. We illustrate this in Figure 8.13. □

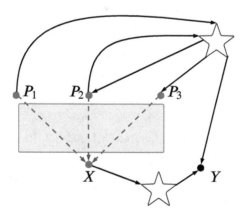

Fig. 8.13: Illustration of Lemma 8.2. By conditioning on both $X$ and its parents ($P_1$, $P_2$, and $P_3$), the interactions within the control box are irrelevant, from the perspective of $Y$. The star shapes represent subgraphs of arbitrary complexity.

Together, these lemmas allow us to capture a powerful result, which allows us to quantify the causal effect of an intervention with reference solely to the original (non-manipulated) model that generated our observational data.

**Theorem 8.2 (The Adjustment Formula).** *Let $X$ be a variable in an acyclic causal model $M$ and let $Z = \oplus \mathrm{par}(X)$ be the random vector consisting of all direct causes (parents) of $X$. Let $Y$ be a variable (or a set of variables) disjoint from $\{X\} \cup \mathrm{par}(X)$. The causal effect of $X$ on $Y$ is then given by the following adjustment formula:*

$$\mathbb{P}(Y = y | \mathrm{do}(X = x)) = \sum_{z \in \mathcal{R}_Z} \mathbb{P}(Y = y | X = x, Z = z)\mathbb{P}(Z = z), \quad (8.4)$$

*where $\mathcal{R}_Z = \mathrm{range}(Z)$ is the set of all possible values of the random vector $Z$.*

When we apply the adjustment formula, we say we are adjusting for, or controlling for, the direct causes of $X$.

We may alternatively express Equation (8.4) as

$$\mathbb{P}(Y = y | \mathrm{do}(X = x)) = \sum_{z \in \mathcal{R}_Z} \frac{\mathbb{P}(X = x, Y = y, Z = z)}{\mathbb{P}(X = x | Z = z)}, \quad (8.5)$$

*the denominators of which are known as the* propensity scores.

*Proof.* By definition,

$$\mathbb{P}(Y = y | \mathrm{do}(X = x)) = \mathbb{P}_m(Y = y | X = x).$$

By applying the law of total probability, we can now write

$$\mathbb{P}_m(Y = y | X = x) = \sum_{z \in \mathcal{R}_Z} \mathbb{P}_m(Y = y | X = x, Z = z)\mathbb{P}_m(Z = z | X = x).$$

However, in the manipulated model, the causal connection between $X$ and $Z$ is broken, so that

$$\mathbb{P}_m(Z = z | X = x) = \mathbb{P}_m(Z = z).$$

Our result follows by applying Lemmas 8.1 and 8.2. In summary:

$$\mathbb{P}(Y = y | \mathrm{do}(X = x)) = \mathbb{P}_m(Y = y | X = x)$$
$$= \sum_{z \in \mathcal{R}_Z} \mathbb{P}_m(Y = y | X = x, Z = z)\mathbb{P}_m(Z = z | X = x)$$
$$= \sum_{z \in \mathcal{R}_Z} \mathbb{P}_m(Y = y | X = x, Z = z)\mathbb{P}_m(Z = z)$$
$$= \sum_{z \in \mathcal{R}_Z} \mathbb{P}(Y = y | X = x, Z = z)\mathbb{P}(Z = z).$$

Equation (8.5) is obtained by multiplying and dividing the summand in Equation (8.4) by the propensity score $\mathbb{P}(X = x | Z = z)$, and then simplifying, using the definition of conditional probability. This completes the proof.  $\square$

Notice the difference between the adjustment formula Equation (8.4), which conditions on an intervention, versus regular conditioning,

$$\mathbb{P}(Y = y | X = x) = \sum_{z \in \mathcal{R}_R} \mathbb{P}(Y = y | X = x, Z = z)\mathbb{P}(Z = z | X = x),$$

which clearly yields a different result.

*Example 8.9.* Let us consider again our rain–sprinkler–pavement Example 8.8. Let us recompute both the conditional probability $\mathbb{P}(W = 1 | S = 1)$ and the causal effect $\mathbb{P}(W = 1 | \mathrm{do}(S = 1))$ again, this time using the appropriate formulas as opposed to a made-up data set.

The conditional probability is given by

$$\begin{aligned}\mathbb{P}(W = 1 | S = 1) &= \mathbb{P}(W = 1 | S = 1, R = 0)\mathbb{P}(R = 0 | S = 1)\\ &\quad + \mathbb{P}(W = 1 | S = 1, R = 1)\mathbb{P}(R = 1 | S = 1)\\ &= (0.10)(7/8) + (0.90)(1/8) = 0.20,\end{aligned}$$

which agrees with our previous result. (We will discuss how we obtained the probabilities $\mathbb{P}(R = 0 | S = 1)$ and $\mathbb{P}(R = 1 | S = 1)$ momentarily.)

The causal effect, on the other hand, is obtained by

$$\begin{aligned}\mathbb{P}(W = 1 | \mathrm{do}(S = 1)) &= \mathbb{P}(W = 1 | S = 1, R = 0)\mathbb{P}(R = 0)\\ &\quad + \mathbb{P}(W = 1 | S = 1, R = 1)\mathbb{P}(R = 1)\\ &= (0.10)(0.70) + (0.90)(0.30) = 0.34,\end{aligned}$$

which again agrees with our previous result.

The conditional probability $\mathbb{P}(R = 1 | S = 1)$ used in the first calculation can be obtained using an application of Bayes' law, which we will discuss in Chapter 10. For now, however, we will simply rely on the result, which states

$$\mathbb{P}(R = 1 | S = 1) = \frac{\mathbb{P}(S = 1 | R = 1)\mathbb{P}(R = 1)}{\mathbb{P}(S = 1)}.$$

Now,

$$\begin{aligned}\mathbb{P}(S = 1) &= \mathbb{P}(S = 1 | R = 1)\mathbb{P}(R = 1) + \mathbb{P}(S = 1 | R = 0)\mathbb{P}(R = 0)\\ &= (0.10)(0.30) + (0.30)(0.70) = 0.24.\end{aligned}$$

Therefore,

$$\mathbb{P}(R = 1 | S = 1) = \frac{\mathbb{P}(S = 1 | R = 1)\mathbb{P}(R = 1)}{\mathbb{P}(S = 1)} = \frac{(0.10)(0.30)}{(0.24)} = 0.125,$$

or $1/8$. Naturally, $\mathbb{P}(R = 0 | S = 1) = 1 - \mathbb{P}(R = 1 | S = 1) = 7/8$.

It is interesting to note that, with Bayes' rule, we see that our formula for the regular conditional probability is equivalent to

$$\mathbb{P}(W = 1|S = 1) = \mathbb{P}(W = 1|S = 1, R = 0)\mathbb{P}(R = 0)\frac{\mathbb{P}(S = 1|R = 0)}{\mathbb{P}(S = 1)}$$

$$+\mathbb{P}(W = 1|S = 1, R = 1)\mathbb{P}(R = 1)\frac{\mathbb{P}(S = 1|R = 1)}{\mathbb{P}(S = 1)},$$

which differs from the formula for the causal effect by the factors

$$\frac{\mathbb{P}(S = 1|R = 0)}{\mathbb{P}(S = 1)} \quad \text{and} \quad \frac{\mathbb{P}(S = 1|R = 1)}{\mathbb{P}(S = 1)}.$$

This provides some additional insight, as it shows that the causal effect $\mathbb{P}(W = 1|\text{do}(S = 1))$ is equivalent to the regular conditional probability $\mathbb{P}(W = 1|S = 1)$ precisely when $R$ and $S$ are independent. This makes a good deal of sense: if $R$ and $S$ are independent, then $R$ is not a parent of $S$, and removing the link between $R$ and $S$ does absolutely nothing.    ▷

*Example 8.10.* Let us return to our single-layer Simpson's paradox Example 8.1, which considers only gender, treatment, and recovery. The DAG for this model is shown in Figure 8.14 (a).

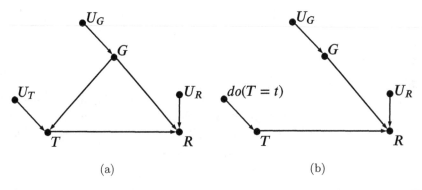

Fig. 8.14: Simpson's paradox example, considering gender $G$, treatment $T$, and recovery $R$. (a) Original model. (b) Manipulated model, showing an intervention on $T$.

The aggregated probabilities in Table 8.2 can be found by conditioning on gender and applying the law of total probability:

$$\mathbb{P}(R = 1|T = A) = \mathbb{P}(R = 1|T = A, G = F)\mathbb{P}(G = F|T = A)$$

$$+\mathbb{P}(R = 1|T = A, G = M)\mathbb{P}(G = M|T = A)$$

$$= (0.70)(0.20) + (0.50)(0.80) = 0.54,$$

where the individual probabilities are given in Table 8.1. Similarly,

$$\mathbb{P}(R = 1|T = B) = \mathbb{P}(R = 1|T = B, G = F)\mathbb{P}(G = F|T = B)$$
$$+\mathbb{P}(R = 1|T = B, G = M)\mathbb{P}(G = M|T = B)$$
$$= (0.65)(0.80) + (0.45)(0.20) = 0.61,$$

which agrees with our aggregated result from Table 8.2.

However, these results are contrary to the more granular results when we look at the data by gender. To resolve this paradox, we can *control for gender* by using the adjustment formula:

$$\mathbb{P}(R = 1|\text{do}(T = A)) = \mathbb{P}(R = 1|T = A, G = F)\mathbb{P}(G = F)$$
$$+\mathbb{P}(R = 1|T = A, G = M)\mathbb{P}(G = M)$$
$$= (0.70)(0.50) + (0.50)(0.50) = 0.60,$$
$$\mathbb{P}(R = 1|\text{do}(T = B)) = \mathbb{P}(R = 1|T = B, G = F)\mathbb{P}(G = F)$$
$$+\mathbb{P}(R = 1|T = B, G = M)\mathbb{P}(G = M)$$
$$= (0.65)(0.50) + (0.45)(0.50) = 0.55.$$

These probabilities reflect the recovery rates when we control for gender by intervening on the treatment, as shown in Figure 8.14 (b). These results show that treatment A is in fact superior to treatment B when we control for gender, a confounding factor.                                         ▷

The adjustment formula can control for confounding variables; however, *the adjustment formula cannot control for our ignorance!* Example 8.10 shows that when we apply the adjustment formula to the causal model shown in Figure 8.14, we obtain the corrected result, *as far as gender is concerned.* However, recall that we changed the story in Example 8.3 by adding an additional layer of granularity: severity of illness. The data now support a modified causal model (shown in Figure 8.3), in which gender is no longer a direct cause of recovery; instead it affects recovery only through its influence on severity and treatment.

In conclusion, applying the adjustment formula at the layer of gender did resolved Simpson's paradox *in the context of the paradoxical result between Tables 8.1 and 8.2.* The adjustment formula did not, however, extend its magic hand into further subdivisions of the data based on severity, which, when taken into account, led us to a different causal structure and conclusion.

### 8.3.3 The Backdoor Criterion

So far, we have seen how the adjustment formula can control for direct causes of a given variable whose causal effect we are trying to measure. As it turns out, there is a more general (and more powerful) alternative than controlling for each of the a variable's parents. The idea is as follows: if we are trying to measure the causal effect that $X$ has on $Y$, instead of controlling for each of the parents of $X$, we can instead control for an alternate set of variables that

1. blocks all spurious paths between $X$ and $Y$,
2. does not affect any directed path between $X$ and $Y$,
3. does not create any new spurious relationships between $X$ and $Y$.

In other words, we seek to control for a minimal set of confounding variables in order to remove any spurious relationship between $X$ and $Y$. To achieve this, we must first identify a certain set of path between the two variables.

**Definition 8.20 (Backdoor/Front-door Path).** *Given a node $X$ in a directed graph $G$, a* backdoor path *of $X$ is any path that contains $X$ as an endpoint and is furthermore connected to $X$ through a backdoor of $X$; i.e., the path connects to $X$ with an arrow that is directed into $X$.*

*Similarly, a* front-door path *of $X$ is any path that is connected to $X$, as an endpoint, through a front door of $X$.*

*When a backdoor/front-door path of $X$ connects to a second endpoint $Y$, we call it a* backdoor/front-door path from $X$ to $Y$.

*Warning! The language "from $X$ to $Y$" does not imply that a backdoor/front-door path is unidirectional; rather the directional implication "from $X$" merely implies that it is a backdoor/front-door path of $X$, but it is free to connect to its secondary endpoint $Y$ as it will.*

An example backdoor path from $X$ to $Y$ is

$$X \leftarrow A \rightarrow B \rightarrow C \leftarrow D \rightarrow E \leftarrow Y;$$

but if we change the direction of the first arrow, $X \rightarrow A \cdots$, we would instead have a front-door path.

Since spurious associations between $X$ and $Y$ must operate through the backdoor paths of $X$, we next make the following useful definition.

**Definition 8.21 (Backdoor Criterion).** *Given an ordered[2] pair of variables $(X, Y)$ in a directed acyclic graph $G$, a set of variables $\mathcal{Z}$ satisfies the* backdoor criterion *(BDC) relative to $(X, Y)$ if*

1. *$\mathcal{Z}$ does not contain any descendants of $X$; i.e., $\mathcal{Z} \cap \mathrm{desc}(X) = \emptyset$; and*
2. *$\mathcal{Z}$ blocks all backdoor paths from $X$ to $Y$.*

Naturally, we should not need to control for each parent of $X$, as long as we can control for a set of variables that satisfies the backdoor criterion, thereby blocking all possible spurious paths between $X$ and $Y$. This is the result of the following theorem.

---

[2] relative to the partial ordering of the set; i.e., $Y \in \mathrm{desc}(X)$.

**Theorem 8.3.** *Let $(X, Y)$ be an ordered pair of variables in a causal model $M = (\mathcal{U}, \mathcal{V}, \mathcal{F}, \mathbb{P})$, and let $\mathcal{Z}$ be a set of variables that satisfies the backdoor criterion relative to $(X, Y)$. Then the causal effect of $X$ on $Y$ is given by the* backdoor adjustment formula

$$\mathbb{P}(Y = y | \mathrm{do}(X = x)) = \sum_{z \in \mathcal{R}_Z} \mathbb{P}(Y = y | X = x, Z = z) \mathbb{P}(Z = z); \quad (8.6)$$

*where $Z = \oplus \mathcal{Z}$ and $\mathcal{R}_Z = \mathrm{range}(Z)$, as usual. When $\mathcal{Z} = \emptyset$ is the empty set, this degenerates to*

$$\mathbb{P}(Y = y | \mathrm{do}(X = x)) = \mathbb{P}(Y = y | X = x), \quad \text{when } \mathcal{Z} = \emptyset. \quad (8.7)$$

*Note 8.4.* Equations (8.4) and (8.6) are identical in form; the only distinction is what constitutes the set $\mathcal{Z}$. ▷

*Note 8.5.* Theorem 8.2 is a special case of Theorem 8.3, as the full set of parents of a variable $X$ automatically satisfy the backdoor criterion relative to $(X, Y)$; i.e., the parents of $X$ must necessarily block all backdoor paths from $X$ to $Y$, as $X$ can only communicate through backdoor channels via its parents. ▷

*Example 8.11.* To illustrate the usefulness of this result, consider Figure 8.15. Using the original adjustment formula, we would have to control

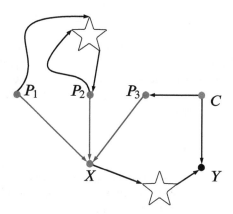

Fig. 8.15: Illustration of Theorem 8.3. Instead of controlling for parents $(P_1, P_2, P_3)$, we can instead control for the single confounding variable $C$.

for each of the three parent variables $P_1$, $P_2$, and $P_3$. However, we notice that there are no backdoor paths from $X$ to $Y$ that travel through $P_1$ or $P_2$; thus, we have cut out unnecessary work for ourselves. It suffices to control for $P_3$ alone, which blocks the backdoor path

$$X \leftarrow P_3 \leftarrow C \rightarrow Y,$$

as it blocks the chain $X \leftarrow P_3 \leftarrow C$. Let's take this a step further, however, and notice that we could, alternatively, control for $C$, as the variable $C$ blocks the fork $P_3 \leftarrow C \rightarrow Y$. It is always attractive to have alternatives. Suppose, for sake of argument, that the parent variable $P_3$ possesses a high cardinality, taking on, say, ten or twenty possible values. If variable $C$, on the other hand, were to be low cardinality, perhaps even binary, we immediately see the savings in terms of the number of enumerations—and hence subdivisions of our data—when carrying out the summation.     ▷

*Example 8.12.* Let us again return to our original Simpson's paradox example, but the double-paradox example illustrated in Example 8.3 that took into account both gender and illness severity. We illustrated the causal structure of these data in Example 8.4; in particular, by the DAG represented in Figure 8.3.

By careful examination of the DAG in Figure 8.3, we immediately see our folly in trying to control for gender alone. In seeking to understand the causal effect of treatment on recovery, i.e., $\mathbb{P}(R = 1|\mathrm{do}(T = t))$, it is now immediately clear that gender $G$ blocks the backdoor path

$$p_1 : T \leftarrow G \rightarrow S \rightarrow R,$$

but it leaves the backdoor path

$$p_2 : T \leftarrow S \rightarrow R$$

unblocked! Controlling for severity alone, on the other hand, blocks both backdoor paths. In fact, we see that either set $\mathcal{Z} = \{S\}$ or set $\mathcal{Z} = \{S, G\}$ accomplish the same goal in blocking both of our backdoor paths. We can therefore obtain our desired result by aggregating by illness severity, and then controlling for severity.     ▷

### 8.3.4 The Front-door Criterion

We saw in Example 8.10 that the backdoor adjustment formula cannot control for our ignorance. For example, if we *only* consider gender in the gender–severity–illness–recovery problem, we will not catch the true culprit: severity. But what if you are in a legal battle with a foe using ignorance as a defense—not negligible ignorance, but the ignorance of certain *unknown unknowns* that we could never possibly control for? Such was the case in the years prior to 1970, when the tobacco industry made such a defense in order to successfully lobby *against* antismoking legislation.

*Example 8.13.* Consider the causal model depicted by the DAG in Figure 8.16. In trying to assess the causal effect that smoking has on lung cancer, the devil's advocates argued successfully (at least, for some time)

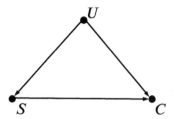

Fig. 8.16: Causal relation between smoking $S$ and lung cancer $C$, with an unobserved confounder $U$ (an *unknown unknown*).

that it was possible that an unobserved confounding variable could be leading to a spurious association between smoking and cancer. If such an idea seems laughable and transparent from our modern perspective, try proving, for example, that gender does *not cause* recovery in our Simpson's paradox example, by using Table 8.1 alone; i.e., without reference to the further subdivision of illness severity (Table 8.4). Come to think of it, how can we prove that any of our results are valid? Isn't there always the possibility of an unseen hand wreaking havoc over our data?

In the case of the tobacco industry, lobbyists argued that it is possible that an as-of-yet undiscovered genotype ($U$ in the diagram) could be causing *both* the propensity of an individual to have a penchant towards smoking *and* the likelihood for that individual to end up with lung cancer later in life. In other words, the same gene that makes a person likely to smoke is also responsible for that individual's later ailments. And since it would be unethical to performed a controlled experiment, it seemed like hope was lost in proving the effect that smoking has on lung cancer.    ▷

The resolution should be obvious: if the backdoor criterion doesn't work, maybe we could try a front-door criterion? Consider the modified causal diagram of Figure 8.17, which shows the addition of a new variable $Z$, representing the amount of tar deposits in an individual's lungs. The idea is to use this intermediate variable with two applications of the backdoor criterion in order to infer the causal effect that smoking has on lung cancer. To achieve this, we next define the front-door analog to Definition 8.21.

**Definition 8.22 (Front-door Criterion).** *Given an ordered pair of variables $(X, Y)$ in a directed acyclic graph $G$, a set of variables $\mathcal{Z}$ satisfies the front-door criterion (BDC) relative to $(X, Y)$ if*

1. *$\mathcal{Z}$ intercepts all directed paths from $X$ to $Y$;*
2. *there are no unblocked backdoor paths from $X$ to any variable in $\mathcal{Z}$;*
3. *all backdoor paths from the variables in $\mathcal{Z}$ to $Y$ are blocked by $X$.*

For example, the single variable $Z$ satisfies the front-door criterion relative to $(S, C)$ in Figure 8.16.

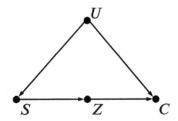

Fig. 8.17: Causal relation between smoking $S$, tar deposits $Z$, and lung cancer $C$, with an unobserved confounder $U$ (an *unknown unknown*).

**Theorem 8.4.** *Let $(X, Y)$ be an ordered pair of variables in a causal model $M = (\mathcal{U}, \mathcal{V}, \mathcal{F}, \mathbb{P})$, and let $\mathcal{Z}$ be a set of variables that satisfies the front-door criterion relative to $(X, Y)$. Then the causal effect of $X$ on $Y$ is given by the* front-door adjustment formula

$$\mathbb{P}(Y = y | \mathrm{do}(X = x)) = \tag{8.8}$$
$$\sum_{z \in \mathcal{R}_Z} \sum_{s \in \mathcal{R}_X} \mathbb{P}(Y = y | X = s, Z = z) \mathbb{P}(X = s) \mathbb{P}(Z = z | X = x).$$

*where $Z = \oplus \mathcal{Z}$ and $\mathcal{R}_Z = \mathrm{range}(Z)$; $\mathcal{R}_X = \mathrm{range}(X)$, as usual.*

*Proof.* The proof follows in two steps. First, since there are no unblocked backdoor paths from $X$ to $\mathcal{Z}$, the degenerate form of Theorem 8.3 implies that

$$\mathbb{P}(Z = z | \mathrm{do}(X = x)) = \mathbb{P}(Z = z | X = x).$$

Second, since all backdoor paths from $\mathcal{Z}$ to $Y$ are blocked by conditioning on $X$, Theorem 8.3 implies that

$$\mathbb{P}(Y = y | \mathrm{do}(Z = z)) = \sum_{s \in \mathcal{R}_X} \mathbb{P}(Y = y | Z = z, X = s) \mathbb{P}(X = s),$$

where we have used a dummy variable $s$ for purpose of summation.

We have thus determined both the causal effect of $X$ on $Z$, as well as the causal effect of $Z$ on $Y$. We can combine these results using the law of total probability in the form

$$\mathbb{P}(Y = y | \mathrm{do}(X = x)) = \sum_{z \in \mathcal{R}_Z} \mathbb{P}(Y = y | \mathrm{do}(Z = z)) \mathbb{P}(Z = z | \mathrm{do}(X = x)).$$

Replacing the terms on the right-hand side with our earlier two formulas yields the result.                                                                 $\square$

Note that the term $\mathbb{P}(Z = z | X = x)$ can be factored out of the inner summation for convenience, so that the right-hand side of Equation (8.8) may be equivalently expressed as

$$\sum_{z \in \mathcal{R}_Z} \left[ \mathbb{P}(Z = z | X = x) \sum_{s \in \mathcal{R}_X} \mathbb{P}(Y = y | X = s, Z = z) \mathbb{P}(X = s) \right].$$

*Example 8.14.* Let us return to the tobacco Example 8.13, and suppose we wish to apply the front-door adjustment formula to account for tar deposits $Z$, as shown in Figure 8.16. In order to provide some numbers to this story, consider the data from a fictitious observational study of 1,000,000 individuals as shown in Table 8.7. We can apply the front-door criterion and our causal story Figure 8.16 in order to determine the causal effect of smoking on cancer. (The logic is that patients who smoke are much more likely to develop tar in their lungs, and the development of tar similarly makes a patient more likely to have lung cancer.)

|  | No tar ($Z = 0$) | Tar ($Z = 1$) |
|---|---|---|
| Nonsmokers ($S = 0$) | 14/665 (2.1%) | 2/35 (5.7%) |
| Smokers ($S = 1$) | 1/15 (6.7%) | 40/285 (14%) |

Table 8.7: Observational data on the relation between smoking and tar deposits on lung cancer. Numbers in thousands.

In order to compute the probability of cancer given the intervention of smoking or non-smoking, we apply the front-door adjustment formula to obtain

$$
\begin{aligned}
\mathbb{P}(C = 1 | \mathrm{do}(S = 0)) = \mathbb{P}(Z = 0 | S = 0) \; & [\mathbb{P}(C = 1 | S = 0, Z = 0)\mathbb{P}(S = 0) \\
& + \mathbb{P}(C = 1 | S = 1, Z = 0)\mathbb{P}(S = 1)] \\
+ \mathbb{P}(Z = 1 | S = 0) \; & [\mathbb{P}(C = 1 | S = 0, Z = 1)\mathbb{P}(S = 0) \\
& + \mathbb{P}(C = 1 | S = 1, Z = 1)\mathbb{P}(S = 1)] \\
\mathbb{P}(C = 1 | \mathrm{do}(S = 1)) = \mathbb{P}(Z = 0 | S = 1) \; & [\mathbb{P}(C = 1 | S = 0, Z = 0)\mathbb{P}(S = 0) \\
& + \mathbb{P}(C = 1 | S = 1, Z = 0)\mathbb{P}(S = 1)] \\
+ \mathbb{P}(Z = 1 | S = 1) \; & [\mathbb{P}(C = 1 | S = 0, Z = 1)\mathbb{P}(S = 0) \\
& + \mathbb{P}(C = 1 | S = 1, Z = 1)\mathbb{P}(S = 1)]
\end{aligned}
$$

Now, from Table 8.7, we have

$$\mathbb{P}(S = 0) = 0.7 \qquad \text{and} \qquad \mathbb{P}(S = 1) = 0.3;$$

i.e., the data were collected from a population comprised of 30% smokers. Furthermore, we have

$$\mathbb{P}(Z = 1 | S = 0) = 5\%$$
$$\mathbb{P}(Z = 1 | S = 1) = 95\%,$$

from which we can easily obtain $\mathbb{P}(Z = 0|S = s)$ for $s = 0, 1$. Combining the above, we find that

$$\mathbb{P}(C = 1|\text{do}(S = 0)) \approx 0.0371$$
$$\mathbb{P}(C = 1|\text{do}(S = 1)) \approx 0.0797,$$

which shows that smokers have double the risk[3] than nonsmokers.

The upshot is we were able to prove the causal link between smoking and cancer using the mediating variable of tar deposit, ruling out the possibility of a spurious association between the two.                                    ▷

## 8.4 Counterfactuals

The English word *counterfactual* is an adjective that means *contrary to fact* or *expressing what has not happened or what isn't the case*. We used the word in this way in our definition of an intervention (Definition 8.18), which makes reference to a "counterfactual reality;" i.e., what would have the overall results looked like had we, contrary to fact, swooped in and intervened to enforce a particular value of a variable. Thus, unlike prediction (i.e., conditioning), we are not limiting the focus of the data, but rather enforcing our intervention across all the data and reporting an overall result. We are, however, missing part of the story.

A *counterfactual* (as a noun) is a particular instance in our counterfactual reality. Whereas intervention is a population-level report (i.e., *what is the probability the pavement would have been wet, had we intervened to make sure the sprinkler was turned on?*), a counterfactual is an individual-level statement limited to a single instance (i.e., *What is the probability that the pavement would have been wet, had we intervened to make sure the sprinkler was turned on, when, in fact, the sprinkler was turned off and and it had rained and the pavement was not wet?*).

A quick glance at Tables 8.5 and 8.6 in Example 8.8 confirms that, in the context of interventions, we moved the entire population of rain and no sprinkler (85% chance of wet pavement) to rain and sprinkler (90% chance of wet pavement). We did not, however, answer the question for particular individuals: what is the probability that the pavement would have been wet under our intervention of sprinkler for the instances of rain and no sprinkler and wet vs. dry pavement?

To answer this, we first craft a precise definition for counterfactuals, and then we discuss the three-step process for evaluating them.

### 8.4.1 Counterfactuals: An Illustration

We begin with an example to illustrate the meaning of a counterfactual statement.

---

[3] using our made-up numbers; in reality, the increased risk is much greater.

*Example 8.15.* Let us revisit the rain–sprinkler–pavement problem of Example 8.8. We have recaptured the data of Tables 8.5 and 8.6 in Table 8.8, separating the factual and counterfactual worlds. In this version, however, we track individuals in the counterfactual world according to who they were in reality. Additionally, the individuals are labeled by their actual real-world values, so we can understand how their conditions have changed under the intervention $do(S = 1)$. In the counterfactual world, we have an added row that shows the overall effect of the intervention. For example, in the counterfactual world, no rain and no sprinkler has a total 49/490 (10%) that end up being wet, as we have ensured that the sprinkler is running in the counterfactual world. (The 10% is from the observed rate of wet pavement for no rain and sprinkler.) This row (49/490, 21/210, 243/270, 27/30) is the same as Table 8.6, except that we are accounting for individuals based on whether or not the sprinkler was running in the real world, even though the sprinkler for all of them was running in our alternate world.

<table>
<tr><td rowspan="6">Real World</td><td colspan="4">No Rain<br>700 (70%)</td><td colspan="4">Rain<br>300 (30%)</td></tr>
<tr><td colspan="2">No Sprinkler<br>490 (70%)</td><td colspan="2">Sprinkler<br>210 (30%)</td><td colspan="2">No Sprinkler<br>270 (90%)</td><td colspan="2">Sprinkler<br>30 (10%)</td></tr>
<tr><td>Dry</td><td>Wet</td><td>Dry</td><td>Wet</td><td>Dry</td><td>Wet</td><td>Dry</td><td>Wet</td></tr>
<tr><td>490</td><td>0</td><td>189</td><td>21</td><td>40.5</td><td>229.5</td><td>3</td><td>27</td></tr>
<tr><td>(100%)</td><td>(0%)</td><td>(90%)</td><td>(10%)</td><td>(15%)</td><td>(85%)</td><td>(10%)</td><td>(90%)</td></tr>
</table>

| | No Rain<br>700 (70%) | | | | Rain<br>300 (30%) | | | |
|---|---|---|---|---|---|---|---|---|
| **Counterfactual World** | No Sprinkler<br>490 (70%) | | Sprinkler<br>210 (30%) | | No Sprinkler<br>270 (90%) | | Sprinkler<br>30 (10%) | |
| | Frac. Wet<br>49/490 (10%) | | Frac. Wet<br>21/210 (10%) | | Frac. Wet<br>243/270 (90%) | | Frac. Wet<br>27/30 (90%) | |
| | Dry | Wet | Dry | Wet | Dry | Wet | Dry | Wet |
| | 49/490 | 0/0 | 0/189 | 21/21 | 13.5/40.5 | 229.5/229.5 | 0/3 | 27/27 |
| | (10%) | (100%) | (0%) | (100%) | (33%) | (100%) | (0%) | (100%) |

Table 8.8: Counterfactuals in the Rain–Sprinkler–Pavement Example. The counterfactual world is defined by the intervention $do(S = 1)$; i.e., the sprinkler is running in our counterfactual world for all cases, though we are organizing individuals according to their position in the real world, highlighting the labels in red when their is risk of conflict.

We can capture the full counterfactuals by examining our final row in the table. These numbers are obtained as follows. We first make the reasonable assumption that if the pavement was wet in the real world, it will also be wet in our counterfactual world. (After all, we are turning the sprinkler *on.*) Further, if the sprinkler was on in the real world, the results should be unchanged—for those individuals—in our counterfactual world. Finally, since we are only concerned with the no sprinkler cases, and since we know that our intervention will leave wet the pavement that was actually wet, and since we also know the total count of wet pavement excepted under our intervention, we can simply attribute the remainder of this count to the cases that corresponded to dry pavement in the real world.

For example, consider the 270 individuals that fall under rain and no sprinkler. Since we have intervened to ensure the sprinkler was running in our counterfactual world, we know that 243 of these individuals (90%) will have wet pavement in our counterfactual world. Moreover, the 229.5 individuals who had wet pavement in the real world, due to the rain, will still have wet pavement in our counterfactual world. The difference is 13.5, which must constitute the number of individuals who have wet pavement in our alternate world, out of the 40.5 individuals who had dry pavement, despite it being raining and their sprinkler being off. We can state this counterfactual as follows: there is a 33% probability that the pavement will be wet, if we intervene to ensure the sprinkler is turned on, for individuals who, in fact, had the sprinkler turned off and had dry pavement on a day in which it had rained.                                            ▷

### 8.4.2 A Crisis of Notation

In Example 8.15, we would like capture our counterfactual statement—*the probability of wet pavement under the sprinkler intervention, if, in fact, the pavement was dry, the sprinkler was off, and it was raining*—in mathematical notation, but we are immediately confronted with a problem. For instance, we would like to state

$$\mathbb{P}(W|\mathrm{do}(S), \neg S, R, \neg W)$$

for this precise case: is the pavement wet if we make sure the sprinkler is on, given that the sprinkler is off, it had rained, and the pavement is not wet. Despite the immediate contradiction (what is the probability that the pavement is wet, given that the pavement is not wet), we also find such notation to fall short under our simple calculus of intervention. For example,

$$\mathbb{P}(W|\mathrm{do}(S), R, \neg S) = \mathbb{P}(W|S, R),$$

that is, our intervention $\mathrm{do}(S)$ *eats* the reality $\neg S$ when we are searching for the correct probabilities to apply.

These conflicts arise as we are referring to events occurring in separate worlds, the real world (from which we gathered observational data) and the alternate counterfactual world (in which we are making certain surgical alterations, such as making sure that our sprinkler was running when, in fact, it was not).

Pearl [2009] resolves this with his *counterfactual notation*, instead writing

$$\mathbb{P}(W_{S=1}|R=1, S=0, W_{S=0}=0),$$

where $W_S$ represents the event of wet pavement, as dependent on the sprinkler's condition. In part, this negates the good work we have done in establishing the *do* operator, as it slides the *do* portion into the subscript. We will advocate for and introduce an alternate notation, which we construct with heavy reference to Pearl's logical construction and definition of counterfactuals.

*Note 8.6.* Pearl treats interventions and counterfactuals as distinct. The logic is that a counterfactual, to Pearl, involves probability statements from conflicting worlds. We take a different view. The intervention is, in fact, a statement about a counterfactual reality, which is composed of many individual counterfactuals. Thus, an intervention is a sort of aggregation over counterfactuals. An intervention is, therefore, a counterfactual statement, as it is a statement about a world that is contrary to our reality. A counterfactual (noun), on the other hand, is simply an atomic instance from that world.                                                                                                   ▷

### 8.4.3 Defining Counterfactuals

Pearl [2009] shows how counterfactuals should be defined and computed using a three step process: abduction, intervention, and prediction. The logic is that since a counterfactual is a statement about an individual, we should first trace back the values of that individual's exogenous variables (abduction), then perform our minimal graph surgery (intervention), and finally determine the modified probability of a variable of interest, using that individual's value for $U = u$ and the modified graph (prediction). Analogous to Pearl's *do* operator, we therefore define an *if* operator, to capture the meaning of an abduction. The *if* operator traces a statement regarding the endogenous variables back to the exogenous ones, thereby defining a new probability distribution over $\mathcal{U}$. We state this definition formally as follows.

**Definition 8.23.** *Let $\mathcal{E}$ be a set of endogenous variables ("the evidence") in a causal model $M = (\mathcal{U}, \mathcal{V}, \mathcal{F}, \mathbb{P})$. Then an* abduction *on $E = \oplus \mathcal{E}$ with value $e$, denoted* if$(E = e)$, *represents the conditional probability distribution*

$$\mathbb{P}_{E=e}(U = u) = \mathbb{P}(U = u | E = e)$$

*over $\mathcal{U}$, such that the probability statement*

$$\mathbb{P}(V = v | \text{if}(E = e)) = \mathbb{P}_{E=e}(V = v) \tag{8.9}$$

*on an endogenous variable $V \in \mathcal{V}$ represents the the resultant probability distribution over $V$ arising from the modified distribution $\mathbb{P}(U = u | E = e)$ over $\mathcal{U}$; i.e., it represents a modification of the probability distribution over $\mathcal{U}$ that is consistent with the evidence $E = e$.*

Thus, our *if* operator is a shorthand for the statement *if, in fact, we observed* ____. The *if* operator then takes a statement about our observation and translates it back to an updated probability distribution over $\mathcal{U}$. Recall that each possible value of $U = \oplus\mathcal{U}$ uniquely determines all values in $\mathcal{V}$. Having captured our broadest understanding of the variable $U$, for an individual with evidence $E = e$, we are now free to alter the graph and make predictions about what would have happened for this given individual had we intervened to do $X$.

Given this definition, we may now define a counterfactual as follows.

**Definition 8.24.** *Let $X$ and $Y$ be endogenous variables in a causal model $M = (\mathcal{U}, \mathcal{V}, \mathcal{F}, \mathbb{P})$, and let $\mathcal{E} \subset \mathcal{V}$ be a set of endogenous variables, possibly and most likely containing $X$ and $Y$. Then a counterfactual is a triple $(\text{if}(E = e), \text{do}(X = x), Y = y)$ consisting of*

1. *an abduction $\text{if}(E = e)$,*
2. *an intervention $\text{do}(X = x)$,*
3. *and a prediction $Y = y$,*

*that defines the quantity*

$$\mathbb{P}(Y = y | \text{do}(X = x), \text{if}(E = e)) = \mathbb{P}_{\substack{E=e \\ M_{X=x}}} (Y = y) \qquad (8.10)$$

*as the probability distribution over $Y$ obtained by propagating the modified probability distribution $\mathbb{P}(U = u | E = e)$ over $\mathcal{U}$ through the modified model $M_{X=x}$ (Definition 8.18), which represents the counterfactual statement "the probability that $Y = y$, if we intervene to do $X = x$, if, in reality, we observed $E = e$".*

*Example 8.16.* So far, we have yet to prescribe the actual structural equations for the rain–sprinkler–pavement problem from Example 8.8, referencing instead the conditional probabilities. These, however, are easy enough to write down:

$$R = f_R(U_R) = \mathbb{I}[U_R < 0.30],$$

$$S = f_S(R, U_S) = \begin{cases} \mathbb{I}[U_S < 0.3] & \text{if } R = 0 \\ \mathbb{I}[U_S < 0.1] & \text{if } R = 1 \end{cases},$$

$$W = f_W(R, W, U_W) = \begin{cases} 0 & \text{if } R = 0 \text{ and } S = 0 \\ \mathbb{I}[U_W < 0.10] & \text{if } R = 0 \text{ and } S = 1 \\ \mathbb{I}[U_W < 0.85] & \text{if } R = 1 \text{ and } S = 0 \\ \mathbb{I}[U_W < 0.90] & \text{if } R = 1 \text{ and } S = 1 \end{cases},$$

where each exogenous variable is an independent and identically distributed uniform random variable $U_R, U_S, U_W \sim \text{Unif}(0, 1)$.

Consider now the counterfactual statement, *the probability that the pavement is wet, if we had intervened to ensure the sprinkler was on, if, in fact,* the sprinkler was off, it was raining, and the pavement was dry, as represented using our counterfactual notation as

$$\mathbb{P}(W|\mathrm{do}(S), \mathrm{if}(R, \neg S, \neg W)).$$

In order to compute this, we first perform our abduction. It is simple to show that our abduction yields

$$U_R|R \sim \mathrm{Unif}(0, 0.30),$$
$$U_S|R, \neg S \sim \mathrm{Unif}(0.10, 1.0),$$
$$U_W|R, \neg S, \neg W \sim \mathrm{Unif}(0.85, 1.0).$$

Moreover, our intervention $\mathrm{do}(S)$ replaces the structural equation $S = f_S(R, U_S)$ with $S = 1$.

Finally, to evaluate our counterfactual, we propagate the reduced probability distribution over $\mathcal{U}$ through the modified model. First, since $U_R|R \sim \mathrm{Unif}(0, 0.30)$, we see that $R = 1$. Second, our intervention replaces the structural equation for $S$ with $S = 1$. Thus, we see from our third structural equation, that

$$W = \mathbb{I}[U_W < 0.90],$$

since $R = 1$ (from abduction) and $S = 1$ (from intervention). Now, our abducted distribution for $U_W$ was $U_W|R, \neg S, \neg W \sim \mathrm{Unif}(0.85, 1.0)$, so we see that

$$\mathbb{P}(W|\mathrm{do}(S), \mathrm{if}(R, \neg S, \neg W)) = \frac{0.90 - 0.85}{1.00 - 0.85} = \frac{1}{3},$$

agreeing with our previous result of Example 8.15.          ▷

## Problems

**8.1.** In Example 8.8, determine the causal effect of *not running the sprinkler* on the wetness of the pavement; i.e., determine

$$\mathbb{P}(W = 1|\mathrm{do}(S = 0)).$$

What is the difference $\mathbb{P}(W = 1|\mathrm{do}(S = 1)) - \mathbb{P}(W = 1|\mathrm{do}(S = 0))$?

# 9

## Time to Get Serious

Thus far, we have discussed many topics of statistics and statistical modeling. The element of time, however, has yet to play a role in our discussion, aside from our brief introduction to stochastic processes in Section 5.4. When we consider IID draws from a random variable, we imagine an infinite urn with an infinite number of marbles that somehow remains unchanged no matter how many (or for how long) we draw. In reality, however, many statistical processes play our over time. We devote the following pages to an introduction to such processes. Our favorite references are Box, *et al.* [2016], Shumway and Stoffer [1999], and Wei [2019]. An additional detailed and comprehensive classic introduction is Hamilton [1994]. Finally, for a data-science introduction to time series, see Nielson [2019].

## 9.1 Basic Concepts

In this section, we first lay out our basic definitions (time series, basic statistics, and stationarity), and then introduce white-noise time series and random walks (Section 9.1.4). The anxious reader, averse to abstraction, may at times glance ahead to Section 9.2 to gander examples.

### 9.1.1 Time Series

We previously introduced a stochastic process as an index collection of random variables defined over a common probability space (see Definition 5.11). A time series is a special case of such processes, which has grown into a field in its own right. We formally define time series as follows.

**Definition 9.1.** *A* time series *is a discrete stochastic process indexed by the integers* $\mathbb{Z}$*, or a subset thereof; i.e., a time series is a collection* $\{X_t\}_{t \in \mathbb{Z}}$ *of random variables defined over a common probability space* $(\Omega, \mathcal{E}, \mathbb{P})$*.*

Moreover, we shall normally consider the indexing set $\mathbb{Z}$ to represent a discretization of time.

Recall that the observed values of any stochastic process are referred to as a *realization* or *sample path* of that process. The phrase *time series* is often used interchangeably to refer to either the process or its realization, as the meaning is typically clear in context.

### 9.1.2 Common Time Series Statistics

We next introduce three basic time series statistics: mean, autocovariance, and autocorrelation.

### Mean

The most basic statistic of a time series $\{X_t\}$ is its mean,

$$\mu_t = \mathbb{E}[X_t].$$

Since each $X_t$ is a separate random variable, the mean of a time series will, in general, vary over time.

### Autocovariance

Since a time series is a sequence of random variables, we are naturally interested in how those random variables are related to each other. The most basic measure of this is autocovariance, which describes the variance of a time series with itself.

**Definition 9.2.** *Given a time series $\{X_t\}$, its* autocovariance function *is defined by the second moment*

$$\gamma_X(s,t) = \mathrm{COV}(X_s, X_t) = \mathbb{E}\left[(X_s - \mu_s)(X_t - \mu_t)\right], \tag{9.1}$$

*where $\mu_t = \mathbb{E}[X_t]$ is the mean at time $t$. When there is no danger of ambiguity, we will drop the subscript and refer to the autocovariance simply as $\gamma(s,t)$.*

Note that for the special case $s = t$, we have the variance of the time series

$$\gamma(t,t) = \mathbb{V}(X_t). \tag{9.2}$$

## Autocorrelation

We are often interested in normalizing the autocovariance to the interval $[-1, 1]$, similar to the normalization of the covariance of two random variables into their correlation. We thus arrive at the following definition.

**Definition 9.3.** *Given a time series* $\{X_t\}$, *its* autocorrelation function (ACF) *is defined by*

$$\rho_X(s, t) = \frac{\gamma(s, t)}{\sqrt{\gamma(s, s)\gamma(t, t)}}, \tag{9.3}$$

*where* $\gamma(s, t)$ *is the autocovariance of the time series. When there is no danger of ambiguity, we will drop the subscript and refer to the* ACF *simply as* $\rho(s, t)$.

Now, the Cauchy–Schwarz inequality can be used to show that

$$\gamma(s, t)^2 \leq \gamma(s, s)\gamma(t, t),$$

which, in turn, implies that the autocorrelation of a time series is bounded between $-1 \leq \rho(s, t) \leq 1$.

Note that the autocorrelation measures the *linear* predictability of the time series' value $X_t$ based on $X_s$. When there is a perfect linear relation, $X_t = \beta_0 + \beta_1 X_s$, the correlation will be perfect $\rho(s, t) = \pm 1$, with the sign depending on the sign of $\beta_1$. Recall that zero correlation $\gamma(s, t) = 0$ implies only that there is no *linear* dependence between $X_s$ and $X_t$, which does not rule out the possibility of a *nonlinear* dependence between the two random variables.

## 9.1.3 Stationarity

An important concept is that of stationarity. Stationarity intuitively means that the process is time invariant, or stationary. There are two flavors of stationarity: strong and weak. Strong stationarity can be thought of as more theoretical, whereas weak stationarity is more practical.

**Definition 9.4.** *A time series* $\{X_t\}$ *is said to be* strongly stationary (or strictly stationary) *if the joint distribution of any subset*

$$\{X_{t_1}, \ldots, X_{t_k}\}$$

*is invariant under any temporal shift* $t_i \to t_i + h$, *for* $h \in \mathbb{Z}$; *i.e., if the joint cumulative distribution functions are invariant:*

$$\mathbb{P}(X_{t_1} \leq c_1, \ldots, X_{t_k} \leq c_k) = \mathbb{P}(X_{t_1+h} \leq c_1, \ldots, X_{t_k+h} \leq c_k),$$

*for any* $h \in \mathbb{Z}$.

See... impractical.

**Definition 9.5.** *A* time series $\{X_t\}$ *is said to be* weakly stationary (or stationary) *if*

1. *its mean is invariant with time; i.e.,* $\mu_t = \mu_s$, *for any* $t, s \in \mathbb{Z}$;
2. *its variance is finite; i.e.,* $\mathbb{V}(X_t) < \infty$ *for all* $t \in \mathbb{Z}$; *and*
3. *its autocovariance function* $\gamma(s,t)$ *depends on s and t only through their absolute difference* $|s - t|$.

See... practical. We will therefore refer to weakly stationary time series simply as *stationary time series,* as weak stationarity is the flavor commonly used in practice. Whenever we require strict stationarity, we will explicitly call it as such.

We can simplify our notation for autocovariance and autocorrelation for stationary time series as follows.

**Definition 9.6.** *The* autocovariance function *for a stationary time series is defined as*

$$\gamma(h) = \gamma(0, h), \tag{9.4}$$

*where* $\gamma(s,t)$ *is the autocovariance function defined in Definition 9.2. Similarly, the* autocorrelation function ACF *of a stationary time series is defined as*

$$\rho(h) = \frac{\gamma(h)}{\gamma(0)}, \tag{9.5}$$

*where* $\gamma(h)$ *is the autocovariance function defined by Equation (9.4).*

These definitions are well defined since, for a stationary time series,

$$\gamma(t, t + h) = \gamma(0, h),$$

since the autocovariance depends on its independent variables only through their difference. Similarly, the autocorrelation for a stationary time series simplifies as

$$\rho(t, t + h) = \frac{\gamma(t, t + h)}{\sqrt{\gamma(t, t)\gamma(t + h, t + h)}} = \frac{\gamma(0, h)}{\gamma(0, 0)}.$$

Equation (9.5) follows.

### 9.1.4 Seedling Examples

We next introduce two cornerstone examples of time series: white noise and the random walk. We will start building more sophisticated examples in Section 9.2, when we introduce moving averages and autoregressive processes.

## White Noise

White noise is a basic building block of more complex time series, and it derives its name from white light, which is a superposition of oscillations over the entire frequency spectrum.

**Definition 9.7.** *A stationary time series* $\{W_t\}$ *is referred to as* white noise *if its mean is zero*

$$\mathbb{E}[W_t] = 0$$

*and if it is uncorrelated; i.e., if its autocovariance is given by*

$$\gamma(t) = \begin{cases} \sigma_w^2 & \text{if } t = 0 \\ 0 & \text{otherwise} \end{cases},$$

*for some constant variance* $\sigma_w^2$. *We denote such a process by* $W_t \sim$ WN$(0, \sigma_w^2)$.

*If, in addition, the random variables* $\{W_t\}$ *are independent and identically distributed, we will refer to the time series as* white independent noise *or* IID noise.

*When the random variables* $\{W_t\}$ *are independent and identically distributed normal random variables,* $W_t \sim$ N$(0, \sigma^2)$, *we refer to the time series as* Gaussian white noise.

*Example 9.1.* In this example, we use a generator function that constructs iterators that output values from a Gaussian white noise sequence. Though this is unnecessary for such a simple example, the simplicity allows us to highlight usage of such function prior to the more complex generators we will construct in the next section.

In Code Block 9.1, we define two functions, `whiteNoise` and `tsPlot`, that return a white-noise generator and that, given a time-series generator as input, produce a set of time-series graphs. We rely on the built-in `statsmodels.tsa.stattools` package to provide a numerical estimator for both ACF and PACF, the latter of which we discuss in Section 9.1.6. The output graphs are shown in Figure 9.1.

We note that the sample ACF (and PACF) both have errors for $h > 0$, as the true values of both of these functions is zero for $h > 0$. The time series itself is comprised of independent and identically distributed draws from a Gaussian distribution.                                                    ▷

## Random Walk

We next present an alternate version of the random walk from Example 5.15.

```python
from statsmodels.tsa.stattools import acf, pacf

# Generator function
def whiteNoise(sigma_w=1):
    while True:
        yield np.random.normal(scale=sigma_w)

# time-series plot function
def tsPlot(Z, n=500, lag=40):
    z = [next(Z) for i in range(n)]
    rho = acf(z, nlags=lag, fft=True)
    phihh = pacf(z, nlags=lag)

    plt.figure(figsize=(8, 9/2))
    plt.subplot(2,1,1)
    plt.plot(z)

    plt.subplot(2,2,3)
    plt.bar(np.arange(lag+1), rho)

    plt.subplot(2,2,4)
    plt.bar(np.arange(lag+1), phihh)

# code
W = whiteNoise()
tsPlot(W)
```

Code Block 9.1: Generator function for Gaussian white noise

**Definition 9.8.** *Let $\{W_t\}$ represent a white-noise process. Then a* random walk with drift *is a time series $\{X_t\}$ defined recursively by the relation*

$$X_t = X_{t-1} + \delta + W_t, \qquad (9.6)$$

*for a parameter $\delta \in \mathbb{R}$, referred to as the* drift.
    *When $\delta = 0$, we refer to the time series $\{X_t\}$ simply as a* random walk.

Note that the drift $\delta$ represents a secular term that builds up over time. The random walk can be integrated as follows.

**Proposition 9.1.** *Given the initial condition $X_0 = 0$, the random walk defined by Equation (9.6) is equivalent to*

$$X_t = \delta t + \sum_{s=1}^{t} W_s. \qquad (9.7)$$

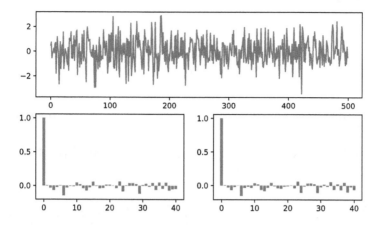

Fig. 9.1: Gaussian white noise (top); plotted with sample ACF (bottom left) and PACF (bottom right).

*Proof.* This follows from mathematical induction: For the basis step, note that our formula yields $X_0 = 0$. For the induction step, assume that Equation (9.7) is true for some $t \geq 0$. Then by applying Equation (9.6) to obtain $X_{t+1}$, we obtain

$$X_{t+1} = X_t + \delta + W_{t+1}$$
$$= \left( \delta t + \sum_{s=1}^{t} W_s \right) + \delta + W_{t+1}$$
$$= \delta(t+1) + \sum_{s=1}^{t+1} W_s,$$

which is equivalent to Equation (9.7) for $t+1$. The result follows by induction. $\square$

**Proposition 9.2.** *The mean and autocovariance of a random walk with drift $\delta$ and variance $\sigma^2$ are given by*

$$\mu_t = \delta t \tag{9.8}$$
$$\gamma(s, t) = \min(s, t)\sigma^2, \tag{9.9}$$

*respectively.*

*Proof.* Equation (9.8) follows directly from Equation (9.7), since

$$\mathbb{E} \left[ \delta t + \sum_{s=1}^{t} W_s \right] = \delta t,$$

which follows since the mean of any white noise process is zero, by definition.

We compute the autocovariance as follows

$$\begin{aligned}
\gamma(s,t) &= \text{COV}(X_s, X_t) \\
&= \mathbb{E}[(X_s - \mu_s)(X_t - \mu_t)] \\
&= \mathbb{E}[X_s X_t - X_s \mu_t - X_t \mu_s + \mu_s \mu_t] \\
&= \mathbb{E}[X_s X_t] - \mu_s \mu_t.
\end{aligned}$$

Next, we substitute Equation (9.7) into this final expression to obtain

$$\begin{aligned}
\mathbb{E}[X_s X_t] &= \mathbb{E}\left[ \left( \delta s + \sum_{i=1}^{s} W_i \right) \left( \delta t + \sum_{j=1}^{t} W_j \right) \right] \\
&= \delta^2 st + \mathbb{E}\left[ \left( \sum_{i=1}^{s} W_i \right) \left( \sum_{j=1}^{t} W_j \right) \right] \\
&= \delta^2 st + \mathbb{E}\left[ \sum_{i=1}^{s} \sum_{j=1}^{t} W_i W_j \right].
\end{aligned}$$

But $\mathbb{E}[W_i W_j] = \sigma^2 \mathbb{I}[i = j]$. And since $\mu_s \mu_t = \delta^2 st$, the result follows. $\qquad\square$

Not only is the random walk nonstationary (as the mean clearly depends on time), but its variance grows linearly with time as well, since

$$\mathbb{V}(X_t) = \sigma^2 t, \tag{9.10}$$

from Equation (9.9). Note that this expression is independent of the drift.

*Example 9.2.* A simple generator function for the random walk is given in Code Block 9.2, with an initial condition $X_0 = 0$. An example output is shown in Figure 9.2.

```
def randomWalk(delta=0, sigma_w=1):
    X = 0
    while True:
        X = X + delta + np.random.normal(scale=sigma_w)
        yield X
```

Code Block 9.2: Generator function for the random walk

Despite the driftless nature of the random walk of Figure 9.2, the time series has wandered quite a bit away from its expected value (the $t$-axis). This is a powerful visual reminder of the linear growth in the walk's variance, as given by Equation (9.10). We also note that the autocorrelation only gradually diminishes with time, whereas the partial autocorrelation, which we will discuss shortly, drops off after lag 1.                     $\triangleright$

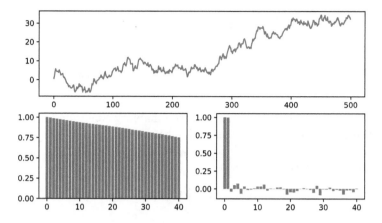

Fig. 9.2: Random walk with zero drift ($\delta = 0$) (top); plotted with sample ACF (bottom left) and PACF (bottom right).

## Linear Processes and Invertibility

We may define time series as arbitrary linear combinations of white noise processes as follows.

**Definition 9.9.** *A* mean-zero linear process *is a time series defined as a linear combination of white noise; i.e., it is a time series that can be expressed by*

$$Z_t = \sum_{j=-\infty}^{\infty} \psi_j W_{t-j}, \qquad where \qquad \sum_{j=-\infty}^{\infty} |\psi_j| < \infty, \qquad (9.11)$$

*for a white noise process $W_t \sim \mathrm{WN}(0, \sigma_w^2)$ and coefficients $\{\psi_j\}$. A linear process with nonzero mean can be established by setting $X_t = Z_t + \mu$, for $\mu \neq 0$.*

*A* causal linear process *is a linear process with $\psi_j = 0$ for $j < 0$; i.e.,*

$$Z_t = \sum_{j=0}^{\infty} \psi_j W_{t-j}, \qquad (9.12)$$

*where we may take $\psi_0 = 1$.*

Note that the values for a general linear process are determined by the past ($j > 0$), the present ($j = 0$), and the future ($j < 0$). For this reason, we call such a process *causal* if its values are *not* influenced by the future; i.e., if $\psi_j = 0$ for $j < 0$. We will only consider causal linear processes in this text, as they are real-world rooted.

**Proposition 9.3.** *The absolute summability condition*

$$\sum_{j=0}^{\infty} |\psi_j| < \infty, \tag{9.13}$$

*is a sufficient condition to prove stationarity for any (causal) linear process Equation (9.12).*

*Proof.* We must show that the absolute summability condition implies each of the three stationarity conditions laid out in Definition 9.5. First, stationarity of the mean is obvious, since $\mathbb{E}[W_t] = 0$, for all $t \in \mathbb{Z}$.

Next, let us consider the autocovariance,

$$\gamma(t, t - h) = \mathbb{E}[Z_t Z_{t-h}].$$

We first note that

$$Z_{t-h} = \sum_{i=0}^{\infty} \psi_i W_{t-h-i} = \sum_{j=h}^{\infty} \psi_{j-h} W_{t-j},$$

which follows by reindexing the summation using $j = i + h$. It follows that

$$Z_t Z_{t-h} = \left( \sum_{i=0}^{\infty} \psi_i W_{t-i} \right) \left( \sum_{j=h}^{\infty} \psi_{j-h} W_{t-j} \right)$$

$$= \sum_{i=0}^{\infty} \sum_{j=h}^{\infty} \psi_i \psi_{j-h} W_{t-i} W_{t-j}$$

Noting that $\mathbb{E}[W_t W_s] = 0$ whenever $t \neq s$, we next take the expectation to uncover

$$\gamma(t, t - h) = \mathbb{E}\left[ \sum_{j=h}^{\infty} \psi_j \psi_{j-h} W_{t-j}^2 \right] = \sigma_w^2 \sum_{j=h}^{\infty} \psi_j \psi_{j-h}, \tag{9.14}$$

which, of course is independent of $t$. This proves the third condition of stationarity, that the autocovariance is time independent.

Finally, we may express the variance from Equation (9.14) as

$$\mathbb{V}(Z_t) = \gamma(0, 0) = \sigma_w^2 \sum_{j=0}^{\infty} = \psi_j^2,$$

which converges since the coefficients $\psi_j$ are absolute summable. This completes the proof. □

**Corollary 9.1.** *The convergence of the complex power series*

$$\psi(z) = \sum_{j=0}^{\infty} \psi_j z^j$$

*on the unit disk $|z| \leq 1$ is a sufficient condition for the stationarity of the linear process Equation (9.12).*

*Proof.* This follows since the complex power series converges on the unit disk $|z| \leq 1$ if and only if the series of coefficients $\psi(1)$ is absolutely convergent. For details regarding the theory of complex variables, see, for example, Brown and Churchill [2013] or Conway [1978]. □

We also make a distinction whenever the white noise process can be written as a linear combination of the time series, as follows.

**Definition 9.10.** *A mean-zero linear process $\{Z_t\}$ is said to be* invertible *if it can be expressed as*

$$W_t = \sum_{j=0}^{\infty} \pi_j Z_{t-j}, \tag{9.15}$$

*where $\sum_{j=0}^{\infty} |\pi_j| < \infty$, for white noise $W_t \sim \mathrm{WN}(0, \sigma_w^2)$.*

Corollary 9.1 has a direct analog for invertible time series: convergence of the complex power series

$$\pi(z) = \sum_{j=0}^{\infty} \pi_j z^j$$

on the unit disk $|z| \leq 1$ implies absolute convergence of the coefficients $\sum_{j=0}^{\infty} |\pi_j| < \infty$.

### 9.1.5 Autoregression

We are often interested in the best linear estimate for a particular value in a time series based on its history.

**Definition 9.11.** *Given a stationary time series $\{X_t\}$, we define the* backward *and* forward *autoregression of length $h$ and lag $k$ as the linear regressions*

$$\hat{X}_t^{\langle h]k} = \alpha_0 + \alpha_1 X_{t-k} + \cdots + \alpha_h X_{t-k-h+1} \tag{9.16}$$

$$\hat{X}_t^{k[h\rangle} = \alpha_0 + \alpha_1 X_{t+k} + \cdots + \alpha_h X_{t+k+h-1}, \tag{9.17}$$

*respectively, for $k \geq 1$. When no lag is specified, the lag is taken as one, and we use an alternate double-subscript notation for the regression coefficients; i.e.,*

$$\hat{X}_t^{\langle h]} = \phi_{h0} + \phi_{h1} X_{t-1} + \cdots + \phi_{hh} X_{t-h} \qquad (9.18)$$

$$\hat{X}_t^{[h\rangle} = \phi_{h0} + \phi_{h1} X_{t+1} + \cdots + \phi_{hh} X_{t+h}. \qquad (9.19)$$

*The regression coefficients of Equations (9.16) and (9.17) are equal due to stationarity; similarly for Equations (9.18) and (9.19). Here, $\phi_{hi}$ is the $i$th regression coefficient when regressing on the $h$ lagging (or leading) values.*

*Note 9.1.* The subscript in Equations (9.16)–(9.19) represents the time at which the prediction is made; e.g., $\hat{X}_t^{\langle h]k}$ is a prediction for the target variable $X_t$. The ordering of the superscript notation visually represents the timeline, as anchored at time $t$:

$$\underbrace{\langle \quad h \quad ]}_{\text{backward window}} \quad \underbrace{k}_{\text{backward lag}} \quad \underbrace{X_t}_{\text{anchor}} \quad \underbrace{k}_{\text{forward lead}} \quad \underbrace{[ \quad h \quad \rangle}_{\text{forward window}} \; .$$

The backward or forward window lengths are encapsulated within $\langle \cdot ]$ or $[\cdot\rangle$, respectively, with the lag sitting on the square-bracket side of the window, closest to the anchor point.    ▷

Without loss of generality, we will consider mean-zero stationary time series. For a non-mean-zero time series $\{X_t\}$, we can always obtain a mean-zero time series by subtracting out the mean

$$Z_t = X_t - \mu.$$

**Proposition 9.4.** *Let $\{Z_t\}$ be a mean-zero stationary time series. Then the regression coefficients for $\hat{Z}_t^{\langle h]}$ and $\hat{Z}_t^{[h\rangle}$ are the solutions to the linear system*

$$\begin{bmatrix} 1 & \rho_1 & \rho_2 & \cdots & \rho_{h-2} & \rho_{h-1} \\ \rho_1 & 1 & \rho_1 & \cdots & \rho_{h-3} & \rho_{h-2} \\ \vdots & \vdots & \vdots & \ddots & \vdots & \vdots \\ \rho_{h-1} & \rho_{h-2} & \rho_{h-3} & \cdots & \rho_1 & 1 \end{bmatrix} \cdot \begin{bmatrix} \phi_{h1} \\ \phi_{h2} \\ \vdots \\ \phi_{hh} \end{bmatrix} = \begin{bmatrix} \rho_1 \\ \rho_2 \\ \vdots \\ \rho_h \end{bmatrix}, \qquad (9.20)$$

*where $\rho_1, \ldots, \rho_h$ are the values of the autocorrelation function.*

*Proof.* We will prove the result for Equation (9.19). A similar calculation yields the identical result for Equation (9.18).

From Equation (7.5), we have

$$\phi_{h1}\mathbb{E}[Z_{t+1}Z_{t+k}] + \phi_{h2}\mathbb{E}[Z_{t+2}Z_{t+k}] + \cdots + \phi_{hh}\mathbb{E}[Z_{t+h}Z_{t+k}] = \mathbb{E}[Z_t Z_{t+k}],$$
$$(9.21)$$

for $k = 1, \ldots, h$. However, since the time series has zero mean, $\mathbb{E}[Z_t] = 0$, the expected products are equivalent to their covariances:

$$\mathbb{E}[Z_t Z_s] = \mathrm{COV}(Z_t, Z_s) = \gamma_{|t-s|},$$

for any $t, s \in \mathbb{Z}$, where the last equality holds due to the stationarity of the time series. Therefore, the regression coefficients satisfy the equations

$$\phi_{h1}\gamma_{k-1} + \phi_{h2}\gamma_{k-2} + \cdots + \phi_{hh}\gamma_{h-k} = \gamma_k,$$

for $k = 1, \ldots, h$. Dividing these equations by $\gamma_0$ yields the system Equation (9.20).

A similar calculation shows that the regression coefficients for $\hat{Z}_t^{\langle h \rangle}$ satisfy the exact same set of equations.    □

As one might expect, the autocorrelations together with the general form of Equation (9.20) allow us to generate the regression coefficients recursively. The following linear algebra definitions, though slightly off the beaten path, will prove useful.

**Definition 9.12.** *The $n$-dimensional identity matrix $I_n$ and exchange matrix $J_n$ are defined componentwise by the relations*

$$[I_n]_{ij} = \mathbb{I}[i = j]$$
$$[J_n]_{ij} = \mathbb{I}[i + j = n + 1],$$

*for $1 \leq i, j \leq n$; i.e., the matrix $J_n$ is the identity matrix with columns in reversed order. For a vector $x \in \mathbb{R}^n$, the quantity $J_n x$ is obtained by reversing the order of the elements of $x$; $[J_n x]_i = x_{n+1-i}$.*

*Given an $n \times n$ matrix $A$, its* counterdiagonal transpose *is the matrix $A_T$ defined by the relation*

$$A_T = J_n A^T J_n,$$

*which is formed by reflecting $A$ about its* counterdiagonal *(the northeast to southwest diagonal). ($A^T$ is the regular matrix transpose.) The matrix $A$ is* symmetric *if $A^T = A$ and* persymmetric *if $A_T = A$. A matrix that is both symmetric and persymmetric is* bisymmetric.

Note that the coefficient matrix in Equation (9.20) is bisymmetric. Also, note that since

$$\det(J_n) = (-1)^{n(n-1)/2},$$

it follows from properties of the determinant that $\det(A_T) = \det(A)$; i.e., the matrix determinant is invariant under the operation of counterdiagonal transpose. We will also make use of the following row and column substitution operations.

**Definition 9.13.** *Let $A \in \mathbb{R}^{n \times n}$ and let $x \in \mathbb{R}^n$. We define the row-substitution (row-sub) and column-substitution (col-sub) operations as*

$$[\mathscr{S}_s(A, x)]_{ij} = \begin{cases} x_j & \text{if } i = s \\ A_{ij} & \text{otherwise} \end{cases}$$

$$[\mathscr{S}^s(A, x)]_{ij} = \begin{cases} x_i & \text{if } j = s \\ A_{ij} & \text{otherwise} \end{cases}$$

*respectively; i.e., $\mathscr{S}_i(A, x)$ is obtained by replacing the ith row of matrix $A$ with vector $x$ and $\mathscr{S}^j(A, x)$ is obtained by replacing the jth column of $A$ with vector $x$. Note we use a subscript or superscript to represent the row or column operation, respectively.*

Notice that for any bisymmetric matrix $A$, we have

$$\mathscr{S}^j(A, x) = \mathscr{S}_{n+1-j}(A, J_n \cdot x)_T,$$

and therefore

$$\det \left[ \mathscr{S}^j(A, x) \right] = \det \left[ \mathscr{S}_{n+1-j}(A, J_n \cdot x) \right]. \tag{9.22}$$

We next derive a recursion relation that allows us to generate the autoregression coefficients for a stationary time series. The following result is due to Durbin [1960].

**Theorem 9.1.** *Let $\{Z_t\}$ be a mean-zero stationary time series. The autoregression coefficients $\phi_{hi}$, for $h \in \mathbb{Z}^+$, $i = 1, \ldots, h$, can be generated using the initial condition*

$$\phi_{11} = \rho_1 \tag{9.23}$$

*along with the following recursion relations*

$$\phi_{h+1,h+1} = \frac{\rho_{h+1} - \sum_{j=1}^{h} \phi_{hj} \rho_{h+1-j}}{1 - \sum_{j=1}^{h} \phi_{hj} \rho_j} \tag{9.24}$$

$$\phi_{h+1,j} = \phi_{hj} - \phi_{h+1,h+1} \phi_{h,h+1-j}, \tag{9.25}$$

*for $j = 1, \ldots, h$.*

*Proof.* Equation (9.23) follows immediately from Equation (9.20) for the case $h = 1$.

To proceed, we define the $h \times h$ bisymmetric matrix $R_h$ and the $h$-dimensional vectors $\phi_h$ and $r_h$ as

$$[R_h]_{ij} = \rho_{i-j}, \qquad [\phi_h]_i = \phi_{hi}, \qquad \text{and} \qquad [r_h]_i = \rho_i,$$

respectively, so that Equation (9.20) is equivalent to the matrix equation

$$R_h \cdot \phi_h = r_h.$$

By Cramer's rule, from linear algebra, the coefficient $\phi_{hj}$ is given by the ratio of determinants

$$\phi_{hj} = \frac{\det(\mathscr{S}^j(R_h, r_h))}{\det(R_h)}.$$

Next, note that $R_h$ is the $n \times n$ submatrix of $R_{h+1}$ comprised of the latter matrix's first $n$ columns and $n$ rows:

$$R_{h+1} = \left[ \begin{array}{cccccc|c} 1 & \rho_1 & \rho_2 & \cdots & \rho_{h-2} & \rho_{h-1} & \rho_h \\ \rho_1 & 1 & \rho_1 & \cdots & \rho_{h-3} & \rho_{h-2} & \rho_{h-1} \\ \vdots & \vdots & \vdots & \ddots & \vdots & \vdots & \vdots \\ \rho_{h-1} & \rho_{h-2} & \rho_{h-3} & \cdots & \rho_1 & 1 & \rho_1 \\ \hline \rho_h & \rho_{h-1} & \rho_{h-2} & \cdots & \rho_2 & \rho_1 & 1 \end{array} \right]$$

To compute the determinant of this matrix, we perform cofactor expansion along the $(h+1)th$ column, starting at the bottom, thereby obtaining

$$\det(R_{h+1}) = \det(R_h) - \sum_{j=1}^{h} \rho_j \det(\mathscr{S}_{h+1-j}(R_h, J_h \cdot r_h)).$$

The index $j$ in the summation coincides with the index of $\rho_j$ in the final column of matrix $R_{h+1}$. Note that the first $h$ columns of the $(h+1)$th row of $R_{h+1}$ can be expressed as $J_h \cdot r_h$, and when eliminating the $(h+1-j)$th row of $R_{h+1}$, we can move $J_h \cdot r_h$ to replace it with precisely $(j-1)$ row swaps, such that the terms in the summation no longer alternate signs. However, from Equation (9.22), we see this is equivalent to

$$\det(R_{h+1}) = \det(R_h) - \sum_{j=1}^{h} \rho_j \det(\mathscr{S}^j(R_h, r_h)).$$

Similarly, by replacing the coefficients (elements of the $(h+1)$th column) with $r_{h+1}$, we obtain

$$\det(\mathscr{S}^{h+1}(R_{h+1}, r_{h+1})) = \rho_{h+1} \det(R_h) - \sum_{j=1}^{h} \rho_{h+1-j} \det(\mathscr{S}^j(R_h, r_h)).$$

But, by Cramer's rule, we can replace

$$\det(\mathscr{S}^j(R_h, r_h)) = \phi_{hj} \det(R_h),$$

so that the factor $\det(R_h)$ cancels out when computing the ratio

$$\phi_{h+1,h+1} = \frac{\det(\mathscr{S}^{h+1}(R_{h+1}, r_{h+1}))}{\det(R_{h+1})}.$$

Equation (9.24) follows.

We leave the proof for Equation (9.25) to the motivated reader.    □

### 9.1.6 Partial Autocorrelation Function

Another useful statistic for stationary time series is defined as follows.

**Definition 9.14.** *Given a stationary time series* $\{X_t\}$, *the partial auto-correlation function* (PACF) *is defined as*

$$P_h = \text{CORR}\left(X_t - \hat{X}_t^{[h-1\rangle}, X_{t+h} - \hat{X}_{t+h}^{\langle h-1]}\right), \qquad (9.26)$$

*for* $h \in \mathbb{Z}^+$, *where* $\hat{X}_t^{[h-1\rangle}$ *and* $\hat{X}_{t+h}^{\langle h-1]}$ *are the regressions of* $X_t$ *and* $X_{t+h}$ *on the intermediary variables* $X_{t+1}, \dots, X_{t+h-1}$, *respectively, as defined by Definition 9.11; i.e.,*

$$\hat{Z}_{t+h}^{\langle h-1]} = \beta_1 Z_{t+h-1} + \beta_2 Z_{t+h-2} + \cdots + \beta_{h-1} Z_{t+1} \qquad (9.27)$$

$$\hat{Z}_t^{[h-1\rangle} = \beta_1 Z_{t+1} + \beta_2 Z_{t+2} + \cdots + \beta_{h-1} Z_{t+h-1}, \qquad (9.28)$$

*where we used the coefficients* $\beta_i = \phi_{h-1,i}$, *for brevity.*

*Note 9.2.* If $\{X_t\}$ is a Gaussian process, the PACF is equivalent to

$$P_h = \mathbb{E}\left[\text{CORR}(X_t, X_{t+h}|X_{t+1}, \dots, X_{t+h-1})\right],$$

where the expectation is taken with respect to the intermediary variables $X_{t+1}, \dots, X_{t+h-1}$. This follows form Proposition 7.2 and the fact that

$$\mathbb{E}[X|Z] = \hat{X},$$

where $\hat{X}$ is the liner regression of $X$ on $Z$, whenever the joint distribution of $X$ and $Z$ is Gaussian. ▷

In order to derive a formula for the partial autocorrelation function, we first require two lemmas.

**Lemma 9.1.** *Let* $\{Z_t\}$ *be a mean-zero stationary time series. Then the variance of the difference* $Z_t - \hat{Z}_t^{[h-1\rangle}$ *is given by*

$$\mathbb{V}\left(Z_t - \hat{Z}_t^{[h-1\rangle}\right) = \gamma_0 - \beta_1 \gamma_1 - \cdots - \beta_{h-1} \gamma_{h-1}, \qquad (9.29)$$

*where* $\beta_i = \phi_{h-1,i}$. *Moreover, due to the stationarity, it follows that*

$$\mathbb{V}\left(Z_t - \hat{Z}_t^{[h-1\rangle}\right) = \mathbb{V}\left(Z_{t+h} - \hat{Z}_{t+h}^{\langle h-1]}\right).$$

*Proof.* Since $\mathbb{E}\left[Z_t - \hat{Z}_t^{[h-1\rangle}\right] = 0$, we have

$$\mathbb{V}\left(Z_t - \hat{Z}_t^{[h-1\rangle}\right) = \mathbb{E}\left[(Z_t - \beta_1 Z_{t+1} - \cdots - \beta_{h-1}Z_{t+h-1})^2\right]$$
$$= \mathbb{E}[Z_t(Z_t - \beta_1 Z_{t+1} - \cdots - \beta_{h-1}Z_{t+h-1})]$$
$$\quad - \beta_1 \mathbb{E}[Z_{t+1}(Z_t - \beta_1 Z_{t+1} - \cdots - \beta_{h-1}Z_{t+h-1})]$$
$$\quad - \cdots - \beta_{h-1}\mathbb{E}[Z_{t+h-1}(Z_t - \beta_1 Z_{t+1} - \cdots - \beta_{h-1}Z_{t+h-1})].$$

However, all but the first term vanishes, due to Equation (9.21), which can be rewritten for $h-1$, replacing $\phi_{h-1,i}$ with $\beta_i$, as

$$\beta_1 \mathbb{E}[Z_{t+1}Z_{t+k}] + \beta_2 \mathbb{E}[Z_{t+2}Z_{t+k}] + \cdots + \beta_{h-1}\mathbb{E}[Z_{t+h-1}Z_{t+k}] = \mathbb{E}[Z_t Z_{t+k}], \tag{9.30}$$

for $k = 1, \ldots, h-1$. Thus,

$$\mathbb{V}\left(Z_t - \hat{Z}_t^{[h-1\rangle}\right) = \mathbb{E}[Z_t(Z_t - \beta_1 Z_{t+1} - \cdots - \beta_{h-1}Z_{t+h-1})].$$

The result follows from the definition of the autocovariance function. $\qquad\square$

**Lemma 9.2.** *Let $\{Z_t\}$ be a mean-zero stationary time series, and let $\hat{Z}_{t+h}^{(h-1]}$ and $\hat{Z}_t^{[h-1\rangle}$ be the regressions defined by Equations (9.27) and (9.28), respectively. Then the covariance of the difference $Z_t - \hat{Z}_t^{[h-1\rangle}$ and $Z_{t+h} - \hat{Z}_{t+h}^{(h-1]}$ is given by*

$$\mathrm{COV}\left(Z_t - \hat{Z}_t^{[h-1\rangle}, Z_{t+h} - \hat{Z}_{t+h}^{(h-1]}\right) = \gamma_h - \beta_1 \gamma_{h-1} - \cdots - \beta_{h-1}\gamma_1. \tag{9.31}$$

*Proof.* The covariance in Equation (9.31) is equivalent to the expression

$$\mathbb{E}\left[(Z_t - \beta_1 Z_{t+1} - \cdots - \beta_{h-1}Z_{t+h-1})(Z_{t+h} - \beta_1 Z_{t+h-1} - \cdots - \beta_{h-1}Z_{t+1})\right].$$

However, recalling again Equation (9.30), this simplifies as

$$\mathbb{E}\left[(Z_t - \beta_1 Z_{t+1} - \cdots - \beta_{h-1}Z_{t+h-1})Z_{t+h}\right].$$

The result follows. $\qquad\square$

**Theorem 9.2.** *The partial autocorrelation function for a mean-zero stationary time series $\{Z_t\}$ at lag $h$ is given by the ratio of determinants*

$$P_h = \frac{\det\begin{bmatrix} 1 & \rho_1 & \rho_2 & \cdots & \rho_{h-2} & \rho_1 \\ \rho_1 & 1 & \rho_1 & \cdots & \rho_{h-3} & \rho_2 \\ \vdots & \vdots & \vdots & \ddots & \vdots & \vdots \\ \rho_{h-1} & \rho_{h-2} & \rho_{h-3} & \cdots & \rho_1 & \rho_h \end{bmatrix}}{\det\begin{bmatrix} 1 & \rho_1 & \rho_2 & \cdots & \rho_{h-2} & \rho_{h-1} \\ \rho_1 & 1 & \rho_1 & \cdots & \rho_{h-3} & \rho_{h-2} \\ \vdots & \vdots & \vdots & \ddots & \vdots & \vdots \\ \rho_{h-1} & \rho_{h-2} & \rho_{h-3} & \cdots & \rho_1 & 1 \end{bmatrix}}. \tag{9.32}$$

*Proof.* From the definition of the partial autocorrelation function,

$$
\begin{aligned}
P_h &= \mathrm{CORR}\left(Z_t - \hat{Z}_t^{[h-1\rangle}, Z_{t+h} - \hat{Z}_{t+h}^{\langle h-1]}\right) \\
&= \frac{\mathrm{COV}\left(Z_t - \hat{Z}_t^{[h-1\rangle}, Z_{t+h} - \hat{Z}_{t+h}^{\langle h-1]}\right)}{\sqrt{\mathbb{V}\left(Z_t - \hat{Z}_t^{[h-1\rangle}\right)\mathbb{V}\left(Z_{t+h} - \hat{Z}_{t+h}^{\langle h-1]}\right)}} \\
&= \frac{\gamma_h - \beta_1\gamma_{h-1} - \cdots - \beta_{h-1}\gamma_1}{\gamma_0 - \beta_1\gamma_1 - \cdots - \beta_{h-1}\gamma_{h-1}},
\end{aligned}
$$

where the last equality follows from Lemmas 9.1 and 9.2. Now, by dividing the numerator and denominator by $\gamma_0$, we obtain

$$
P_h = \frac{\rho_h - \beta_1\rho_{h-1} - \cdots - \beta_{h-1}\rho_1}{1 - \beta_1\rho_1 - \cdots - \beta_{h-1}\rho_{h-1}}.
$$

However, noting that $\beta_j = \phi_{h-1,j}$, we see that this expression is equivalent to the expression for $\phi_{hh}$, by replacing $h+1$ with $h$ in Equation (9.24).

We conclude that the partial autocorrelation with lag $h$, $P_h$, is equivalent to the final regression coefficient of the $h$th order autoregression, so that

$$
P_h = \phi_{hh}.
$$

Since $\phi_{hh}$ is standard notation for the partial autocorrelation function, we will use it in place of $P_h$. □

### 9.1.7 Sample Estimates

**Definition 9.15.** Let $(x_1, \ldots, x_n)$ be a finite sample from a stationary time series $\{X_t\}$. Then the sample mean is defined by

$$
\overline{x} = \frac{1}{n}\sum_{t=1}^{n} x_t.
$$

The sample autocovariance function is defined by

$$
\hat{\gamma}(h) = \frac{1}{n}\sum_{t=1}^{n-h}(x_{t+h} - \overline{x})(x_t - \overline{x}), \tag{9.33}
$$

with $\hat{\gamma}(-h) = \hat{\gamma}(h)$ for $h = 0, \ldots, n-1$.

Similarly, the sample autocorrelation function is defined by

$$
\hat{\rho}(h) = \frac{\hat{\gamma}(h)}{\hat{\gamma}(0)}. \tag{9.34}
$$

## 9.2 Stationary Time Series

In this section, we introduce an important class of stationary time series, known as the autoregressive moving average process. We will begin by discussing each of its components—autoregressive processes and moving average processes—in turn.

### Backshift and Finite-Difference Operators

To facilitate our discussion, we introduce the following operator.

**Definition 9.16.** *Let $\{X_t\}$ be a time series. We define the* backshift operator $B$ *by the relation*

$$BX_t = X_{t-1}, \tag{9.35}$$

*where we define the integer powers of $B$ in the natural way:*

$$B^h X_t = X_{t-h},$$

*for $h \in \mathbb{Z}$. In particular, we will refer to $B^{-1}$ as the* forward-shift operator.

   In addition, we will be interested in discussing *finite differences*; the first-order compatriot of which being the difference $(1 - B)X_t = X_t - X_{t-1}$. Higher order differences are easily generalized with the following operator.

**Definition 9.17 (Finite Differences).** *The* difference of order $d$ *is the operator*

$$\nabla^d = (1 - B)^d, \tag{9.36}$$

*where the superscript is dropped for $d = 1$.*

### 9.2.1 Moving Average Processes

#### Definition

The simplest way of constructing a stationary time series is by computing a moving average over the trailing values from a white noise process $\{W_t\}$. We define such a process as follows.

**Definition 9.18.** *A* moving-average (MA) *process of order $q$, denoted* MA$(q)$, *is a time series of the form*

$$Z_t = W_t + \theta_1 W_{t-1} + \cdots + \theta_q W_{t-q}, \tag{9.37}$$

*for constant coefficients $\theta_1, \ldots, \theta_q$, where $\theta_q \neq 0$, and white noise process $W_t \sim \mathrm{WN}(0, \sigma_w^t)$.*
   *The associated* moving-average polynomial *is defined as*

$$\theta(z) = 1 + \theta_1 z + \cdots + \theta_q z^q, \tag{9.38}$$

*for* $z \in \mathbb{C}$, *and the* moving-average operator *is defined as*

$$\theta(B) = 1 + \theta_1 B + \cdots + \theta_q B^q, \tag{9.39}$$

*where* $B$ *is the backshift operator (Definition 9.16).*

Given our operator notation Equation (9.39), we may equivalently express a MA($q$) process in the form

$$Z_t = \theta(B) W_t. \tag{9.40}$$

**Autocorrelation and Partial Autocorrelation**

By comparing with our definition of a causal linear process Equation (9.12), we immediately have the following.

**Proposition 9.5.** *The* MA($q$) *process given by Equation (9.40) is always stationary, but invertible if and only if*

$$\theta(z) \neq 0 \quad for \quad |z| \le 1;$$

*i.e., a moving-average process is invertible if and only if the roots of the moving-average polynomial defined by Equation (9.38) lie outside of the unit circle on the complex plane.*

*Proof.* Since a MA process is a finite linear process, stationarity follows from Proposition 9.3, as there are only a finite number of coefficients; i.e.,

$$\sum_{j=0}^{\infty} |\psi_j| = \sum_{j=0}^{q} |\theta_j| < \infty.$$

By inverting the polynomial $\theta(z)$, we may express Equation (9.40) as

$$\theta(B)^{-1} Z_t = W_t.$$

From the theory of complex variables (e.g., see Corollary 9.1), the power series $\pi(z) = \theta(z)^{-1}$ must converge for $|z| \le 1$ for the coefficients to be absolutely summable; i.e., for $\sum_{j=0}^{\infty} |\pi_j| < \infty$. This is only true if $\theta(z)$ does not have any zeros on the unit disk, thus completing the result. □

**Proposition 9.6.** *The* ACF *of a* MA($q$) *process is given by*

$$\rho_h = \begin{cases} \dfrac{-\theta_h + \theta_1 \theta_{h+1} + \cdots + \theta_{q-h} \theta_q}{1 + \theta_1^2 + \cdots + \theta_q^2} & for\ h = 1, \ldots, q \\ 0 & for\ h > q \end{cases} \tag{9.41}$$

*The* PACF *of a* MA($q$) *process tails off as a mixture of exponential decay and/or damped sine waves, depending on the roots of the moving-average polynomial; damped oscillations occurring if and only if the polynomial possesses complex roots.*

*Proof.* We begin by recalling that a MA($q$) process is a linear combination of the trailing $q$ terms of a white-noise process. The autocovariance between $Z_t$ and $Z_{t+h}$ can therefore be expressed as

$$\gamma_h = \text{COV}(Z_t, Z_{t-h})$$
$$= \text{COV}(W_t + \theta_1 W_{t-1} + \cdots + \theta_q W_{t-q},$$
$$W_{t-h} + \theta_1 W_{t-h-1} + \cdots + \theta_q W_{t-h-q}).$$

Now, by definition, $\text{COV}(W_t, W_s) = 0$ whenever $t \neq s$. It follows that, when $h > q$, there is absolutely no overlap between $Z_t$ and $Z_{t-h}$, so that $\text{CORR}(Z_t, Z_{t-h}) = 0$ for $h > q$.

For the case $h \leq q$, on the other hand, there are precisely $q - h + 1$ overlapping terms, so that (if we define $\theta_0 = 1$),

$$\gamma_h = \text{COV}\left(\sum_{i=0}^{q-h} \theta_{h+i} W_{t-h-i}, \sum_{j=0}^{q-h} \theta_i W_{t-h-j}\right)$$

$$= \sum_{i=0}^{q-h} \theta_{h+i} \theta_i \text{COV}(W_{t-h-i}, W_{t-h-i})$$

$$= \sum_{i=0}^{q-h} \theta_{h+i} \theta_i \sigma_w^2.$$

Now, it is clear that the variance is given by

$$\gamma_0 = \sigma_w^2 \sum_{i=0}^{q-h} \theta_i^2,$$

and Equation (9.41) follows.

$\square$

## Examples

*Example 9.3.* We can easily construct a simple *generator function* that represents a MA($q$) process, as shown in Code Block 9.3. We use the `queue.Queue`[1] class to construct a queue that remembers the trailing $q$ value of the white-noise process $W_t$.

Think of a queue object as a list with $O(1)$ get and put operations, and a maximum length. The put method adds an element to the end of the queue, whereas the get method pops the first element at the beginning of the queue. For this reason, we reverse the list of $\theta$ coefficients, as the queue, at any point, represents the values

---

[1] See `https://docs.python.org/3/library/queue.html` for more details.

```
1  from queue import Queue
2  def ma(theta=[], sigma_w=1, burn_length=100, max_iter=1e6):
3      q, t = len(theta), 0
4      assert q > 0
5      W = Queue(maxsize=q)
6      theta = list(reversed(theta))
7
8      while not W.full():
9          W.put(np.random.normal(scale=sigma_w))
10
11     while t < max_iter:
12         t += 1
13         assert q==0 or W.full()
14         W_t = np.random.normal(scale=sigma_w)
15         Z_t = np.dot(theta, W.queue) + W_t
16         W.get()
17         W.put(W_t)
18         if t > burn_length:
19             yield Z_t
```

Code Block 9.3: Generator function for MA($q$) process

$$[W_{t-q}, \ldots, W_{t-1}],$$

whereas our input parameter is

$$\theta = [\theta_1, \ldots, \theta_q].$$

```
1  Z = ma(theta=[0.9])
2  z = [next(Z) for i in range(100)] # generate a list of values
3  tsPlot(Z) # Input is the iterator, not the values
```

Code Block 9.4: Using the generator to construct values from a time series

The generator function itself returns an iterator object, that can be used to construct an arbitrary sequence from our MA($q$) process, as shown in Code Block 9.4. (We reused **tsPlot** function from Code Block 9.1.) The **burn_length** parameter discards the first so-many terms, which is relevant for the AR($p$) process, which we will discuss in the next section. The **max_iter** parameter prevents the sort of infinite loops that would otherwise occur when summing the iterator object.                          ▷

*Example 9.4.* The MA(1) process is given by

$$Z_t = W_t + \theta W_{t-1}.$$

We use the generator function in Code Block 9.3 and the `tsPlot` function we defined in Code Block 9.1 to plot the MA(1) process for $\theta = \pm 0.9$, as shown in Figures 9.3 and 9.4. Compare with the white noise shown in Figure 9.1.

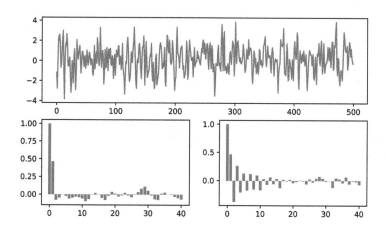

Fig. 9.3: MA(1) with $\theta = 0.9$ (top); plotted with sample ACF (bottom left) and PACF (bottom right).

For each of these, the ACF vanishes after lag 1, except for error from the numerical estimator. The PACF functions trail off    ▷

*Example 9.5.* Consider the MA(2) process

$$Z_t = W_t - W_{t-1} + 0.5W_{t-2},$$

corresponding to the moving-average polynomial $\theta(z) = 1 - z + 0.5z^2$ with complex roots $z = 1 \pm i$. Notice the damped oscillations in the PACF shown in Figure 9.5.    ▷

### 9.2.2 Autoregressive Processes

#### Definition

In Section 9.1.5, we discussed the linear autoregression of time series, as a regression Equation (9.18) of the variable $X_t$ on its $h$ lagging values $X_{t-1}, \ldots, X_{t-h}$. In this section, we show how this regression equation can be used to construct a special class of time series known as *autoregressive processes*, which we define as follows.

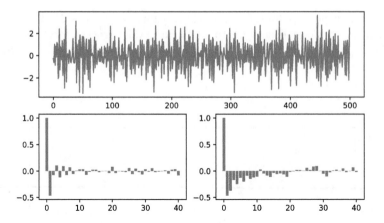

Fig. 9.4: MA(1) with $\theta = -0.9$ (top); plotted with sample ACF (bottom left) and PACF (bottom right).

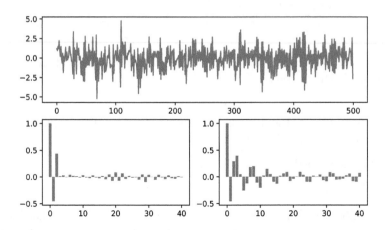

Fig. 9.5: MA(2) process $(1 - B + 0.5B^2)Z_t = W_t$ (top); plotted with sample ACF (bottom left) and PACF (bottom right).

**Definition 9.19.** *An* autoregressive (AR) *process of order $p$, denoted* AR$(p)$, *is a time series of the form*

$$Z_t = \phi_1 Z_{t-1} + \phi_2 Z_{t-2} + \cdots + \phi_p Z_{t-p} + W_t, \tag{9.42}$$

*for constant coefficients $\phi_1, \ldots, \phi_p$, where $\phi_p \neq 0$, and white noise process $W_t \sim \mathrm{WN}(0, \sigma_w^t)$.*

*The associated* autoregressive polynomial *is defined as*

$$\phi(z) = 1 - \phi_1 z - \cdots - \phi_q z^q, \tag{9.43}$$

*for $z \in \mathbb{C}$, and the* autoregressive operator *is defined as*

$$\phi(B) = 1 - \phi_1 B - \cdots - \phi_p B^p, \tag{9.44}$$

*where $B$ is the backshift operator (Definition 9.16).*

Given our operator notation Equation (9.44), we may equivalently express an AR$(p)$ process in the form

$$\phi(B) Z_t = W_t. \tag{9.45}$$

**Autocorrelation and Partial Autocorrelation**

By comparing with our definition of an invertible process Equation (9.15), we immediately have the following.

**Proposition 9.7.** *The* AR$(p)$ *process given by Equation (9.45) is always invertible, but stationary if and only if*

$$\phi(z) \neq 0 \qquad for \qquad |z| \leq 1;$$

*i.e., an autoregressive process is stationary if and only if the roots of the autoregressive polynomial defined by Equation (9.43) lie outside of the unit circle on the complex plane.*

*Proof.* Invertibility follows from the fact that an AR$(p)$ process, as defined by Equation (9.42), is already in the form of Equation (9.15), with a finite set of coefficients.

The stationarity condition follows again from Corollary 9.1. Note that Equation (9.45) is equivalent to the causal linear process

$$Z_t = \phi(B)^{-1} W_t.$$

However, we have seen that this process is invertible only if $\psi(z) = \phi(z)^{-1}$ converges on the unit disk $|z| \leq 1$, which can only be true if $\phi(z) \neq 0$ on the unit disk, thus completing the result.    $\square$

A simple counterexample, showing that not all autoregressive processes are stationary, is the random walk of Equation (9.6), which has a single unity root and time dependent mean and autocovariance (Proposition 9.2).

**Proposition 9.8.** *The ACF of an AR(p) process tails off as a mixture of exponential decay and/or damped sine waves, depending on the roots of the moving-average polynomial; damped oscillations occurring if and only if the polynomial possesses complex roots.*
  *The PACF of an AR(p) process vanishes after lag p; i.e.,*

$$\phi_{hh} = 0 \qquad for \qquad h > p.$$

*Proof.* To show the second part, we begin by considering the autoregression Equation (9.18) as applied to the AR(p) process Equation (9.42). It is immediately clear that the regression coefficients coincide with the coefficients of Equation (9.42), when defined:

$$\phi_{hi} = \phi_i, \qquad for \qquad i \leq p,$$

and $\phi_{hi} = 0$ otherwise. In particular, when $h > p$, it follows that

$$Z_t - \hat{Z}_t^{\langle h-1]} = Z_t - \hat{Z}_t^{\langle p]} = W_t,$$

so that

$$\phi_{hh} = \mathrm{CORR}\left(Z_t - \hat{Z}_t^{\langle h-1]}, Z_{t+h} - \hat{Z}_{t+h}^{\langle h-1]}\right)$$
$$= \mathrm{CORR}\left(Z_t - \hat{Z}_t^{\langle p]}, Z_{t+h} - \hat{Z}_{t+h}^{\langle p]}\right)$$
$$= \mathrm{CORR}(W_t, W_{t+h}) = 0,$$

for $h > p$. We conclude that the PACF of an AR(p) process vanishes for $h > p$.
  The ACF follows a certain recurrence relation. To see this, let us multiply Equation (9.42) by $Z_{t-h}$, for $h > 0$, to obtain

$$Z_t Z_{t-h} = \phi_1 Z_{t-1} Z_{t-h} + \cdots + \phi_p Z_{t-p} Z_{t-h} + W_t Z_{t-h}.$$

Taking expectation, and noting that $\mathbb{E}[W_t Z_{t-h}] = 0$ for $h > 0$, we obtain

$$\gamma_h = \phi_1 \gamma_{h-1} + \cdots + \phi_p \gamma_{h-p}.$$

Dividing by $\gamma_0$ yields the relation

$$\rho_h = \phi_1 \rho_{h-1} + \cdots + \phi_p \rho_{h-p}, \tag{9.46}$$

for $h > 0$.                                                          □

## Examples

We leave it as an exercise for the reader to write a generator function for an $\text{AR}(p)$ process that is similar to Code Block 9.3.

*Example 9.6.* Consider the $\text{AR}(1)$ process

$$Z_t = \phi Z + W_t,$$

for $\phi = \pm 0.9$. These are shown in Figures 9.6 and 9.7

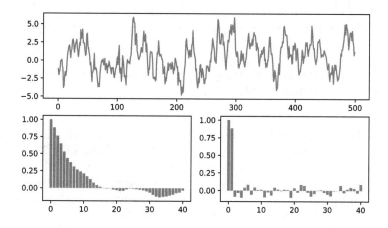

Fig. 9.6: $\text{AR}(1)$ process $\phi = 0.9$ (top); plotted with sample ACF (bottom left) and PACF (bottom right).

$\triangleright$

### 9.2.3 ARMA Processes

#### The Dual Nature of Autoregressive and Moving-Average Processes

Autoregressive and moving-average processes are flip sides of the same coin. A $\text{MA}(q)$ process, as defined in Equation (9.37), is a finite version of a causal stationary linear process Equation (9.12). Similarly, an $\text{AR}(p)$ process is a finite version of the general invertible process of Equation (9.15). It follows that an invertible $\text{MA}(q)$ process, one for which $\theta(z)$ has no roots in the unit disk of the complex plane, can be expressed as an infinite autoregressive process. Similarly, a stationary $\text{AR}(p)$ process, one for which $\phi(z)$ has no

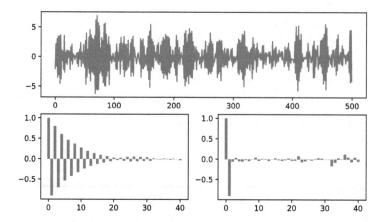

Fig. 9.7: AR(1) process $\phi = -0.9$ (top); plotted with sample ACF (bottom left) and PACF (bottom right).

roots in the unit disk of the complex plane, can be expressed as an infinite moving-average process (shown for the case AR(1) in Exercise 9.2).

This duality also reflects in the dual nature of the ACF and PACF. A MA($q$) process has a finite ACF, but an infinite PACF, reflecting its equivalence to an infinite autoregressive process. Similarly, an AR($p$) process has a finite PACF, but an infinite ACF, also reflecting its dual nature.

It follows that, thus far, we are leaving a class of stationary time series off the table, those that exhibit both elements of autoregressive and elements of moving-average processes. We turn to this topic next.

**Autoregressive Moving-Average Processes**

**Definition 9.20.** *An* autoregressive moving-average (ARMA) *process of orders $p, q$, denoted* ARMA($p, q$), *is a time series of the form*

$$Z_t = \phi_1 Z_{t-1} + \cdots + \phi_p Z_{t-p} + W_t + \theta_1 W_{t-1} + \cdots + \theta_q W_{t-q}, \quad (9.47)$$

*for constant coefficients $\phi_1, \ldots, \phi_p; \theta_1, \ldots, \theta_q$, where $\phi_p \neq 0$ and $\theta_q \neq 0$, and white noise process $W_t \sim \mathrm{WN}(0, \sigma_w^t)$.*

Given the operators Equations (9.39) and (9.44), we may equivalently express an ARMA($p, q$) process in the form

$$\phi(B)Z_t = \theta(B)W_t. \quad (9.48)$$

By combining Proposition 9.5 and 9.7, we arrive at the follwoing.

**Proposition 9.9.** *The* ARMA $(p, q)$ *process given by Equation* (9.47) *is*

1. *stationary if and only if* $\phi(z) \neq 0$ *for* $|z| \leq 1$, *and*
2. *invertible if and only if* $\theta(z) \neq 0$ *for* $|z| \leq 1$.

*Proof.* Stationarity follows as long as we can express Equation (9.48) as

$$Z_t = \phi(B)^{-1}\theta(B)W_t,$$

so that $\psi(B) = \phi(B)^{-1}\theta(B)$, with $||\psi||_2 < \infty$. Invertibility follows if we can express it as

$$\theta(B)^{-1}\phi(B)Z_t = W_t,$$

so that $\pi(B) = \theta(B)^{-1}\phi(B)$. The condition on the roots of the associated polynomials follows from complex-variable theory. □

### Autocorrelation and Partial Autocorrelation

**Proposition 9.10.** *The* ACF *of an* ARMA $(p, q)$ *depends on both sets of coefficients,* $\theta$ *and* $\phi$, *for lag* $h \leq q$, *and decays like an* AR $(p)$ *process for lag* $h > q$.
   *The* PACF *of an* ARMA $(p, q)$ *decays like a* MA $(q)$ *process for lag* $h > p$.

*Proof.* Multiplying both sides of Equation (9.47) by $Z_{t-h}$, we obtain

$$\begin{aligned}
Z_t Z_{t-h} = {} & \phi_1 Z_{t-1}Z_{t-h} + \cdots + \phi_p Z_{t-p}Z_{t-h} \\
& + W_t Z_{t-h} + \theta_1 W_{t-1}Z_{t-h} + \cdots + \theta_q W_{t-q}Z_{t-h},
\end{aligned}$$

so that

$$\begin{aligned}
\gamma_h = {} & \phi_1 \gamma_{h-1} + \cdots + \phi_p \gamma_{h-p} \\
& + \mathbb{E}[Z_{t-h}W_t] + \theta_1 \mathbb{E}[Z_{t-h}W_{t-1}] + \cdots + \theta_q \mathbb{E}[Z_{t-h}W_{t-q}].
\end{aligned}$$

Recall that an invertible ARMA model $\phi(B)Z_t = \theta(B)W_t$ can always be written in the form $Z_t = \psi(B)W_t$, where $\psi(B) = \phi(B)^{-1}\theta(B)$ and where $\phi(B)^{-1}$ an infinite series. Using this relation, we can show that

$$\mathbb{E}[Z_{t-h}W_{t-k}] = \begin{cases} 0 & \text{for } k < h \\ \psi_{k-h}\sigma_w^2 & \text{for } k \geq h \end{cases}$$

Note that the condition $\mathbb{E}[Z_{t-h}W_{t-k}] = 0$ for $k < h$ follows from causality: only past values of the shocks can affect the time series; i.e., when $k < h$, $t - k$ represents a time after $t - h$, so that the effect of $W_{t-k}$ cannot be felt by $Z_{t-h}$. Putting this together, we have

$$\gamma_h = \phi_1 \gamma_{h-1} + \cdots + \phi_p \gamma_{h-p} + \sigma_w^2 \left( \theta_h \psi_0 + \theta_{h+1}\psi_1 + \cdots + \theta_q \psi_{q-h} \right).$$

Note that there are $q - h + 1$ terms in the final term. In particular, note that This implies that

$$\gamma_h = \phi_1 \gamma_{h-1} + \cdots + \phi_p \gamma_{h-p},$$

for $h > q$. Dividing by $\gamma_0$, we obtain the recurrence relation

$$\rho_h = \phi_1 \rho_{h-1} + \cdots + \phi_p \rho_{h-p} \qquad (9.49)$$

for the ACF. Next, let $z_1, \ldots, z_r$ represent the roots of the polynomial $\phi(z) = 1 - \phi_1 z - \cdots - \phi_p z^p$, with multiplicities $m_1, \ldots, m_r$, such that $m_1 + \cdots + m_r = p$. It can be shown that the general solution to Equation (9.49) is given by

$$\rho_h = z_1^{-h} P_1(h) + z_2^{-h} P_2(h) + \cdots + z_r^{-h} P_r(h), \qquad (9.50)$$

for $h \geq q$, where $P_j$ is a polynomial of order $m_j - 1$. In particular, if each root is a single root, than the functions $P_1(h), \ldots, P_r(h)$ are simply constant coefficients. For details of the proof, see, for example, Shumway and Stoffer [1999] or Wei [2019].

For a stationary process, each root of $\phi(z)$ must lie outside of the unit circle, so that $|z_j| \geq 1$. If the roots are real, the solution Equation (9.50) will decay exponentially with $h$. If the roots are complex-conjugate pairs, the solution will exhibit damped oscillations; i.e., sinusoidal waves with an exponentially decaying envelope. We conclude that the ACF for an ARMA$(p, q)$ process decays similarly to an AR$(p)$ process for $h > q$.    □

A summary of results for AR$(p)$, MA$(q)$, and ARMA$(p, q)$ models provided in Table 9.1[2].

## Examples

*Example 9.7.* We can easily extend Code Block 9.3 to accommodate a general ARMA$(p, q)$ process, as shown in Code Block 9.5. We note that a `Queue` object with a `maxsize=0` has infinite length, so we must take care to handle the edge cases $p = 0$ or $q = 0$ whenever we interact with our queues. Other than that, it functions in the same way as our previous code.

The ARMA$(1, 1)$ process defined by

$$Z_t = 0.25 Z_{t-1} + W_t + 0.5 W_{t-1}$$

is shown in Figure 9.8. Similarly, the ARMA$(2, 6)$ process defined by

$$Z_t = 1.5 Z_{t-1} - 0.75 Z_{t-2}$$
$$+ W_t - 0.6 W_{t-1} + 0.05 W_{t-2} + 0.15 W_{t-3} + 0.025 W_{t-4} - 0.025 W_{t-6}$$

---

[2] Note that both Box, *et al.* [2016] and Wei [2019] claim that the ACF and PACF for the ARMA process tail off after $q - p$ and $p - q$ lags, respectively. This claim is related to the solution Equation (9.50) to the linear difference equation Equation (9.49), and a similar equation for the PACF, though neither author, in our opinion, provides a convincing justification for this claim; we invite the interested reader to explore further.

| | AR$(p)$ | MA$(q)$ | ARMA$(p, q)$ |
|---|---|---|---|
| model | $\phi(B)Z_t = W_t$ | $Z_t = \theta(B)W_t$ | $\phi(B)Z_t = \theta(B)W_t$ |
| stationarity condition | $\phi(z) \neq 0$ for $|z| \leq 1$ | always | $\phi(z) \neq 0$ for $|z| \leq 1$ |
| invertibility condition | always | $\theta(z) \neq 0$ for $|z| \leq 1$ | $\theta(z) \neq 0$ for $|z| \leq 1$ |
| ACF | tails off | cuts off after lag $q$ | tails off after lag $q - p$ |
| PACF | cuts off after lag $p$ | tails off | tails off after lag $p - q$ |

Table 9.1: Summary of stationary models. *Tails off* means that the series tails off as exponential decay or dampened sinusoidal waves.

is depicted in Figure 9.9. Note that it is no longer as straightforward to infer the values of $p$ and $q$ from the ACF and PACF plots. Note that we create this plot using our generator Code Block 9.5, instantiated using the command

```
Z = arma(phi = [1.5, −0.75], theta = [−0.6, 0.05, 0.15, 0.025, 0, −0.025]).
```

As a final example, consider the ARMA$(3, 2)$ process

$$Z_t = 2Z_{t-1} - 1.5Z_{t-2} + 0.375Z_{t-3} + W_t - W_{t-1} + 0.5W_{t-2},$$

which is shown in Figure 9.10. We will use this model to illustrate an integrated process in the next section.

▷

### 9.2.4 Inferring Random Shocks

Many authors liken the white-noise process $W_t$ to a sequence of *random shocks* that ultimately build up and determine the values of the time series $Z_t$. These shocks are, however, unobserved, though they may be inferred from the primary time series data, given a sufficient sample size.

To proceed, note that, for a given ARMA$(p, q)$ process, the terms of Equation (9.47) may be rearranged to express the current shock $W_t$ as a function of the current time-series value $Z_t$ and the trailing history $Z_{t-1}, \ldots, Z_{t-p}, W_{t-1}, \ldots, W_{t-q}$, via the relation

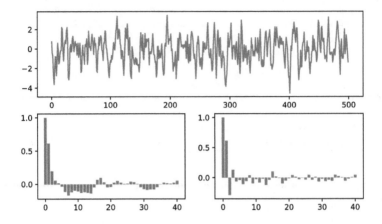

Fig. 9.8: ARMA(1, 1) process (top); plotted with sample ACF (bottom left) and PACF (bottom right).

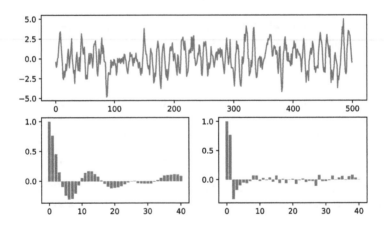

Fig. 9.9: ARMA(2, 6) process (top); plotted with sample ACF (bottom left) and PACF (bottom right).

```
 1  def arma(phi=[], theta=[], sigma_w=1, burn_length=100,
        max_iter=1e6):
 2      p, q, t = len(phi), len(theta), 0
 3      Z = Queue(maxsize=p)
 4      W = Queue(maxsize=q)
 5      theta, phi = list(reversed(theta)), list(reversed(phi))
 6      while (p > 0) and (not Z.full()):
 7          Z.put(np.random.normal(scale=sigma_w))
 8      while (q > 0) and (not W.full()):
 9          W.put(np.random.normal(scale=sigma_w))
10
11      while t < max_iter:
12          t += 1
13          assert p==0 or Z.full()
14          assert q==0 or W.full()
15          W_t = np.random.normal(scale=sigma_w)
16          Z_t = np.dot(phi, Z.queue) + np.dot(theta, W.queue) + W_t
17
18          if p > 0:
19              Z.get()
20              Z.put(Z_t)
21          if q > 0:
22              W.get()
23              W.put(W_t)
24          if t > burn_length:
25              yield Z_t
```

Code Block 9.5: Generator function for an ARMA$(p, q)$ process

$$W_t = \phi(B)Z_t + (1 - \theta(B))W_t$$
$$= Z_t - \phi_1 Z_{t-1} - \cdots - \phi_p Z_{t-p}$$
$$-\theta_1 W_{t-1} - \cdots - \theta_q W_{t-q}.$$

By knowing the previous values of the shocks $W_{t-1}, \ldots, W_{t-q}$ and the previous and current value of the time series $Z_t, Z_{t-1}, \ldots, Z_{t-p}$, we can therefore determine the current shock $W_t$ from the time-series equations.

Since we do not know any of the values of the series $W_t$, we can initialize our estimates by setting the initial $q$ values of $W_t$ to their expected value: $\hat{W}_t = 0$ for $t = p-q+1, \ldots, p$. We may then infer $\hat{W}_{p+1}$ using the $p$ values $Z_1, \ldots, Z_p$, and proceed according to Algorithm 9.1.

Note that, in general, stationary time series may be expressed solely in terms of their current and previous shocks, according to

$$Z_t = W_t + \sum_{t=1}^{\infty} \psi_j W_{t-j},$$

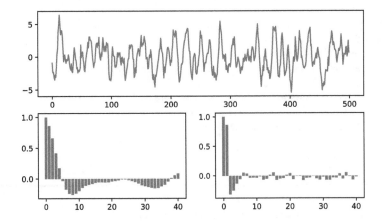

Fig. 9.10: ARMA$(3, 2)$ process (top); plotted with sample ACF (bottom left) and PACF (bottom right).

where the series $\sum \psi_j$ is absolutely convergent. The absolute convergence property therefore implies that impact of the shocks must decay to zero as time goes by. Our errors induced by our incorrect estimates $\hat{W}_t = 0$ for $t \leq p$ will therefore decay to zero as well. We applied this algorithm to an ARMA$(p, q)$ process and compared our estimated shocks $\hat{W}_t$ with the actual, true shocks $W_t$. The result is shown in Figure 9.11. (See Exercise 9.4 for additional details.) Note that our estimates converge fairly rapidly to the underlying true values.

---

**Algorithm 9.1:** Estimating the white-noise variables for a stationary process.

---

**Input:** time-series data $Z_t$, for $t = 1, \ldots, n$;
          an ARMA$(p, q)$ model with known parameters $\theta$, $\phi$, $\sigma_w^2$
**Output:** Estimates $\hat{W}_t$ of the underlying random shocks.
1 Set $\hat{W}_t = 0$, for $t = p - q + 1, \ldots, p$
2 **for** $t = p + 1, \ldots, n$ **do**
3   |  Set $\hat{W}_t = \phi(B)(Z_t) + (1 - \theta(B))\hat{W}_t$
4 **end**
5 **return** $\{\hat{W}_t\}$

---

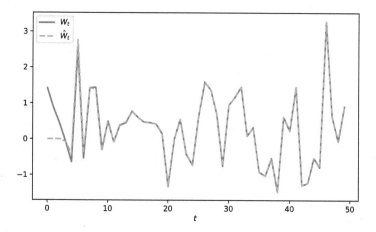

Fig. 9.11: Random shocks $W_t$ and their estimates $\hat{W}_t$ for an ARMA$(3,2)$ process.

## 9.3 Nonstationary Time Series

In this section, we discuss two important nonstationary generalizations of the classic ARMA model. The first is an *integrated* version of an ARMA process, where integration refers to the finite-difference counterpart to the classic integration of a function from calculus. The second is a seasonal version of the model, that can be used to model seasonal effects; i.e., those effects that reoccur with a certain frequency.

### 9.3.1 ARIMA Process

One of our earliest examples of a time series was that of a nonstationary time series: the random walk. However, by considering the definition of this time series (Equation (9.6)), it is clear that the differenced time series

$$D_t = X_t - X_{t-1}$$

is stationary. (In fact, this differenced series is simply a white-noise process.) This observation leads us to an obvious question: can we construct a class of nonstationary time series that can be "differentiated" into a stationary time series? This motivates the following definition.

**Definition 9.21.** *An* autoregressive integrated moving-average process, *denoted* ARIMA$(p, d, q)$, *is a time series* $\{Z_t\}$ *that satisfies the difference equation*

$$\phi(B)\nabla^d Z_t = \theta(B)W_t, \qquad (9.51)$$

*where the functions $\theta$ and $\phi$ are defined in Equations (9.39) and (9.44), respectively, and where the* finite-difference operator $\nabla$ *is defined by* $\nabla = 1 - B$.

Equation (9.51) implies that when the series $\{Z_t\}$ is an ARIMA$(p, d, q)$ process, its $d$th finite difference

$$D_t = \nabla^d Z_t = (1 - B)^d Z_t$$

is an ARMA$(p, q)$ process. Now, instead of expanding out Equation (9.51) to obtain a single complicated recurrence relation for the ARIMA series $Z_t$, our next proposition shows how we can start with an ARMA process and integrate to achieve the same result.

**Proposition 9.11.** *Let $\{D_t\}$ be an* ARMA$(p, q)$ *process. Then the process $\{Z_t\}$, defined by*

$$Z_t = D_t - \sum_{j=1}^{d} (-1)^j \binom{d}{j} Z_{t-j}. \tag{9.52}$$

*for $d \in \mathbb{Z}_+$, is an* ARIMA$(p, d, q)$ *process and is referred to as the $d$th integral of the process $D_t$.*

*Proof.* We see from Equation (9.51) that if $Z_t$ is an ARIMA$(p, d, q)$ process, then $D_t = \nabla^d Z_t$ is an ARMA$(p, q)$ process. The binomial expansion of $\nabla^d$, however, yields

$$D_t = \nabla^d Z_t = (1 - B)^d Z_t = \sum_{j=0}^{d} (-1)^j \binom{d}{j} B^j Z_t.$$

By recognizing $B^j Z_t = Z_{t-j}$ and subtracting all but the $j = 0$th term from both sides, we obtain our result.    □

In order to provide some insight into Equation (9.52), let us consider the cases $d = 1$ and $d = 2$. The first integral $X_t$ of an ARMA$(p, q)$ process $D_t$ is defined by

$$X_t = D_t + X_{t-1},$$

which itself is equivalent to a summation of all historic values of $D_t$: $X_t = \sum_{j=0}^{\infty} D_{t-j}$. Now, if we take the first integral of the series $X_t$, we obtain

$$Y_t = X_t + Y_{t-1}.$$

From this it must follow, also, that $Y_{t-1} = X_{t-1} + Y_{t-2}$. By substituting these last two equations into the equation for $X_t$, we obtain

$$\underbrace{(Y_t - Y_{t-1})}_{X_t} = D_t + \underbrace{(Y_{t-1} - Y_{t-2})}_{X_{t-1}}.$$

By rearranging, we find

$$Y_t - 2Y_{t-1} + Y_{t-2} = D_t,$$

equivalent to Equation (9.52) for the case $d = 2$. Higher-order integrals follow in kind.

Equation (9.52) shows that to determine the next term of an ARIMA process, we need only the current term of the base ARMA process and the trailing $d$ terms of the ARIMA process. This is the key observation for modifying Code Block 9.5 for ARIMA models. A generator function for a general ARIMA$(p, d, q)$ process is shown in Code Block 9.6. We use `scipy.special.comb` in order to compute the binomial coefficients. The coefficients for the queue of $Z$ values is calculated on line 8; this is equivalent to removing the first value (`range(1,d+1)` instead of `range(d)`), reversing (which alternates each of the signs), and then multiplying by $-1$ (which alternates the signs again). We also simplified some of the queue operations with the `put` function.

The once integrated process derived from the ARMA$(3, 2)$ stationary process of Figure 9.10 is shown in Figure 9.12. This is constructed with the line

$$Z = \texttt{arima}(\texttt{d} = 1, \texttt{phi} = [2, -1.5, 0.375], \texttt{theta} = [-1, 0.5]),$$

as usual. Notice the spike in $\phi_{11}$ in the PACF, and the slow decay of the ACF. These are characteristic features of integrated / nonstationary models which hint at their want for differencing.

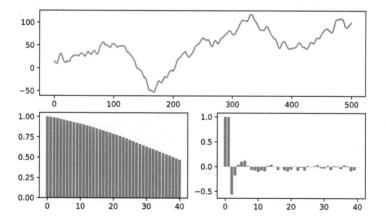

Fig. 9.12: ARIMA$(3, 1, 2)$ process (top); plotted with sample ACF (bottom left) and PACF (bottom right).

```
1  from scipy.special import comb
2  def put(Q, value=0, sigma_w = 1, fill_random=False):
3      if Q.maxsize == 0:
4          return
5      if fill_random:
6          while not Q.full():
7              Q.put(np.random.normal(scale=sigma_w))
8      if value:
9          Q.get()
10         Q.put(value)
11
12 def arima(d=0, phi=[], theta=[], sigma_w=1, burn_length=100,
       max_iter=1e6):
13     p, q, t = len(phi), len(theta), 0
14     D = Queue(maxsize=p)
15     W = Queue(maxsize=q)
16     Z = Queue(maxsize=d)
17     theta, phi = list(reversed(theta)), list(reversed(phi))
18     z_coeff = [(-1)**i * comb(d, i, exact=True) for i in range(d)]
19     put(D, sigma_w=sigma_w, fill_random=True)
20     put(W, sigma_w=sigma_w, fill_random=True)
21     put(Z, sigma_w=sigma_w, fill_random=True)
22
23     while t < max_iter:
24         t += 1
25         assert p==0 or D.full()
26         assert q==0 or W.full()
27         assert d==0 or Z.full()
28         W_t = np.random.normal(scale=sigma_w)
29         D_t = np.dot(phi, D.queue) + np.dot(theta, W.queue) + W_t
30         Z_t = D_t + np.dot(z_coeff, Z.queue)
31
32         put(D, D_t)
33         put(W, W_t)
34         put(Z, Z_t)
35         if t >= burn_length:
36             yield Z_t
```

Code Block 9.6: Generator function for ARIMA($p, d, q$) process

## 9.3.2 Seasonality

Another situation commonly encountered in nonstationary time series is that of *seasonality*, or periodic deviations from the set behavior. Seasonal effects could exist at a variety of frequencies: weekly (e.g., users of a mobile app might have higher engagement during the weekend), monthly (e.g., pension accounts might deposit funds into the stock market at the end/beginning of each month), quarterly (e.g., company quarterly reports), annually (e.g., holiday effects), or at other frequencies.

A simple approach of including seasonality is to decompose a time series into three components: a baseline trend component $B_t$, a season component $S_t$, and a random component $W_t$, so that

$$Z_t = B_t + S_t + W_t.$$

The baseline trend could be linear or polynomial in time, and seasonal, for example, could be a Fourier series representation. Coefficients may then be inferred by regression. This approach, however, often yields poor results, as it is too rigid to accurately model the time-series data. A more modern approach, which is the topic of this section, is to incorporate seasonality directly into the time-series model.

### Inspirations: A Purely Seasonal Model

To get a handle on seasonal effects, we begin with a model that is purely seasonal. Fundamentally, seasonal effects occur when data that differ by an integer multiple of a given period $s$ behave in a similar fashion. If a given time series exhibits periodicity at lag $s$, it is therefore natural to assume that the operator $B^s$, defined by $B^s Z_t = Z_{t-s}$, plays an important role in capturing the seasonal effect. In particular, let us consider the following *purely seasonal autoregressive moving-average process*

$$\Phi(B^s)S_t = \Theta(B^s)W_t, \tag{9.53}$$

where $W_t \sim \mathrm{WN}(0, \sigma_w^2)$ is a white-noise process and the polynomials $\Phi$ and $\Theta$ are order $P$ and $Q$ polynomials defined analogously to the polynomials $\phi$ and $\theta$ defined in Equations (9.39) and (9.44), so that the operators

$$\Phi(B^s) = 1 - B^s - B^{2s} - \cdots - B^{Ps}, \tag{9.54}$$
$$\Theta(B^s) = 1 + B^s + B^{2s} + \cdots + B^{Qs} \tag{9.55}$$

constitute a *seasonal autoregressive operator* and a *seasonal moving average operator*, respectively.

To illustrate this, we consider the purely seasonal moving average process

$$S_t = 0.95S_{t-12} + W_t,$$

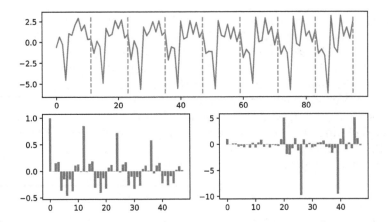

Fig. 9.13: A purely seasonal moving average process with lag 12 and coefficient $\Phi_1 = 0.9$. Increments of period $s = 12$ are illustrated with the vertical dashed green lines.

with $W_t \sim \mathrm{WN}(0, 0.25)$, as shown in Figure 9.13. Notice the slowly decaying peaks in the ACF at lags $12k$, for $k = 1, 2, 3, \ldots$. Theoretically, there is a spike in the PACF at lag 12; however, our numerical simulations failed to clearly detect this peak. We conclude that the periodic peaks in the ACF are the clearest signal in identifying the periodicity of a seasonal effect.

### SARIMA Models

In order to define our general seasonal autoregressive integrated moving-average process, we next require two additions to the purely seasonal model discussed in the previous paragraph: integrating the seasonal effect and incorporating nonseasonal trends. To achieve the former, we consider the *seasonal finite-difference operator* $\nabla_s = 1 - B^s$, so that we may express a purely seasonal ARIMA model by the equation

$$\Phi(B^s)\nabla_s^D S_t = \Theta(B^s)W_t.$$

This is completely analogous to the basic ARIMA model, except that the backshift operator $B$ is replaced by its seasonal counterpart $B^s$. To achieve the latter (i.e., to integrate nonseasonal effects into our model), we have the following.

**Definition 9.22.** *The* seasonal autoregressive moving-average (SARIMA) *process of trend order* $(p, d, q)$ *and seasonal order* $(P, D, Q)$ *and seasonal* period (periodicity) $s$, *denoted* SARIMA $(p, d, q) \times (P, D, Q)_s$, *is the time series* $Z_t$ *defined by equation*

$$\Phi(B^s)\phi(B)\nabla_s^D\nabla^d Z_t = \Theta(B^s)\theta(B)W_t, \qquad (9.56)$$

*where* $W_t \sim \text{WN}(0, \sigma_w^2)$.

We can add an autoregressive seasonal component to the ARIMA$(3, 1, 2)$ model of Figure 9.12, to obtain

$$(1 - 0.9B^{12})(1 - 2B + 1.5B^2 - 0.375B^3)(1 - B)Z_t = (1 - B + 0.5B^2)W_t.$$

Expanding, we find this model is equivalent to

$$\begin{aligned}
Z_t &= 3Z_{t-1} - 3.5Z_{t-2} + 1.875Z_{t-3} - 0.375Z_{t-4} \\
&\quad + 0.9Z_{t-12} - 2.7Z_{t-13} + 3.15Z_{t-14} - 1.6875Z_{t-15} + 0.3375Z_{t-16} \\
&\quad + W_t - W_{t-1} + 0.5W_{t-2}.
\end{aligned}$$

The autoregressive component has complex roots $2, 1 \pm i/\sqrt{3}$, and the moving-average component has complex roots $1 \pm i$. We can use our existing `arima` class to simulate this time series by using the command

```
Z = arima(d = 1,
    phi = [2, −1.5, 0.375, 0, 0, 0, 0, 0, 0, 0, 0, 0.9, −1.8, 1.35, −0.3375],
    theta = [−1, 0.5], burn_length = 0, sigma_w = 1).
```

(The $\phi$ coefficients are obtained by expanding $(1 - 0.9B^{12})(1 - 2B + 1.5B^2 - 0.375B^3)$.) A simulation from this model is shown in Figure 9.14.

Note the general trends of an integrated process: a slowly decaying ACF and a peak in the PACF at $\phi_{11}$ that immediately cuts off.

## 9.4 Model Identification, Estimation, and Forecasting

So far, we have explored both stationary and nonstationary time series, showing how we can build and simulate common time series from their definitions. We next discuss the problem from the other direction: given a time series, how do we form an appropriate model and, having decided upon such a model, how do we estimate the coefficients? We will conclude this section with a discussion on forecasting.

### 9.4.1 Model Identification

So far, we have discussed many types of time series. In this section, we will discuss how to identify which particular model represents a good fit to a given problem; i.e., given an actual real-life time series, how can we go about determining the parameters $p$, $d$, and $q$ to fit a general ARIMA$(p, d, q)$ model to the data? Naturally, we will make use of the time series along with its sample ACF and sample PACF in order to determine an appropriate model for our data.

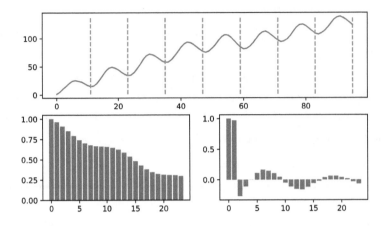

Fig. 9.14: A SARIMA$(3, 1, 2) \times (1, 0, 0)_{12}$ model.

## Overview

Our task is to determine the hyperparameters $p$, $d$, and $q$ in order to fit a reasonable ARIMA model to a given set of time-series data. A good rule of thumb is that we require at least $n = 50$ observations to fit a good model, and that we consider the sample ACF and PACF only up to a lag of $n/4$. Assuming we have a sufficient set of observations, we may proceed to the model identification stage, which we outline in this section.

Model identification is a process and not an exact science; it requires us to roll up our sleeves and don our detectives' caps in order to uncover ideal candidates for appropriate models based on our data set and its properties. At a high level, the process proceeds according to the following steps:

1. Exploratory analysis
2. Transformation
3. Differencing
4. Fitting a stationary model.

We next discuss each of these topics in turn.

## Exploratory Analysis

We begin by plotting our data. By visual inspection of the time series data, we often gain a good feel for which ingredients will be needed to produce an appropriate model. In particular, we examine the data to get a handle on several key questions:

1. Does the time-series plot exhibit any seasonality?

2. Does the series contain a trend?
3. Does the series contain any outliers?
4. Is the variance constant, or does it vary over time?
5. Does the series exhibit any other unusual nonstationary phenomena?

## Variance-stabilizing Transforms

A common approach to remedy nonconstant variance is the *Box-Cox power transform*, which typically takes the form of a logarithmic transformation on the dependent variable $Z_t$. When such transformations are necessary, they should, as a rule of thumb, be performed prior to differencing, as they sometimes require positive values of the series.

In particular, it is common for the variance in a nonstationary process to change based on its expected value; i.e.,

$$\mathbb{V}(Z_t) = cf(\mu_t),$$

for some function $f$. A common remedy is the Box-Cox power transform, which we define as follows.

**Definition 9.23.** *The* Box-Cox power transform *is the transformation $T$ defined by*

$$T(Z_t) = \frac{Z_t^\lambda - 1}{\lambda}. \tag{9.57}$$

Common values of the parameter $\lambda$ are shown in Table 9.2 along with their corresponding transformation. Note that the logarithmic transformation for the case $\lambda = 0$ is defined by taking the limit of Equation (9.57) as $\lambda \to 0$, which, of course, as everybody knows, yields the natural logarithm. For details of how Equation (9.57) may stabilize the variance, see, for example, Box, *et al.* [2016] or Wei [2019].

| $\lambda$ | $T(Z_t)$ |
|:---:|:---:|
| $-1$ | $\dfrac{1}{Z_t}$ |
| $-0.5$ | $\dfrac{1}{\sqrt{Z_t}}$ |
| $0$ | $\ln(Z_t)$ |
| $0.5$ | $\sqrt{Z_t}$ |
| $1$ | $Z_t$ |

Table 9.2: Common forms of the Box-Cox power transformation.

If such a transformation is necessary, it is typically performed before further analysis. It is therefore common for some authors to refer to the transformed series as the original series.

### Finite Differencing

Our next task is to deduce a minimum necessary degree $d$ of differencing, such that our transformed series $X_t = \nabla^d Z_t$ is a stationary ARMA$(p, q)$ series, for some value of $p$ and $q$. Typically, the degree of differencing $d$ will be either 0, 1, or 2. In order to determine whether or not differencing (or an additional degree of differencing) is required, we examine the sample ACF and sample PACF, and invoke the following rule of thumb.

**Proposition 9.12 (Rule of Thumb).** *A strong indicator of a nonstationary time series, which therefore requires at least one degree of differencing* $(1 - B)Z_t$, *is the concurrence of a slowly decaying sample* ACF *and a corresponding* PACF *that cuts off sharply after lag 1.*

For example, see Figures 9.12 and 9.14. In practice, it is usually sufficient to inspect the first 20 or so sample ACF coefficients, to determine whether or not they rapidly decay. To see why this rule of thumb is true, consider again Equation (9.50). For a stationary time series, each root of the function $\phi(z)$ lies outside the unit circle, so that the ACF decays exponentially. For a nonstationary ARIMA model, on the other hand, the polynomial coefficient $\varphi(z) = \phi(z)(1 - B)^d$ of $Z_t$ in the ARIMA time series $\varphi(B)Z_t = \theta(B)W_t$ has precisely $d$ roots of unity. If we consider a root $z_0 = 1 + \delta$, where $\delta$ is a small positive coefficient, we see that

$$\rho_h \approx a(1 - h\delta),$$

for large $h$ as $\delta \to 0^+$, for some coefficient $a$, illustrating the slow decay of the ACF.

To understand the statement regarding PACF, consider the following. First, note that the PACF and ACF of lag 1 are equivalent, since

$$\phi_{11} = \text{COV}(X_t, X_{t+1}) = \rho_1,$$

so that $\phi_{11}$ should be close to 1, since the ACF slowly decays. For lags $h > 1$, the PACF conditions on the intermediary values. Again assuming a root $z_0 = 1 - \delta$, so that $\rho_h \approx 1 - h\delta$, we have

$$\rho_1^2 \approx 1 - 2\delta + \delta^2$$
$$\rho_2 = 1 - 2\delta.$$

Now, from the recurrence relation Equation (9.24) for the PACF, we have

$$\phi_{22} = \frac{\rho_2 - \rho_1^2}{1 - \rho_1^2} \approx \frac{\delta}{2 - \delta},$$

which goes to zero as $\delta \to 0^+$. This shows that the correlation between values of the time series is broken when conditioning on at least one intermediary value.

By considering the ACF and PACF of the differenced series $(1 - B)^d Z_t$ for $d = 0, 1, 2, \ldots$, we can then choose the minimum degree of differencing $d$ that breaks these patterns.

### The Peril of Overdifferencing

It is important not to difference our time series more times than required; though some authors have argued these perils are much less than the perils of *underdifferencing*. Nonetheless, overdifferencing may lead to a tendency to overfit our model. Take, for example, the random walk, our first encounter with a nonstationary time series. The differenced equation is, of course, simple white noise. However, if we apply finite differencing to a white-noise process, we actually obtain a MA(1) process! (See Exercise 9.5.) Thus, by applying two rounds of differencing, we would incorrectly identify the random walk as an ARIMA$(0, 2, 1)$ process, as opposed to its simpler and correct representation as an ARIMA$(0, 1, 0)$ process.

### Fitting a Stationary Model

Now that we have applied a sufficient degree of differencing, our next task is to fit a stationary ARMA model to the resultant time series. In order to fit the parameters $p$ and $q$, we examine the ACF and PACF of the differenced time series $X_t = (1 - B)^d Z_t$, and compare them with our theoretical results from Table 9.1. In practice, we typically find that we can fit a reasonable model with orders of $p$ and $q$ less than or equal to three.

### 9.4.2 Estimation

Given a stationary ARMA$(p, q)$ time series with known (or suspected) values of the hyperparameters $p$ and $q$, we next turn to the problem of estimating the model's $p + q + 1$ parameters $\xi = (\phi_1, \ldots, \phi_p, \theta_1, \ldots, \theta_q, \sigma_w^2)$. We will discuss two methods for achieving such a result: method of moments and maximum likelihood. The latter takes several forms, as it is more common to solve for an approximation of the likelihood function than the true likelihood itself.

### Method of Moments for Autoregressive Processes

Our first approach involves the method of moments, which we previously discussed in the form of Satterthwaite's approximation (Definition 3.17). Formally, we define the approach as follows.

**Definition 9.24.** *The* method of moments *is a method for estimating a model's parameters that involves computing the sample moments (mean, variance, etc.) and equating them to their theoretical counterparts, expressed as a function of the model's parameters.*

In particular, for time-series data, we consider the sample mean $\overline{Z}$, sample variance $\hat{\gamma}_0$, and the sample autocorrelations $\hat{\rho}_j$. Note that we may use the estimate $\hat{\mu} = \overline{Z}$, and assume, without loss of generality, that our time series has zero mean.

In particular, the method of moments is commonly used with an AR($p$) process, as follows. The method of moments estimates for such a process are based on the following result.

**Proposition 9.13.** *An* AR*($p$) process satisfies the following system of equations*

$$\rho_1 = \phi_1 + \phi_2\rho_1 + \cdots + \phi_p\rho_{p-1}$$
$$\rho_2 = \phi_1\rho_1 + \phi_2 + \cdots + \phi_p\rho_{p-2}$$
$$\vdots \qquad \vdots$$
$$\rho_p = \phi_1\rho_{p-1} + \phi_2\rho_{p-2} + \cdots + \phi_p, \qquad (9.58)$$

*which are known as the* Yule–Walker equations.

*Proof.* The $p$ Yule–Walker equations are equivalent to Equation (9.46), explicitly expressed for $h = 1, \ldots, p$. $\qquad\qquad\square$

If we let

$$\rho = \begin{bmatrix} \rho_1 \\ \rho_2 \\ \vdots \\ \rho_p \end{bmatrix}, \qquad \phi = \begin{bmatrix} \phi_1 \\ \phi_2 \\ \vdots \\ \phi_p \end{bmatrix}, \qquad R = \begin{bmatrix} 1 & \rho_1 & \rho_2 & \cdots & \rho_{p-1} \\ \rho_1 & 1 & \rho_1 & \cdots & \rho_{p-2} \\ \rho_2 & \rho_2 & 1 & \cdots & \rho_{p-3} \\ \vdots & \vdots & \vdots & \ddots & \vdots \\ \rho_{p-1} & \rho_{p-2} & \rho_{p-3} & \cdots & 1 \end{bmatrix},$$

we see that the Yule–Walker equations may be expressed in matrix form as

$$\rho = R\phi.$$

The *Yule–Walker estimates* $\hat{\phi}_1, \ldots, \hat{\phi}_p$ for the autoregression parameters $\phi_1, \ldots, \phi_p$ are given by inverting this relation and substituting the sample autocorrelations for $\rho_1, \ldots, \rho_p$:

$$\hat{\phi} = \hat{R}^{-1}\hat{\rho}. \qquad (9.59)$$

In order to estimate the variance $\sigma_w^2$ of the white noise process, we consider

$$\gamma_0 = \mathbb{E}[Z_t Z_t]$$
$$= \mathbb{E}[Z_t(\phi_1 Z_{t-1} + \cdots + \phi_p Z_{t-p} + W_t)]$$
$$= \phi_1\gamma_1 + \cdots + \phi_p\gamma_p + \sigma_w^2,$$

which may be rearranged to obtain

$$\sigma_w^2 = \gamma_0 \left( 1 - \sum_{j=1}^{p} \phi_j \rho_j \right). \tag{9.60}$$

By substituting the sample variance $\hat{\gamma}_0$ for $\gamma_0$ and the sample autocorrelations $\hat{\rho}_j$ for their theoretical counterparts $\rho_j$, we obtain an estimate $\hat{\sigma}_w^2$ for the variance of the white noise, completing our estimate for the AR($p$) parameters.

Asymptotically, we have

$$\sqrt{n} \left( \hat{\phi} - \phi \right) \to \mathrm{N} \left( 0, \sigma_w^2 R \right)$$

in distribution and $\hat{\sigma}_w^2 \to \sigma_w^2$ in probability as $n \to \infty$.

## Maximum Likelihood Estimation

For a stationary ARMA($p, q$) process, we wish to express the likelihood function $L(\xi; Z)$ for the parameters

$$\xi = (\phi_1, \ldots, \phi_p, \theta_1, \ldots, \theta_q, \sigma_w^2),$$

so that we may then apply standard techniques, such as the Newton–Raphson method or the method of scoring (see Section 5.1), in order to determine the value $\hat{\xi}$ that maximizes the likelihood function. The exact likelihood function for a general ARMA process, however, is quite complicated; see Newbold [1974] for details. We will instead consider two approximations, known as conditional and unconditional likelihoods, where conditioning refers to the option of conditioning the likelihood on an assumed set of initial conditions.

*Conditional Maximum Likelihood*

In order to derive the likelihood equation, we rewrite Equation (9.47) in terms of $W_t$ as

$$W_t = Z_t - \phi_1 Z_{t-1} - \cdots - \phi_p Z_{t-p} - \theta_1 W_{t-1} - \cdots - \theta_q W_{t-q}. \tag{9.61}$$

Assuming the random variables $W_t$ constitute an IID Gaussian white-noise process, we may express the joint density of $w = (w_1, \ldots, w_n)$ as

$$p(w|\theta, \phi, \sigma_w^2) = (2\pi\sigma_w^2)^{-n/2} \exp \left[ \frac{-1}{2\sigma_w^2} \sum_{t=1}^{n} w_t^2 \right]. \tag{9.62}$$

The problem with this expression is that the values of the variables $w_1, \ldots, w_n$ cannot be solved using Equation (9.61) due to the lack of initial conditions. However, if we *assume* prior values

$$Z_* = \langle Z_{1-p}, \ldots, Z_{-1}, Z_0 \rangle \tag{9.63}$$

$$W_* = \langle W_{1-q}, \ldots, W_{-1}, W_0 \rangle, \tag{9.64}$$

we may then successively calculate $w_1, \ldots, w_n$, using Equation (9.61) with the initial conditions Equations (9.63) and (9.64), to obtain

$$w_t(\phi, \theta | w_*, z_*, z),$$

where $z = (z_1, \ldots, z_n)$ represents the observed values of the time series. If we define the conditional sum of squares

$$S_*(\phi, \theta) = \sum_{t=1}^{n} w_t^2(\phi, \theta | w_*, z_*, z), \tag{9.65}$$

we may then express the conditional log-likelihood of the density Equation (9.62) as

$$l_*(\phi, \theta, \sigma_w^2) = -\frac{n}{2} \ln(2\pi\sigma_w^2) - \frac{S_*(\phi, \theta)}{2\sigma_w^2}. \tag{9.66}$$

Note, in particular, that the dependence on $l_*(\phi, \theta, \sigma_w^2)$ on $\phi$ and $\theta$ is only through the sum-of-squares term $S_*(\phi, \theta)$. It is therefore sufficient to minimize $S_*(\phi, \theta)$ to determine the maximum-likelihood estimates of $\phi$ and $\theta$.

One approach to assuming $w_*$ and $z_*$ is to replace these initial conditions with their expected values; i.e., $z_* = \bar{z}$ and $w_* = 0$.

Alternatively, if we have a sufficient amount of data, we may assume that $w_p = w_{p-1} = \cdots = w_{p+1-q} = 0$ and then calculate $w_t$ for $t \geq (p+1)$ using the difference formula of Equation (9.61) (see Algorithm 9.1). We then replace the sum of squares with

$$S_*(\phi, \theta) = \sum_{t=p+1}^{n} w_t^2(\phi, \theta | z). \tag{9.67}$$

After estimating the parameters, we then estimate the variance of the white-noise process using

$$\hat{\sigma}_w^2 = \frac{S_*(\hat{\phi}, \hat{\theta})}{df},$$

where $df$ is the number of degrees of freedom, given by the difference $df = (n-p) - (p+q+1)$ between the number of data points used to estimate $S_*$ and the number of parameters estimated.

*Unconditional Maximum Likelihood*

Box, *et al.* [2016] proposed an interesting alternative approach to the problem of initial conditions, which has become known as the *unconditional maximum likelihood estimate.* For an additional description, see also Wei [2019].

The idea, in short, is to replace the conditional sum of squares, given by Equation (9.65), in the log-likelihood equation Equation (9.66), with an unconditional sum of squares

$$S(\phi, \theta) = \sum_{t=-\infty}^{n} \mathbb{E}\left[W_t | \phi, \theta, z\right]^2 \approx \sum_{t=-m}^{n} \mathbb{E}\left[W_t | \phi, \theta, z\right]^2,$$

for some sufficiently large $m > 0$. Here, $\mathbb{E}[W_t | \phi, \theta, z]$ represents the conditional expected values for $W_t$, given the parameters $\phi$ and $\theta$ and the observed time-series values $z$. For $t \leq 0$, we rewrite the forward form of the ARMA$(p, q)$ model (Equation (9.47)) in its equivalent *backward form*:

$$\phi(F)Z_t = \theta(F)W_t,$$

where $F$ is the *forwardshift operator*, defined analogously to Definition 9.16:

$$FZ_t = Z_{t+1}.$$

This allows us to *backcast* our time series to a sequence of historic values $Z_0, Z_{-1}, Z_{-2}, \ldots$.

### 9.4.3 Forecasting

The point of time series analysis is, of course, not simply to build a model that describes variables we have already observed, but also, and most importantly, to predict, or *forecast*, future values from the given series of study. We take on this topic presently; for additional references, see the usual suspects Shumway and Stoffer [1999] and Wei [2019], as well as Brockwell and Davis [2016]. Our goal is to predict both the expected value as well as the variance—from which we may obtain error bars—of future unseen values of a given time series model.

### Bootstrap Forecasts

The simplest way to approximate a time-series forecast and its error bounds is using the bootstrap. From our observed history $Z_1, \ldots, Z_n$, we may compute $d$ finite differences to obtain the differenced series $D_t$, from whence we may infer the values of the white-noise process $W_t$ using Algorithm 9.1. We may then use the trailing $d$ values of $Z$, $p$ values of $D$, and $q$ values of $W$ as initial conditions for our time-series model, from which we may run $n_{\text{boot}}$ simulations using a random variable generator for the future values of $W_t$. The steps for this process are given in Algorithm 9.2.

In order to implement this approach in Python, we require the ability to simulate the future of a time series with a *prescribed* set of initial conditions.

---

**Algorithm 9.2:** Generating bootstrap forecasts.

---

**Input:** time-series data $Z_t$, for $t = 1, \ldots, n$;
an ARIMA$(p, d, q)$ model with known parameters $d$, $\theta$, $\phi$, $\sigma_w^2$;
number $n_{\text{boot}}$ of bootstrap samples.

**Output:** Forecasts $\hat{Z}_t$ for $t = n + 1, \ldots, n + m$.

1 Compute the differenced time series $X_t = (1 - B)^d Z_t$

2 Use Algorithm 9.1 to estimate the shocks $\hat{W}_t$ for $t = n - q + 1, \ldots, n$

3 **for** $b = 1, \ldots, n_{\text{boot}}$ **do**

4     Set $\tilde{Z}_t^b = Z_t$ for $t = 1, \ldots, n$

5     Set $V_t = \hat{W}_t$ for $t = 1, \ldots, n$

6     Generate a sequence of random shocks $V_t \sim \text{WN}(0, \sigma_w^2)$ for
    $t = n + 1, \ldots, n + m$

7     Calculate the bootstrap sample $\{\tilde{Z}_t^b\}$ for $t = n + 1, \ldots, n + m$ using
    the ARIMA equation $\phi(B)(1 - B)^d \tilde{Z}_t = \theta(B) V_t$.

8 **end**

9 **return** $\{\{\tilde{Z}_t^b\}_{t=n+1}^{n+m}\}_{b=1}^{n_{\text{boot}}}$

---

```
1   def arima(d=0, phi=[], theta=[], sigma_w=1, burn_length=100,
        max_iter=1e6, D0=[], W0=[], Z0=[]):
2       ### Modify lines 19-21
3       if len(D0)>0 or len(W0)>0:
4           burn_length = 0
5           assert len(D0) == p
6           assert len(Z0) == d
7           assert len(W0) == q
8           for d0 in D0:
9               D.put(d0)
10          for w0 in W0:
11              W.put(w0)
12          for z0 in Z0:
13              Z.put(z0)
14      else:
15          put(D, sigma_w=sigma_w, fill_random=True)
16          put(W, sigma_w=sigma_w, fill_random=True)
17          put(Z, sigma_w=sigma_w, fill_random=True)
18      ###
```

Code Block 9.7: Modified section of generator function for ARIMA$(p, d, q)$ process; modified from Code Block 9.6.

This requires a small modification to our ARIMA generator function (Code Block 9.6), which is shown in Code Block 9.7.

We implement Algorithms 9.1 and 9.2 in Code Block 9.8. Note that we set the first column of the matrix z_boots to equal the final value of the observed time series, as to connect our forecasts and their error bars with the anchoring point when we produce our graphs.

```
1   phi, theta = [2, -1.5, 0.375], [-1, 0.5]
2   n, m, n_boot = 100, 40, 1000
3   Z = arima(phi=phi, theta=theta)
4   z = [next(Z) for i in range(n)]
5   w_hat = np.zeros(n)
6   for i in range(3, n):
7       w_hat[i] = z[i] - 2*z[i-1] + 1.5*z[i-2] - 0.375*z[i-3] +
            w_hat[i-1] - 0.5*w_hat[i-2]
8   Z0, D0, W0 = [], z[-3:], w_hat[-2:]
9   z = z + [next(Z) for i in range(m)] # actual future-values
10  z_boots = np.zeros((n_boot, m+1))
11  z_boots[:, 0] = z[n-1]
12  for i in range(n_boot):
13      Z_b = arima(phi=phi, theta=theta, Z0=Z0, D0=D0, W0=W0)
14      z_boots[i, 1:] = [next(Z_b) for i in range(m)]
```

Code Block 9.8: Bootstrap forecasts with error bounds for an ARMA$(3, 2)$ process; results shown in Figure 9.15.

The result is shown in Figure 9.15 (top). The solid blue line represents the actual time series, and the red line represents the average bootstrap (solid red) and its 95% error bounds (dashed red), for the bootstrap forecasts based at time $n = 100$. The similar result for an ARIMA$(3, 1, 2)$ process is also shown in Figure 9.15 (bottom). We leave it as an exercise for the reader to modify Code Block 9.8 to produce a bootstrap forecast for the integrated model (see Exercise 9.6).

From these bootstrap simulations, we note that the confidence intervals for a stationary ARMA process level out at a constant, whereas the confidence intervals for the integrated ARIMA process grow linearly over time. Moreover, the forecasts for the stationary ARMA process quickly decay back to the process mean. We will establish a theoretical basis for these observations over the coming pages.

In generating our bootstrap forecasts and, in general, in the other forecasting techniques we will soon discuss, we have assumed that the model parameters $\phi_1, \ldots, \phi_p$, $\theta_1, \ldots, \theta_q$, and $\sigma_w^2$ are known. One advantage to the bootstrap forecast is that when we include error bounds in our modeling parameters, we can still produce bootstrap forecasts by simply sampling the

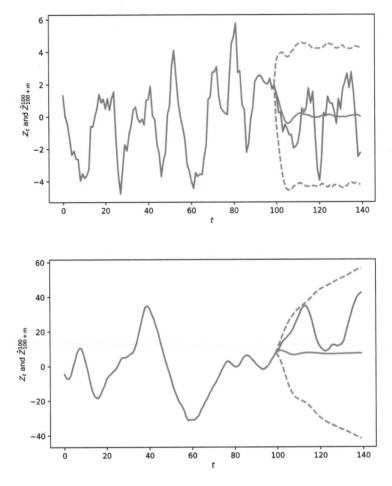

Fig. 9.15: An ARMA$(3, 2)$ process (top) and an ARIMA$(3, 1, 2)$ process (bottom), observed up to time $n = 140$ (blue), with forecasts (red) and 95% error bounds (red dashed) anchored at time $n = 100$.

joint distribution from our parameter space for each individual trace. Thus, the bootstrap forecast, with minor modification, constitutes a straightforward and elegant way to incorporate the errors of our parameter estimates into our forecasts.

## Best Linear Predictors

In the context of forecasting, we wish to use a time series' entire observed history, $x_1, \ldots, x_n$, to predict future values. Our prediction window, in the context of Equation (9.16), is therefore coincident with our forecast origin $t = n$. We are therefore interested in determining the *m-step ahead predictor based at forecast origin n*

$$\hat{X}^n_{n+m} = \mathbb{E}\left[X_{n+m}|X_1,\ldots,X_n\right] \tag{9.68}$$

for $m \geq 1$. When this quantity is expressed as a linear regression on the values $X_1,\ldots,X_n$, it is also referred to as the *best linear predictor* and may be expressed as

$$\hat{X}^n_{n+m} = \phi^m_{n0} + \phi^m_{n1}X_n + \cdots + \phi^m_{nn}X_1, \tag{9.69}$$

where $\phi^m_{n0} = 0$ for a mean-zero process. For the case of $m = 1$ (i.e., the *best one-step ahead predictor*), we drop the superscript so that $\phi_{ni} = \phi^1_{ni}$, consistent with Equation (9.18).

For theoretical reasons, it is also useful to define the $m$-step ahead predictor at forecast origin $n$ *based on the infinite past* as the quantity

$$\tilde{X}^n_{n+m} = \mathbb{E}\left[X_{n+m}|X_n, X_{n-1},\ldots\right]. \tag{9.70}$$

In general, $\hat{X}^n_{n+m} \neq \tilde{X}^n_{n+m}$; for large sample sizes, however, we may regard the two predictors as approximately indistinguishable.

Note that Equations (9.68) and (9.70) may be expressed formally in terms of the backward autoregression (Equation (9.16)) as

$$\hat{X}^n_{n+m} = \hat{X}^{\langle n\rangle m}_{n+m} \quad \text{and} \quad \tilde{X}^n_{n+m} = \hat{X}^{\langle \infty\rangle m}_{n+m},$$

respectively.

## Long-range Results for Stationary Processes

For a mean-zero stationary process, we may write

$$Z_{n+m} = \sum_{j=0}^{\infty} \psi_j W_{n+m-j}, \tag{9.71}$$

where we have defined $\psi_0 = 1$. Taking the conditional expectation based on the infinite past anchored at point $t = n$, we have

$$\tilde{Z}^n_{n+m} = \sum_{j=m}^{\infty} \psi_j w_{n+m-j}, \tag{9.72}$$

where $w_n, w_{n-1}, \ldots$ represent the observed (inferred) values of the white-noise process up to time $t = n$. Since the coefficients $\psi_j$ constitute an absolutely convergent series, we conclude that

$$\lim_{m\to\infty} \tilde{Z}^n_{n+m} = 0.$$

Moreover, by combining Equations (9.71) and (9.72), we see that the prediction errors $P_{n+m}^n$ must satisfy

$$P_{n+m}^n = \mathbb{E}\left[(Z_{n+m} - \tilde{Z}_{n+m}^n)^2\right] = \sigma_w^2 \sum_{j=0}^{m-1} \psi_j^2,$$

from whence it follows that

$$\lim_{m \to \infty} P_{n+m}^n = \gamma_0.$$

These theoretical results are illustrated in the Figure 9.15 (top), where our forecasts decay to zero and the error bounds decay to a constant.

This makes a great deal of intuitive sense: since the autocorrelations of a stationary process decay exponentially fast, the ability to predict the future is limited to a finite horizon, after which our predictions relax to the process mean and variance.

It can further be shown that the error bounds for a nonstationary ARIMA$(p, d, q)$ process grow as $P_{n+m}^n \sim O(m^d)$ as $m \to \infty$, similar to the behavior of the $d = 1$ model we examined in Figure 9.15 (bottom).

**Recursive Relation for Invertible Processes**

For an invertible process, we may write

$$W_{n+m} = \sum_{j=0}^{\infty} \pi_j Z_{n+m-j},$$

where we take $\pi_0 = 1$. Now, by recalling that

$$\mathbb{E}[W_{n+m}|X_n, X_{n-1}, \ldots] = 0,$$

we see that by taking the conditional expected value of both sides based on the infinite past anchored at time $t = n$ and rearranging, we obtain

$$\tilde{Z}_{n+m}^n = -\sum_{j=1}^{m-1} \tilde{Z}_{n+m-j}^n - \sum_{j=m}^{\infty} \pi_j Z_{n+m-j}. \qquad (9.73)$$

This equation may be applied recursively for $m = 1, 2, \ldots$ to obtain estimates $\tilde{Z}_{n+1}^n, \tilde{Z}_{n+2}^n, \ldots$. Invertibility implies that the weights $\pi_j$ form a convergent series. In practice, they decay sufficiently rapidly, such that the preceding equation may be approximating using only a finite lookback window.

## The Prediction Equations

We next discuss how one may calculate the coefficients of Equation (9.69) by minimizing the squared error. We will then show how these resulting prediction equations can be expressed in matrix format.

**Proposition 9.14.** *Given time-series data $x_1, \ldots, x_n$, up to the forecast origin $n$, the coefficients $\alpha_j$ of the best linear predictor Equation (9.68) that minimize the expected square error*

$$P_{n+m}^n = \mathbb{E}\left[\left(X_{n+m} - \hat{X}_{n+m}^n\right)^2\right] \tag{9.74}$$

*are obtained by solving the following* prediction equations

$$\mathbb{E}\left[X_{n+m} - \hat{X}_{n+m}^n\right] = 0 \tag{9.75}$$

$$\mathbb{E}\left[\left(X_{n+m} - \hat{X}_{n+m}^n\right) X_k\right] = 0, \tag{9.76}$$

*where $k = 1, \ldots, n$, for the coefficients $\alpha_0, \alpha_1, \ldots, \alpha_n$.*

*Proof.* Equations (9.75) and (9.76) are obtained by differentiating the expected squared-error loss Equation (9.74) with respect to the coefficients $\alpha_0, \ldots, \alpha_n$. (See Exercise 9.7.)    □

For a stationary process, by substituting Equation (9.69), we can show that Equations (9.75) and (9.76) are equivalent to

$$\phi_{n0}^m = \mu\left(1 - \sum_{i=1}^{n} \phi_{ni}^m\right),$$

$$\gamma_{n+m-k} = \phi_{n1}^m \gamma_{n-k} + \cdots + \phi_{nn}^m \gamma_{k-1},$$

for $k = 1, \ldots, n$. In matrix notation, we may express this second equation as

$$\Gamma_n \phi_n^m = \gamma_n^m, \tag{9.77}$$

where we have defined the $n \times n$ matrix $\Gamma_n$ and the $n \times 1$ vectors $\phi_n^m$ and $\gamma_n^m$ as

$$[\Gamma_n]_{ij} = \gamma_{i-j}, \text{ for } i, j = 1, \ldots, n,$$
$$\phi_n^m = \langle \phi_{n1}^m, \ldots, \phi_{nn}^m \rangle,$$
$$\gamma_n^m = \langle \gamma_m, \ldots, \gamma_{m+n-1} \rangle,$$

respectively. Moreover, one can show that the prediction error is given by

$$P_{n+m}^n = \gamma_0 - (\gamma_n^m)^T \Gamma_n^{-1} \gamma_n^m. \tag{9.78}$$

## The Durbin–Levinson Algorithm

For large $n$, the system of equations Equation (9.77) may be difficult and time consuming to solve. We will conclude this chapter with an introduction to two recursive approaches, that allows one to use a sequence of one-step predictors to generate forecasts recursively. We refer the reader to Brockwell and Davis [2016] for a discussion of these methods based on the theory of projection operators.

**Proposition 9.15 (Durbin–Levinson algorithm).** *The one-step ahead prediction coefficients $\phi_{nj}$ for a mean-zero stationary process can be computed recursively using the equations*

$$\phi_{11} = \rho_1 \tag{9.79}$$

$$\phi_{nn} = \frac{\rho_n - \sum_{j=1}^{n-1} \phi_{n-1,j}\rho_{n-j}}{1 - \sum_{j=1}^{n-1} \phi_{n-1,j}\rho_j} \; for \; n \geq 2 \tag{9.80}$$

$$\phi_{nj} = \phi_{n-1,j} - \phi_{nn}\phi_{n-1,n-j} \; for \; n \geq 2, j = 1, \ldots, n-1, \tag{9.81}$$

$$P_1^0 = \gamma_0 \tag{9.82}$$

$$P_{n+1}^n = P_n^{n-1}(1 - \phi_{nn}^2) \; for \; n \geq 1. \tag{9.83}$$

Note that Equations (9.79)–(9.81) constitute a restatement of Theorem 9.1.

## The Innovations Algorithm

Finally, we state a result due to Brockwell and Davis [1992]. An *innovation* is the one-step prediction error, defined by

$$U_n = X_n - \hat{X}_n^{n-1}.$$

Now, it can be shown that the innovations are related to the one-step predictions via a lower-triangular matrix $\Theta$ with null diagonals:

$$
\begin{bmatrix} \hat{X}_1^0 \\ \hat{X}_2^1 \\ \vdots \\ \hat{X}_n^{n-1} \end{bmatrix}
=
\begin{bmatrix}
0 & 0 & 0 & \cdots & 0 \\
\theta_{11} & 0 & 0 & \cdots & 0 \\
\theta_{22} & \theta_{21} & 0 & \cdots & 0 \\
\vdots & \vdots & \vdots & \ddots & \vdots \\
\theta_{n-1,n-1} & \theta_{n-1,n-2} & \theta_{n-1,n-3} & \cdots & 0
\end{bmatrix}
\cdot
\begin{bmatrix} U_1 \\ U_2 \\ U_3 \\ \vdots \\ U_n \end{bmatrix},
$$

such that

$$\hat{X}_{n+1}^n = \sum_{j=1}^{n} \theta_{nj}U_{n+1-j},$$

for $n = 1, 2, \ldots$. For additional discussion, see Brockwell and Davis [2016] or Shumway and Stoffer [1999].

**Proposition 9.16 (Innovations algorithm).** *Given a time series with finite second moments, the coefficients $\theta_{nj}$ may be computed recursively via the relations*

$$\theta_{n,n-j} = (P_{j+1}^j)^{-1}\left[\gamma(n+1,j+1) - \sum_{i=0}^{j-1}\theta_{j,j-1}\theta_{n,n-i}P_{i+1}^i\right], \quad (9.84)$$

$$P_1^0 = \gamma(1,1), \tag{9.85}$$

$$P_{n+1}^n = \gamma(n+1,n+1) - \sum_{j=0}^{n-1}\theta_{n,n-j}^2 P_{j+1}^j. \tag{9.86}$$

Moreover, for an $m$-step predictor, the innovations algorithm can be used to show

$$\hat{X}_{n+m}^n = \sum_{j=m}^{n+m-1}\theta_{n+m-1,j}U_{n+m-j}, \tag{9.87}$$

$$P_{n+m}^n = \gamma(n+m,n+m) - \sum_{j=h}^{n+m-1}\theta_{n+m-1,j}^2 P_{n+m-j}^{n+m-j-1}. \tag{9.88}$$

For details, see references.

Finally, we note that while the Durbin–Levinson algorithm generates the coefficients to express the one-step predictors in the form

$$\hat{X}_{n+1}^n = \sum_{j=1}^{n}\phi_{nj}X_{n+1-j},$$

the innovations algorithm represents the analogous formula,

$$\hat{X}_{n+1}^n = \sum_{j=1}^{n}\theta_{nj}\left(X_{n+1-j} - \hat{X}_{n+1-j}^{n-j}\right),$$

expressed in terms of linear combinations of the innovations. This is useful as the innovations, unlike the individual predictors, are uncorrelated. Moreover, the innovations algorithm has a tidy representation for the stationary ARMA$(p,q)$ process.

## Problems

**9.1.** Prove Equation (9.25).

**9.2.** By exploiting the recursive nature of the AR(1) process

$$Z_t = \phi Z_{t-1} + W_t,$$

show that it is equivalent to

$$Z_t = \phi^k Z_{t-k} + \sum_{j=0}^{k-1} \phi^j W_{t-j},$$

for any $k \in \mathbb{Z}_+$. Conclude that the AR(1) process is a stationary linear process (Equation (9.12)) if and only if $|\phi| < 1$, and show that this is equivalent to the condition $\phi(z) = 1 - \phi z \neq 0$ for $|z| \leq 1$.

**9.3.** Write a generator function for an AR($p$) process, following a similar structure as in Code Block 9.3.

**9.4.** Reproduce Figure 9.11. By simulating the ARMA(3, 2) process

$$Z_t = 2Z_{t-1} - 1.5Z_{t-2} + 0.375Z_{t-3} + W_t - W_{t-1} + 0.5W_{t-2};$$

i.e., using

$$\phi(B) = 1 - 2B + 1.5B^2 - 0.375B^3$$
$$\theta(B) = 1 - B + 0.5B^2.$$

Repeat for an ARIMA(3, $d$, 2) process using the same polynomials $\phi(B)$ and $\theta(B)$ and $d = 1, 2$. What do you observe?

**9.5. Overdifferencing** In this exercise, we will illustrate the danger of *overdifferencing* when fitting an ARIMA model.
(a) Show that the differenced series for a random walk is a white noise process. Explain why this implies that the random walk is equivalent to an ARIMA(0, 1, 0) process.
(b) Show the the differenced series for white noise is a MA(1) process. Conclude that by overdifferencing a random walk, one might incorrectly identify a random walk as an ARIMA(0, 2, 1) process, thereby overfitting the model.

**9.6.** Modify Code Block 9.8 to obtain bootstrap forecasts for a simulated ARIMA(3, 1, 2) model.

**9.7.** Show that by differentiating the expected squared-error loss Equation (9.74), one obtains the prediction equations Equations (9.75) and (9.76).

# Night at the Casino

Bayesian inference constitutes a radical departure from the frequentist point of view, which we have exclusively dealt with up until now.

Our main go-to reference for Bayesian inference is Gelman, *et al.* [2014]. For a discussion of Bayesian inference from a graph-theory point of view, see Barber [2012]. Finally, an important reference on probabilistic programming, i.e., how we should go about performing Bayesian inference using simulation, is Davidson-Pilon [2016]. Another excellent source that covers many classic topics in statistics from a Bayesian viewpoint is McElreath [2016].

## 10.1 Bayesian Inference

### 10.1.1 Bayes' Law

#### Bayesian versus Frequentist Points of View

Bayesian inference represents a paradigm shift in how we think of and view probabilities. From the *frequentist* point of view, probability represents the long-run frequency of the occurrence of an event under identical circumstances. This definition of probability is itself often a counterfactual, as the unending supply of IID events is never fully realized in our universe. In some contexts this is okay. When flipping a fair coin, for example, we can easily imagine an unending sequence of flips, and we understand that a 50% probability of heads means that the frequency of heads will approach 50% as we continue to flip the coin more and more. But what about a 30% probability of rain on Tuesday? Next Tuesday can only happen once, and it will either rain or it will not rain, so are we to interpret a "30% chance of rain" as a statement as to what is happening in many, many parallel universes, completely identical except for Tuesday's precipitation?

From the *Bayesian* point of view, probability represents our belief in the likelihood of the occurrence of an event. From this perspective, a 30%

chance of rain makes perfect sense: it is a statement about our belief whether or not it will rain on Tuesday.

At first glance, the distinction between the two definitions can seem superficial: are we not just repackaging the same statement? However, on closer examination, the Bayesian viewpoint represents a dramatic departure from the frequentist point of view. In particular, the Bayesian viewpoint offers two key advantages:

1. Our belief is not only limited to a point estimate, but also can be represented by a distribution;
2. Our belief is capable of changing; i.e., our understanding of a natural phenomenon should improve with experience.

We will explore the implications and precise meaning of these statements throughout the remainder of the chapter.

### Bayes' Law

We begin with the classical statement of Bayes' law, which relates the conditional probability of two events.

**Proposition 10.1.** *Let $(\Omega, \mathcal{E}, \mathbb{P})$ be a probability space (see Definition 1.3). Then, for any $A, B \in \mathcal{E}$, we have*

$$\mathbb{P}(A|B) = \frac{\mathbb{P}(B|A)\mathbb{P}(A)}{\mathbb{P}(B)}. \tag{10.1}$$

*Proof.* This follows immediately from the definition of conditional probability Definition 1.5, since

$$\mathbb{P}(A|B) = \frac{\mathbb{P}(AB)}{\mathbb{P}(B)} = \frac{\mathbb{P}(A)\mathbb{P}(B|A)}{\mathbb{P}(B)},$$

which is equivalent to Equation (10.1). □

*Example 10.1.* Consider a medical diagnostic test for a rare condition, which affects 0.1% of the population, that is 99% sensitive and 99% specific. Recall that 99% sensitive means that the true positive rate is 99%; i.e., 99% of those afflicted with the condition will be tested as positive. Similarly, 99% specific means that the true negative rate is 99%, or that 99% of those not afflicted will correctly be identified with a negative test result. Suppose now that Joe takes the diagnostic test, which results in a positive result. What is the probability that Joe has the condition?

Let $C$ be the event that a person has the condition, and let $+$ and $-$ represent positive and negative test results, respectively. In order to find the conditional probability of $C$, given Joe's positive test result, we must apply Bayes' law:

$$\mathbb{P}(C|+) = \frac{\mathbb{P}(+|C)\mathbb{P}(C)}{\mathbb{P}(+)}.$$

Since the test is 99% sensitive, we know that $\mathbb{P}(+|C) = 0.99$. Moreover, we know that the overall prevalence of the condition is 0.1%, so that $\mathbb{P}(+) = 0.001$. The denominator represents the probability of a positive result (without conditioning on whether or not a person has the condition), which is obtained from the law of total probability:

$$\mathbb{P}(+) = \mathbb{P}(+|C)\mathbb{P}(C)+\mathbb{P}(+|\neg C)\mathbb{P}(\neg C) = 0.99(0.001)+0.01(0.999) = 0.01098.$$

Combining the above, we find that

$$\mathbb{P}(C|+) = \frac{0.00099}{0.01098} \approx 9.016\%.$$

If this result seems counterintuitive, consider a small population of 10,000. Since 0.1% of the population is afflicted with the condition, we expect 10 people to have the condition out of the population. This represents the sum of the first column in Table 10.1. Similarly, 9,990 people will not have the condition, which is the sum of the second column. Now, since the test is 99%

|  | $Y = 1$ | $Y = 0$ |
|---|---|---|
| $\hat{Y} = 1$ | 10 | 100 |
| $\hat{Y} = 0$ | 0 | 9890 |

Table 10.1: Confusion matrix for Example 10.1.

sensitive, we might expect all 10 of the actual cases to be correctly identified. Since the test is 99% specific, we expect 99% of the actual negatives to be correctly identified, leading the the confusion matrix shown in Table 10.1. Thus, out of the positive test results, we expect 10 to be true positives, whereas 100 will be false positives, leading to a 9% probability of having the condition, given a positive result.    ▷

*Example 10.2.* A famous problem in Bayesian probability is the *Monty Hall problem*, as introduced in Selvin [1975a] and Selvin [1975b]. A game-show host presents a contestant with a choice of three doors. Behind one is a valuable prize (a new car, cash, etc.), and behind the other two are dud prizes (a donkey, dirty socks, etc.). The contestant selects one of the three doors—say, door number one. The host then reveals one of the duds from one of the other two doors. (*"It's a good thing you didn't choose door number three, because behind door number three is a used statistics book!"*) The contestant is then asked if he would like to stick to his or her original choice or switch to the other unopened door. Once the final choice is locked in, the doors are opened and the prizes revealed.

As it turns out, it is always better to switch to the unopened door. In fact, the probability of winning the prize is 2/3 if the contestant switches, but only 1/3 if the contestant sticks with their initial choice. But how could this be? We might expect that there should be exactly a 1/3 probability of the prize being behind any of the three doors. However, by the contestant opening one of the doors (specifically, a non-prize, non-selected door), we are receiving additional information which shifts the distribution in probabilities.

To see this, let

1. event $A$ represent the event in which the prize is behind door number one;
2. event $B$ represent the event in which the prize is behind door number two; and
3. event $C$ represent the event in which the prize is behind door number three.

Our prior belief is that of equal a priori probability:

$$\mathbb{P}(A) = \mathbb{P}(B) = \mathbb{P}(C) = \frac{1}{3}.$$

Since the selection of door by the contestant is arbitrary, let us, without loss of generality, suppose that the contestant selects door number one. Given the contestant's initial choice, we know that the host must either reveal the contents behind door number two or door number three, which we will call events $OB$ and $OC$, respectively. ($O$ for open.)

The probability that the host opens door number three (given the contestant's initial selection of door number one) is given by

$$\mathbb{P}(OC) = \mathbb{P}(OC|A)p(A) + \mathbb{P}(OC|B)p(B) + \mathbb{P}(OC|C)p(C)$$
$$= (1/2)(1/3) + (1)(1/3) + (0)(1/3) = 1/2.$$

Note that $\mathbb{P}(OC|A) = 1/2$, since if the prize is behind door number one, the host will arbitrarily decide which of the remaining two doors to open. Also, $\mathbb{P}(OC|B) = 1$, as if the prize is behind door number two, the host is not allowed to reveal this, and is therefore forced to choose door number three. Similarly, $p(OC|C) = 0$, as the host cannot open door number three if this door leads to the prize. A similar calculation shows that $p(OB) = 1/2$.

So far, nothing is surprising. The host has an equal probability of opening doors two or three, due to the symmetry. But once the host's decision is revealed, something interesting happens.

Suppose the host opens door number three. The *posterior probability* that the prize is behind door number one is now given by Bayes' law:

$$\mathbb{P}(A|OC) = \frac{\mathbb{P}(OC|A)\mathbb{P}(A)}{\mathbb{P}(OC)}.$$

Now, the prior probability is $p(A) = 1/3$, and the probability that door three was chosen by the host is $\mathbb{P}(OC) = 1/2$. Also, recall that $\mathbb{P}(OC|A) = 1/2$; if the prize is behind door number one, the host has a 50% probability of selecting door three. But if we put this all together, we find

$$\mathbb{P}(A|OC) = \frac{(1/2)(1/3)}{1/2} = 1/3.$$

The probability of the prize being behind door number one is now only $1/3$. Similarly, $\mathbb{P}(B|OC) = 2/3$. Therefore, the contestant can double their chances at winning the prize by switching doors.                                    ▷

*Example 10.3.* In 1966, Ed Thorp shocked the gambling world by publishing the details at how a player could *beat the dealer* in the game of blackjack; see Thorp [1966]. His discovery was fueled by Bayesian statistics and MIT's IBM 704 computer. A lovely accounting of his life, how he used mathematics to beat the house at both blackjack and roulette, and his later years using mathematics to *beat the market*, is available in his autobiography Thorp [2017].

In the game of *blackjack*, the dealer deals two cards to each player, and two to herself (one face-up and one face-down). Each player in turn has the opportunity to receive a new card (*hit*) or finalize their hand (*stand*). Each card has a value: aces count as a 1 or 10, at the player's choice; 10, Jack, Queen, and King are each worth 10 points; and number cards are worth their face value. A player wins if their total is 21, or closer to 21 than the dealer. The catch is if a player total exceeds 21, it is called a *bust* and the player loses immediately. The other catch is that the dealer must hit until she reaches a total of 16, after which she stands. The house favor is created as players must select their cards first, risking a bust and immediate loss (regardless of whether the dealer busts).

Playing the Baldwin strategy, devised by four mathematicians during their time in the army, the dealer has only a 0.21% advantage over the player: practically 50-50 odds. Ed Thorp, however, realized that those odds shift dramatically based on the cards that were played. In Thorp's words:

> I realized that the odds as the game progressed actually de-
> pended on which cards were still left in the deck and that the edge
> would shift as play continued, sometimes favoring the casino and
> sometimes the player.

Thorp realized that the Baldwin strategy assumed that the probability of drawing a card was always the same. But, as the game progressed, one gains additional information, and is able to understand more precisely how the probability shifts, given the observed data. (A Bayesian update!) In particular, the more ten cards played, the better the odds for the dealer; whereas the more low-value cards (2–6) that are played, the better the odds for the player. The reason is that the dealer relies on low-value cards in order not to bust when it is their turn to draw, as they are required to hit until they reach at least 16.

Thorp devised various card counting strategies to track whether or not the deck was "ten-rich" or "ten-poor." When the odds shifted in his favor, Thorp increased his bet to between $2 and $10; otherwise he bet $1. Eventually the casinos realized they could no longer offer black-jack tables that used a single deck of cards. But Thorp did alright until they figured out his game. ▷

### 10.1.2 Bayesian Inference

*Statistical inference* is the practice of drawing conclusions from data about unobservable quantities or parameters. The frequentist's approach to inference is through hypothesis testing (Definition 3.1); a null hypothesis is made about a population parameter, a test is conducted, and a binary conclusion is drawn that renders the null hypothesis either accepted or rejected. Bayesian inference, on the other hand, allows us to quantify our *belief* in the range of values of a parameter through a distribution, and it further allows for the possibility of that distribution to change with experience. We make this leap by extending Bayes' law to an equivalent statement about probability distributions.

**Proposition 10.2.** *Given an* IID *set* $\mathcal{D} = \{Y_1, \ldots, Y_n\}$ *of values from a distribution that depends on a parameter* $\theta$,

$$p(\theta|\mathcal{D}) = \frac{p(\mathcal{D}|\theta)p(\theta)}{p(\mathcal{D})}, \tag{10.2}$$

*where* $p(\cdot)$ *represents the appropriate (conditional) probability density or mass function.*

In order to more fully understand Equation (10.2), let's unpack its various components.

**Definition 10.1.** *In Equation* (10.2), *the quantity*

- $p(\theta)$ *represents the* prior distribution, *which quantifies our belief of the possible values of the parameter* $\theta$ *prior to having seen the data;*
- $p(\mathcal{D}|\theta)$ *represents the* likelihood *of the data (Definition 5.1); i.e.,*

$$p(\mathcal{D}|\theta) = \prod_{i=1}^{n} p(y_i|\theta),$$

*which is interpreted as a function of* $\theta$, *and where* $p(Y|\theta)$ *is referred to as the* sampling distribution, *as it is the distribution that generates our sample of data;*

- $p(\mathcal{D})$ *represents the* data distribution *or* marginal distribution *for our observed data, which is obtained by marginalizing the likelihood function over the parameter $\theta$ using the prior distribution; i.e.,*

$$p(\mathcal{D}) = \int_{\theta \in \Theta} p(\mathcal{D}|\theta)p(\theta)d\theta,$$

*when $\theta$ is a continuous variable, or*

$$p(\mathcal{D}) = \sum p(\mathcal{D}|\theta_i)p(\theta_i),$$

*when $\theta$ is a discrete variable, and*
- $p(\theta|\mathcal{D})$ *represents the* posterior distribution, *which quantifies our new belief of the possible values of the parameter $\theta$ having accounted for the observed data.*

*The process of converting a prior belief to a posterior belief via Equation* (10.2) *is known as a* Bayesian update.

Since the marginal distribution $p(\mathcal{D})$ does not depend on the parameter $\theta$, Equation (10.2) is often expressed in its *unnormalized form*

$$p(\theta|\mathcal{D}) \propto p(\mathcal{D}|\theta)p(\theta). \tag{10.3}$$

It is understood that the right-hand side must be normalized to make it a proper probability distribution for the parameter $\theta$.

*Example 10.4.* Suppose we wish to determine the probability of a positive outcome for a binary event, such as flipping a coin. Let $Y \sim \text{Bern}(\theta)$ represent the outcome of the coin flip, where $\theta$ represents the probability of heads. If we have no a priori knowledge or expectation as to how the coin is weighted, we might take the uniform distribution

$$\theta \sim \text{Unif}(0, 1)$$

as our prior distribution, so that $p(\theta) = 1$ for $0 < \theta < 1$, or, equivalently,

$$p(\theta) = \text{Unif}(\theta; 0, 1).$$

Now suppose we perform a sequence of ten Bernoulli trials, resulting in three successes; i.e., we measure three heads out of ten coin flips.

The likelihood function is given by the binomial distribution as

$$p(\mathcal{D}|\theta) = \prod_{i=1}^{10} \text{Bern}(y_i; \theta) = \prod_{i=1}^{10} \theta^{y_i}(1-\theta)^{1-y_i} = \theta^3(1-\theta)^7,$$

where $y_i$ is the result of the $i$th Bernoulli trial. Since our prior distribution is uniform, the posterior is proportional to our likelihood:

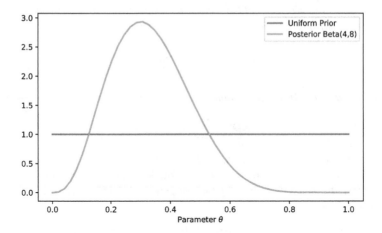

Fig. 10.1: Bayesian update for Example 10.4.

$$p(\theta|\mathcal{D}) \propto \theta^3(1-\theta)^7.$$

We recognize this distribution, however, as the kernel of the beta distribution (see Definition 2.17). We conclude that

$$p(\theta|\mathcal{D}) = \mathrm{Beta}(\theta; 4, 8);$$

i.e., the posterior distribution is a beta distribution with parameters $\alpha = 4$ and $\beta = 8$. (Compare the penultimate equation with Equation (2.49).) The prior and posterior distributions are plotted in Figure 10.1.    ▷

By a careful analysis of the preceding example, we can confirm that the posterior distribution is independent of the order of the precise observations. Moreover, we could select half of the observations, generate a posterior, and use that posterior as the prior for the second half of the observations, and this should lead to the same result that we obtained by considering all observations at once. (This is obvious, in general, given the definition of the likelihood function.)

This, however, leads us to an interesting observation in the context of our previous example: if the prior distribution for a sequence of Bernoulli trials is a beta distribution, so too should be the posterior. We can verify this as follows.

*Example 10.5.* Consider again the case of $n$ Bernoulli trials, resulting in a total of $k$ successful outcomes. Suppose, instead of choosing a uniform, we select a Beta prior; i.e., suppose we take

$$p(\theta) = \mathrm{Beta}(\theta; \alpha, \beta),$$

for some $\alpha, \beta > 0$. From Equation (2.49), the beta distribution is given by

$$p(\theta) = \frac{1}{B(\alpha, \beta)} \theta^{\alpha-1}(1-\theta)^{\beta-1},$$

where $B(\alpha, \beta)$ is the beta function (Equation (2.48)). The likelihood is given by

$$p(\mathcal{D}|\theta) = \prod_{i=1}^{n} \mathrm{Bern}(y_i; \theta) = \theta^k(1-\theta)^{n-k}.$$

Combining the previous two equations, we find that the posterior is given by

$$p(\theta|\mathcal{D}) \propto \theta^{\alpha+k-1}(1-\theta)^{\beta+n-k-1},$$

from which we conclude that

$$p(\theta|\mathcal{D}) = \mathrm{Beta}(\theta; \alpha+k, \beta+n-k). \tag{10.4}$$

Thus, when the prior distribution is a beta distribution, so too is the posterior. Moreover, the Bayesian update is equivalent to the following arithmetic operations

$$\alpha' = \alpha + k$$
$$\beta' = \beta + n - k.$$

Moreover, the expected value of $\theta$, following the observations is given by Equation (2.50) as

$$\mathbb{E}[\theta|\mathcal{D}] = \frac{\alpha + k}{\alpha + \beta + n}. \tag{10.5}$$

For example, consider a prior $p(\theta) \sim \mathrm{Beta}(2, 2)$, following which we observe $n = 10$ trials with $k = 3$ successes, so that the posterior is given by $\mathrm{Beta}(5, 9)$. The prior and posterior distributions are plotted in Figure 10.2.

Two points are in order. The first is that we can view the parameters $\alpha$ and $\beta$ of the prior distribution as a collection of *virtual observations*, where $\alpha$ constitutes the number of virtual successes and $\beta$ the number of virtual failures. The ratio $\mu = \alpha/(\alpha + \beta)$ represents our prior belief regarding the expected probability of success, whereas the magnitude $\alpha + \beta$ reflects our confidence; i.e., prior values $(3, 7)$ are easier to overcome than $(30, 70)$, as the latter represents a more firm belief that the true rate should be 30%. Second, the variance given by Equation (2.51) is equivalent to

$$\mathbb{V}(\theta) = \frac{\mu(1-\mu)}{\alpha + \beta + 1}. \tag{10.6}$$

Since the value of $\alpha$ and $\beta$ change with added data, and since $\alpha' + \beta' = \alpha + \beta + n$, this indicates that, even if our observed actual results agree exactly with our prior expected value, the variance of our belief decreases with added data. That is, the act of observing more data shrinks the variance of the probability distribution, making our belief more certain.    $\triangleright$

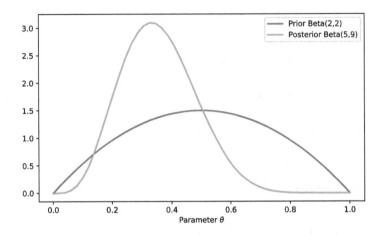

Fig. 10.2: Bayesian update for Example 10.5.

### 10.1.3 Conjugate Priors and Predictive Posteriors

#### Conjugate Priors

In Example 10.5, we examined a case where the prior and posterior distributions belonged to the same functional family of distributions. This is not unique to Bernoulli trials, but occurs for a variety of different distributions. When it does, we say that we are using the *conjugate prior* of the likelihood function. Formally, we have the following.

**Definition 10.2.** *Let $p(Y|\theta)$ be a parametric family of distributions. Then the parameteric family of distributions $p(\theta|\phi)$ is called a* conjugate prior *to the original family if, for any $Y$ and $\phi$, there exists a $\phi'$ such that*

$$p(Y|\theta)p(\theta|\phi) \propto p(\theta|\phi');$$

*i.e., if the posterior distribution $p(\theta|Y)$ is a member of the same family of distributions as the prior $p(\theta)$.*

*The parameter(s) $\phi$ of the prior distribution is referred to as the* hyperparameter(s) *for the system.*

We have already seen an example of this in Example 10.5. When the distribution $p(Y|\theta) = \text{Bern}(Y;\theta)$ and the prior distribution $p(\theta|\alpha,\beta) = \text{Beta}(\theta;\alpha,\beta)$, then so too is the posterior $p(\theta|Y) = \text{Beta}(\theta;\alpha+Y,\beta+1-Y)$.

#### Posterior Predictive Distribution

The parameter $\theta$ is never directly observable, as it is a modeling parameter. Once we have collected a set of data $\mathcal{D}$ and calculated our posterior distribution $p(\theta|\mathcal{D})$, it is natural to ask what we can do with this. In particular, what do we think about the *next* outcome $Y$, as sampled from

$$Y \sim p(Y|\theta),$$
$$\theta \sim p(\theta|\mathcal{D}).$$

We answer this by marginalizing over $\theta$, relative to the posterior distribution, to obtain the following.

**Definition 10.3.** *Given a parametric distribution $p(Y|\theta)$, a set of data $\mathcal{D}$, and our corresponding posterior distribution $p(\theta|\mathcal{D})$, the* posterior predictive distribution *is the distribution of the random variable $Y$ obtained by marginalizing over $\theta$; i.e.,*

$$p(Y|\mathcal{D}) = \int_{\theta \in \Theta} p(Y|\theta)p(\theta|\mathcal{D}) \, d\theta. \tag{10.7}$$

*Similarly, given the prior distribution $p(\theta)$, the* prior predictive distribution *is defined by*

$$p(Y) = \int_{\theta \in \Theta} p(Y|\theta)p(\theta) \, d\theta. \tag{10.8}$$

*Integrals are replaced with summations when the parameter $\theta$ is discrete.*

Note that the prior predictive distribution is similar to the marginal distribution in Definition 10.1, except that it is defined using the single probability distribution $p(Y|\theta)$, as opposed to the likelihood. Nevertheless, some authors use the term interchangeably.

*Note 10.1.* When a conjugate prior is used, $p(\theta)$ and $p(\theta|\mathcal{D})$ share the same functional form; i.e., the act of Bayesian update is a transformation on the hyperparameter $\phi \to \phi'$. As such, both the prior and posterior predictive distributions have the same form, except evaluated at a different value of the hyperparameter(s). For this reason, in the context of conjugate priors, we shall refer to the *predictive distribution*, as the distinction between prior and posterior is controlled by the choice of hyperparameter.    ▷

*Example 10.6.* Determine the prior predictive distribution in Example 10.5, where the prior distribution is $p(\theta) = \text{Beta}(\theta; \alpha, \beta)$. (Note this is equivalent to calculating the posterior predictive distribution, if we replace $\alpha, \beta \to \alpha', \beta'$.

We proceed by applying Equation (10.8), from which we obtain

$$p(Y) = \int_0^1 p(Y|\theta)p(\theta)\,d\theta$$

$$= \int_0^1 \frac{\theta^Y(1-\theta)^{1-Y}}{B(\alpha,\beta)}\theta^{\alpha-1}(1-\theta)^{\beta-1}\,d\theta$$

$$= \int_0^1 \frac{\theta^{\alpha+Y-1}(1-\theta)^{\beta-Y}}{B(\alpha,\beta)}\,d\theta$$

$$= \frac{B(\alpha+Y,\beta+1-Y)}{B(\alpha,\beta)}$$

$$= \text{BetaBin}(Y;1,\alpha,\beta).$$

Thus the prior (or posterior) predictive distribution for $Y \sim \text{Bern}(\theta)$ and $\theta \sim \text{Beta}(\alpha,\beta)$ is the beta-binomial distribution (Equation (2.59)). The penultimate equality follows since the preceding integral contains the kernel of the $\text{Beta}(\alpha+Y,\beta+1-Y)$ distribution. This can further be extended as a predictive distribution for the total number of successes from the next $n$ flips, as shown in Exercise 10.2.                    ▷

**Variance Reduction**

When calculating a Bayesian update for the probability parameter $\theta$ of a Bernoulli random variable, we saw that the variance *decreases* inversely proportionally to the data size. The variance is expressed by Equation (10.6), where the sum $\alpha + \beta$ increases by the number of observations (Equation (10.4)). In other words, if Equation (10.6) represents the variance of our prior distribution $p(\theta)$, then the variance of the posterior is obtained by using the same equation, with $\mu$ replaced by Equation (10.5) and with $\alpha + \beta$ in the denominator replaced with $\alpha' + \beta' = \alpha + \beta + n$.

The idea of a Bayesian update reducing the variance of our belief in the value of a parameter (which would therefore reduce our confidence intervals) is not unique to Bernoulli random variables, but a staple of Bayesian inference. The result also makes intuitive sense: our certainty should improve the more data we collect. In the limit as $n \to \infty$, the posterior distribution will collapse to a delta function centered at a particular point.

In order to capture the expected variance reduction mathematically, we next revist Equations (1.19) and (2.58), both of which are based on the conditional expectation

$$\mathbb{E}[\theta|\mathcal{D}],$$

which is the expected value of $\theta$ relative to the posterior distribution. The posterior distribution itself is, however, dependent on the particular set of data $\mathcal{D}$ that was realized, which occurs relative to the marginal distribution $p(\mathcal{D})$. We may therefore consider the expectation and variance of the above quantity relative to all possible sets of data we might observe. Equation (1.19) implies that

$$\mathbb{E}[\theta] = \mathbb{E}\left[\mathbb{E}[\theta|\mathcal{D}]\right].$$

The outer expectation is taken relative to the data $\mathcal{D}$. This equation states that the expectation of the prior distribution is equivalent to the average posterior expectation, over all possible realizations of the data consistent with the data distribution $p(\mathcal{D})$.

Similarly, Equation (2.58) implies that

$$\mathbb{V}(\theta) = \mathbb{E}\left[\mathbb{V}(\theta|\mathcal{D})\right] + \mathbb{V}\left(\mathbb{E}[\theta|\mathcal{D}]\right). \tag{10.9}$$

This states that, on average, the variance in the posterior is *less* than the variance in the prior by an amount equal to the variance of the possible posterior means. Moreover, if we think of our data size as variable, and we let $\mathcal{D}_n$ represent an IID sample of $n$ observations, then we find that

$$\lim_{n\to\infty} \mathbb{V}\left(\mathbb{E}[\theta|\mathcal{D}_n]\right) = \mathbb{V}(\theta),$$

which implies that $\mathbb{V}(\theta|\mathcal{D}_n) \to 0$ as $n \to \infty$; i.e., the variance of the posterior vanishes in the limit of unlimited data.

## 10.1.4 Conjugate Priors and Exponential Dispersion Models

In general, any member of the family of exponential dispersion models has a conjugate prior, as we show in our next proposition. In our next section, we will then derive some common conjugate families.

**Proposition 10.3.** *Let $p(Y|\theta) \sim$ EDM be an exponential dispersion model (Definition 7.10), for a given cumulant function $\kappa(\theta)$ and fixed dispersion parameter $\phi$. Then two-parameter family defined by*

$$p(\theta|\mu,\psi) = b(\mu,\psi)\exp\left[\frac{\theta\mu - \kappa(\theta)}{\psi}\right] \tag{10.10}$$

*constitutes a conjugate prior to the family $p(Y|\theta)$, where the function $b(\mu,\psi)$ is the normalizing constant given by*

$$b(\mu,\psi) = \left(\int \exp\left[\frac{\theta\mu - \kappa(\theta)}{\psi}\right]d\theta\right)^{-1}, \tag{10.11}$$

*where the integration is taken over the domain of the parameter $\theta$.*

*Moreover, given a set of data $\mathcal{D} = \{y_i\}_{i=1}^n$, the Bayesian update in the hyperparameter $(\mu,\psi)$ is given by*

$$\mu' = \lambda\overline{y}. + (1-\lambda)\mu, \tag{10.12}$$
$$\psi' = (1-\lambda)\psi, \tag{10.13}$$

*where $y. = \sum_{i=1}^n y_i$ is the sum, $\overline{y}. = (1/n)\sum_{i=1}^n y_i$ is the sample mean, and the parameter $\lambda$, known as the* credibility, *is given by*

$$\lambda = \frac{n}{n + \eta}, \tag{10.14}$$

where $\eta = \phi/\psi$ and $n = |\mathcal{D}|$.

*Proof.* From Equation (7.53), the likelihood function from a set of IID $Y_i \sim$ EDM is given by

$$p(\mathcal{D}|\theta) = \left(\prod_{i=1}^{n} a(y_i, \phi)\right) \exp\left[\frac{\theta \sum_{i=1}^{n} y_i - n\kappa(\theta)}{\phi}\right].$$

Note that the coefficient is independent of the variable $\theta$. From this expression, it is clear that if the prior distribution is of the form

$$p(\theta|\nu, \eta) = b(\nu/\eta, \phi/\eta) \exp\left[\frac{\theta\nu - \eta\kappa(\theta)}{\phi}\right], \tag{10.15}$$

then so too is the posterior, with updated hyperparameters given by

$$\nu' = \nu + \sum_{i=1}^{n} y_i, \tag{10.16}$$

$$\eta' = \eta + n. \tag{10.17}$$

Now, we can rearrange the argument of the exponential function in Equation (10.15) to obtain

$$\frac{\theta\nu/\eta - \kappa(\theta)}{\phi/\eta}.$$

Making the substitution $\mu = \nu/\eta$ and $\psi = \phi/\eta$ yields the expression Equation (10.10). The normalizing function Equation (10.11) follows as the function Equation (10.10) must be a proper density for the parameter $\theta$.

To derive Equation (10.12), we may apply Equations (10.15) and (10.16) to obtain

$$\mu' = \frac{\nu'}{\eta'} = \frac{\nu + y.}{\eta + n} = \frac{\eta}{\eta + n}\frac{\nu}{\eta} + \frac{n}{\eta + n}\frac{y.}{n} = (1 - \lambda)\mu + \lambda\overline{y}..$$

Similarly,

$$\psi' = \frac{\phi}{\eta'} = \frac{\eta}{\eta + n}\frac{\phi}{\eta} = (1 - \lambda)\psi,$$

which is equivalent to Equation (10.13). □

*Note 10.2.* Equations (10.12) and (10.13) take the same form as found in credibility theory, a topic we will discuss toward the end of the chapter. The parameter $\lambda$ may be regarded as a "credibility factor" expressing the fraction of the observed weight to the total observed and prior weights. This latter expression is exactly in the appropriate form for the credibility estimator (Theorem 10.3), which can be interpreted as a *linear* Bayes' estimator for risk premiums.    ▷

*Note 10.3.* We are at present unaware of a closed-form expression for the mean and variance of the family of conjugate priors given by Equation (10.10). The moment-generating function may be expressed as

$$M(t) = \frac{b\left(\mu, \psi\right))}{b\left(\mu + t\psi, \psi\right)},$$

but it is unclear how we could use this to determine an expression for the moments. Though one can show that

$$\mathbb{E}[\Theta] = -\psi \frac{b_\mu\left(\mu, \psi\right)}{b\left(\mu, \psi\right)}$$

and

$$\mathbb{V}(\Theta) = \psi^2 \left[ \frac{b_\mu\left(\mu, \psi\right)^2}{b\left(\mu, \psi\right)^2} - \frac{b_{\mu\mu}\left(\mu, \psi\right)}{b\left(\mu, \psi\right)} \right],$$

where $b_\mu(\cdot, \cdot)$ is the partial derivative of the normalizing coefficient with respect to its first argument. Since we cannot say much about $b$, it is unclear how we might uncover additional expressions from these results.    ▷

As it turns out, the parameter $\mu$ has a simple interpretation as the expected value of $Y$ when marginalizing over $\theta$.

**Proposition 10.4.** *Given the hierarchical model $Y \sim \text{EDM}(Y; \Theta)$ and $\Theta \sim p(\Theta; \mu, \psi)$, where $p(\Theta; \mu, \psi)$ is given by Equation (10.10), we have the result*

$$\mathbb{E}[Y] = \mu. \tag{10.18}$$

*Moreover, if we define*

$$\sigma^2 = \mathbb{E}\left[\mathbb{V}(Y|\Theta)\right], \tag{10.19}$$
$$\tau^2 = \mathbb{V}\left(\mathbb{E}[Y|\Theta]\right), \tag{10.20}$$

*we have that*

$$\frac{\sigma^2}{\tau^2} = \frac{\phi}{\psi} = \eta. \tag{10.21}$$

Note that, given Equation (10.21), the credibility factor Equation (10.14) may be expressed as

$$\lambda = \frac{n}{n + \sigma^2/\tau^2}.$$

The variance in means $\tau^2$ and the conditional variance $\sigma^2(\Theta)$ of the EDM are illustrated in Figure 10.3, with $\sigma^2 = \mathbb{E}[\sigma^2(\Theta)]$.

*Proof.* Recall from Theorem 7.7 the expression

$$\mu(\Theta) = \kappa'(\Theta).$$

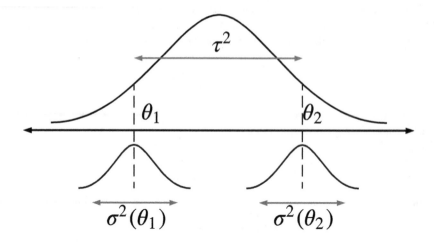

Fig. 10.3: Visualization of the variance of an EDM, $\sigma^2(\Theta)$, versus the variance of the prior $\tau^2$.

where we have defined $\mu(\Theta) = \mathbb{E}[Y|\Theta]$ as the expected value of $Y$, given the parameter $\Theta$. Now, from Equation (1.19), we have

$$\mathbb{E}[Y] = \mathbb{E}[\mathbb{E}[Y|\Theta]] = \mathbb{E}[\kappa'(\Theta)].$$

However, note that

$$\mu - \mathbb{E}[\kappa'(\Theta)] = \int b(\mu, \psi) \left[\mu - \kappa'(\theta)\right] \exp\left[\frac{\theta\mu - \kappa(\theta)}{\psi}\right] d\theta,$$

which integrates to zero, as the density of an EDM vanishes on its boundaries (typically $\pm\infty$; but in some cases $-\infty$ and 0). This implies that $\mathbb{E}[\kappa'(\Theta)] = \mu$, completing the proof of Equation (10.18).

For the second part of the proposition, we again recall from Theorem 7.7 that

$$\mathbb{V}(Y|\Theta) = \phi\kappa''(\Theta),$$

so that

$$\sigma^2 = \phi\mathbb{E}[\kappa''(\Theta)].$$

Moreover,

$$\tau^2 = \mathbb{V}(\mu(\Theta)) = \mathbb{E}\left[(\kappa'(\theta) - \mu)^2\right].$$

Now, not only does Equation (10.10) vanish at the limits of the domain of $\theta$, but so too does its derivative

$$p'(\theta|\mu, \psi) = b(\mu, \psi) \left[\frac{\mu - \kappa'(\theta)}{\psi}\right] \exp\left(\frac{\mu\theta - \kappa(\theta)}{\psi}\right).$$

This implies that the integral of $p''(\theta|\mu, \psi)$ must also vanish; i.e.,

$$\int b(\mu, \psi) \left\{ \left[ \frac{\mu - \kappa'(\theta)}{\psi} \right]^2 - \frac{\kappa''(\theta)}{\psi} \right\} \exp\left( \frac{\mu\theta - \kappa(\theta)}{\psi} \right) d\theta = 0.$$

However, this is equivalent to the expression

$$\frac{\mathbb{E}\left[(\kappa'(\Theta) - \mu)^2]\right]}{\psi^2} = \frac{\mathbb{E}[\kappa''(\Theta)]}{\psi},$$

or, alternatively, $\phi\tau^2 = \psi\sigma^2$, which completes the proof. $\qquad\square$

It follows that we can interpret the parameter $\psi$ as the ratio of the variance $\tau^2$ in expected means $\mu(\Theta)$ to the expected variance $\sigma^2$. We may visualize these two variances as shown in Figure 10.3. As we continue to make Bayesian updates, the variance in the prior/posterior will collapse to zero, so that $\tau^2 \to 0$ and $\psi \to 0$ as $n \to \infty$, whereas the expected variance $\sigma^2$ will approach a limiting constant: $\sigma^2 \to \sigma^2(\hat{\theta})$. Finally, note that the variance in the posterior predictive distribution is given by $\mathbb{V}(Y) = \sigma^2 + \tau^2$, due to the law of total variance, as given by Equation (2.58).

### 10.1.5 A Catalogue of Conjugate Families

We have already computed the conjugate prior and posterior predictive distributions for a Bernoulli random variable, in Examples 10.5 and 10.6. Generalization to a binomial random variable is straightforward (see Exercise 10.2). Moreover, we proved that *any* member of the exponential dispersion family similarly has a conjugate prior. We now turn to a few specific examples: a number of common conjugate families are recorded in Tables 10.2 and 10.3 along with their posterior predictive distributions, when available. We will derive a few of these, and leave the rest as an exercise for the reader.

### Poisson–Gamma

The gamma distribution is the conjugate prior for the Poisson distribution. To see this, let

$$p(Y|\lambda) = \text{Poiss}(Y; \lambda) = \frac{e^{-\lambda}\lambda^y}{y!},$$

$$p(\lambda) = \text{Gamma}(\lambda; \alpha, \beta) = \frac{1}{\Gamma(\alpha)\beta^\alpha}\lambda^{\alpha-1}e^{-\lambda/\beta}.$$

The likelihood of a set of data $\mathcal{D} = \{Y_i\}_{i=1}^n$ is therefore given by

$$p(\mathcal{D}|\lambda) = \frac{e^{-n\lambda}\lambda^{\sum y_i}}{\prod y_i!}.$$

| sampling distribution | unknown parameter | conjugate prior | Bayesian update | predictive distribution |
|---|---|---|---|---|
| $\text{Bern}(\theta)$ | $\theta$ | $\text{Beta}(\alpha, \beta)$ | $\alpha' = \alpha + \sum Y_i$ $\beta' = \beta + n - \sum Y_i$ | $\text{BetaBin}(Y; 1, \alpha, \beta)$ |
| $\text{Binom}(n_i; \theta)$ | $\theta$ | $\text{Beta}(\alpha, \beta)$ | $\alpha' = \alpha + \sum Y_i$ $\beta' = \beta + \sum n_i - \sum Y_i$ | $\text{BetaBin}(Y; n, \alpha, \beta)$ |
| $\text{Poiss}(\lambda)$ | $\lambda$ | $\text{Gamma}(\alpha, \beta)$ | $\alpha' = \alpha + \sum Y_i$ $(\beta')^{-1} = \beta^{-1} + n$ | $\text{NBD}(Y; \alpha, (1+\beta)^{-1})$ |
| $\text{Geom}(\theta)$ | $\theta$ | $\text{Beta}(\alpha, \beta)$ | $\alpha' = \alpha + n$ $\beta' = \beta + \sum Y_i$ | $\text{BetaGeom}(Y; \alpha, \beta)$ |
| $\text{NBD}(r, \theta)$ | $\theta$ | $\text{Beta}(\alpha, \beta)$ | $\alpha' = \alpha + \sum Y_i$ $\beta' = \beta + rn$ | $\text{BetaNegBin}(Y; \alpha, \beta)$ |
| $\text{Multi}(\theta)$ | $\theta$ | $\text{Dir}(\alpha)$ | $\alpha' = \alpha + \sum Y_i$ | $\text{DirMulti}(Y; n, \alpha)$ |

Table 10.2: Common conjugate families for discrete distributions.

| sampling distribution | unknown parameter | conjugate prior | Bayesian update | predictive distribution |
|---|---|---|---|---|
| $N(\mu, \sigma^2)$ | $\mu$ | $N(\nu, \tau^2)$ | $\nu' = \left(\frac{1}{\tau^2} + \frac{n}{\sigma^2}\right)^{-1}\left(\frac{\nu}{\tau^2} + \frac{\sum Y_i}{\sigma^2}\right)$, $(\tau')^2 = \left(\frac{1}{\tau^2} + \frac{n}{\sigma^2}\right)^{-1}$ | $N(Y; \nu, \tau^2 + \sigma^2)$ |
| $\mathrm{Exp}(\lambda^{-1})$ | rate $\lambda$ | $\mathrm{Gamma}(\alpha, \beta)$ | $\alpha' = \alpha + n$, $(\beta')^{-1} = \beta^{-1} + \sum Y_i$ | $\mathrm{Lomax}(Y; \alpha, \beta^{-1})$ |
| $\mathrm{Gamma}(\alpha, \beta)$ | $\beta$ | $\mathrm{Gamma}(\alpha_0, \beta_0)$ | $\alpha_0' = \alpha_0 + n\alpha$, $(\beta_0')^{-1} = \beta_0^{-1} + \sum Y_i$ | — |
| $\mathrm{Pareto}(m, k)$ | $k$ | $\mathrm{Gamma}(\alpha, \beta)$ | $\alpha' = \alpha + n$, $(\beta')^{-1} = \beta^{-1} + \sum \ln(Y_i/m)$ | — |

Table 10.3: Common conjugate families for continuous distributions.

The posterior distribution is therefore proportional to

$$p(\lambda|\mathcal{D}) \propto e^{-n\lambda}\lambda^{\sum y_i}\lambda^{\alpha-1}e^{-\lambda/\beta} = \lambda^{\alpha+\sum y_i-1}e^{-\lambda(\beta^{-1}+n)},$$

which we recognize as the kernel of a Gamma$(\alpha + \sum y_i, (\beta^{-1} + n)^{-1})$ distribution.

The predictive distribution is then obtained by marginalizing the sampling distribution over the gamma prior / posterior:

$$
\begin{aligned}
p(y|\alpha, \beta) &= \int_0^\infty p(y|\lambda)p(\lambda|\alpha, \beta)\, d\lambda \\
&= \int_0^\infty \frac{\lambda^{\alpha+y-1}e^{-\lambda(1+1/\beta)}}{y!\Gamma(\alpha)\beta^\alpha}\, d\lambda \\
&= \frac{\Gamma(\alpha+y)(1+1/\beta)^{-(\alpha+y)}}{y!\Gamma(\alpha)\beta^\alpha},
\end{aligned}
$$

where the final equality follows by recognizing the kernel of the Gamma$(\lambda; \alpha + y, (1+1/\beta)^{-1})$ distribution. This, however, is equivalent to

$$p(y|\alpha, \beta) = \frac{\Gamma(\alpha+y)}{\Gamma(y+1)\Gamma(\alpha)}\frac{1}{(1+\beta)^\alpha}\frac{1}{(1+1/\beta)^y}.$$

Setting $r = \alpha$ and $p = (1+\beta)^{-1}$, we recognize this as a negative-binomial distribution; i.e.,

$$p(y|\alpha, \beta) = \text{NBD}(y; \alpha, (1+\beta)^{-1}).$$

As mentioned in Note 10.1, this represents both the prior and posterior predictive distribution, depending on the choice of hyperparameters $(\alpha, \beta)$.

## Geometric–Beta

The beta distribution is the conjugate prior of the geometric distribution. To see this, let

$$p(Y|\theta) = \text{Geom}(Y; \theta) = \theta(1-\theta)^y,$$
$$p(\theta) = \text{Beta}(\theta; \alpha, \beta) = \frac{1}{B(\alpha, \beta)}\theta^{\alpha-1}(1-\theta)^{\beta-1}.$$

The likelihood of a set of data $\mathcal{D} = \{Y_i\}_{i=1}^n$ is therefore given by

$$p(\mathcal{D}|\theta) = \theta^n(1-\theta)^{\sum Y_i}.$$

The posterior distribution is therefore proportional to

$$p(\theta|\mathcal{D}) \propto \theta^n(1-\theta)^{\sum y_i}\theta^{\alpha-1}(1-\theta)^{\beta-1} = \theta^{\alpha+n-1}(1-\theta)^{\beta+\sum y_i-1},$$

which we recognize as the kernel of a Beta$(\alpha + n, \beta + \sum y_i)$ distribution.

**Exponential–Gamma**

The gamma distribution is the conjugate prior to the exponential distribution, when expressed in terms of its rate parameter $\lambda$. To see this, let

$$p(Y|\lambda) = \mathrm{Exp}(Y; \lambda^{-1}) = \lambda e^{-\lambda y},$$

$$p(\lambda) = \mathrm{Gamma}(\lambda; \alpha, \beta) = \frac{1}{\Gamma(\alpha)\beta^\alpha} \lambda^{\alpha-1} e^{-\lambda/\beta}.$$

The likelihood of a set of data $\mathcal{D} = \{Y_i\}_{i=1}^n$ is therefore given by

$$p(\mathcal{D}|\lambda) = \lambda^n e^{-\lambda \sum y_i}.$$

The posterior distribution is therefore proportional to

$$p(\lambda|\mathcal{D}) \propto \lambda^n e^{-\lambda \sum y_i} \lambda^{\alpha-1} e^{-\lambda/\beta} = \lambda^{\alpha+n-1} e^{-\lambda(\beta^{-1}+\sum y_i)},$$

which we recognize as the kernel of a $\mathrm{Gamma}(\alpha + n, (\beta^{-1} + \sum y_i)^{-1})$ distribution.

The predictive distribution is obtained by marginalizing

$$
\begin{aligned}
p(Y) &= \int_0^\infty p(Y|\lambda)p(\lambda) \, d\lambda \\
&= \int_0^\infty \frac{\lambda e^{-\lambda y} \lambda^{\alpha-1} e^{-\lambda/\beta}}{\Gamma(\alpha)\beta^\alpha} \, d\lambda \\
&= \frac{\Gamma(\alpha+1)}{\Gamma(\alpha)\beta^\alpha(\beta^{-1}+Y)^{\alpha+1}}.
\end{aligned}
$$

The final equation follows by recognizing the kernel of the $\mathrm{Gamma}(\alpha + 1, (\beta^{-1} + Y)^{-1})$ distribution in the integrand of the preceding equation. However, this is equivalent to the *Lomax distribution*, or *Pareto Type II distribution*:

$$\mathrm{Lomax}(x; \alpha, \lambda) = \frac{\alpha}{\lambda}\left[1 + \frac{x}{\lambda}\right]^{-(1+\alpha)} = \frac{\alpha\lambda^\alpha}{(x+\lambda)^{\alpha+1}}, \tag{10.22}$$

with support $x \in [0, \infty)$. (Note that the Pareto distribution is a shifted Lomax distribution.) That is, $p(Y) = \mathrm{Lomax}(Y; \alpha, \beta^{-1})$.

**EDM Form**

Finally, we can calculate the parameters $\mu$, $\sigma^2$, $\tau^2$, and $\psi$ for several common conjugate priors, using Equations (10.18)–(10.21). The results are shown in Table 10.4. Note that in each case, the variance of the predictive distribution is equal to the sum of the variances $\sigma^2$ and $\tau^2$. If we reparameterize the conjugate prior distribution using the hyperparameters $(\mu, \psi)$ in place of $(\alpha, \beta)$, the Bayesian update formula for each of these families is given in the form of Equations (10.12) and (10.13). Note that the $(\mu, \psi)$ parameterization is not unique to the prior, but unique to the conjugate pair.

| family | $\mu = \mathbb{E}[Y]$ | $\sigma^2 = \mathbb{E}[\mathbb{V}(Y|\Theta)]$ | $\tau^2 = \mathbb{V}(\mathbb{E}[Y|\Theta])$ | $\psi = \tau^2/\sigma^2$ |
|---|---|---|---|---|
| binomial beta | $\dfrac{n\alpha}{\alpha+\beta}$ | $\dfrac{n\alpha\beta}{(\alpha+\beta)(\alpha+\beta+1)}$ | $\dfrac{n^2\alpha\beta}{(\alpha+\beta)^2(\alpha+\beta+1)}$ | $\dfrac{n}{\alpha+\beta}$ |
| Poisson gamma | $\alpha\beta$ | $\alpha\beta$ | $\alpha\beta^2$ | $\beta$ |
| geom beta | $\dfrac{\beta}{\alpha-1}$ | $\dfrac{\beta(\alpha+\beta-1)}{(\alpha-2)(\alpha-1)}$ | $\dfrac{\beta(\alpha+\beta-1)}{(\alpha-2)(\alpha-1)^2}$ | $\dfrac{1}{\alpha-1}$ |
| exp gamma | $\dfrac{1}{\beta(\alpha-1)}$ | $\dfrac{1}{\beta^2(\alpha-1)(\alpha-2)}$ | $\dfrac{1}{\beta^2(\alpha-1)^2(\alpha-2)}$ | $\dfrac{1}{\alpha-1}$ |

Table 10.4: Each prior may be parameterized by $(\mu, \psi)$, so that the Bayesian update is in the form of Equations (10.12) and (10.13).

### 10.1.6 The Bayes–Kalman Filter (or Recursive Bayes)

The variance-reduction principal (recall Equation (10.9)) implies that as we continue to add data to a Bayesian update, the variance of the posterior distribution shrinks, ultimately approaching zero. In the context of stochastic processes (Definition 5.11), however, this is not always—or even typically—a desired trait. For example, suppose a marketing campaign has been running for 30 days, and each day the observed conversion rate has been constistantly 10%. Then, all of the sudden, the conversion rate drops to 2%, and remains at 2%. (This could be due to unknown or hidden factors, such as a sudden change in market pressures or auction conditions.) Now, if we are simply running a binomial–beta Bayesian update each day, it will take *an additional 30 days* just for the 10% and 2% measurements to be equally weighted, whereas a human would notice the dramatic change in performance almost immediately. This is because our Bayesian update is simply keeping a tab of volume $n$ and conversions $k$, following the update of Equation (10.5). As the total volume grows, the overall variance shrinks, and it becomes harder and harder to "unlearn" historic behavior.

The cause of this dilemma is the IID assumption in the Bayesian update rule of Proposition 10.2. As long as our data are IID, we can perform the Bayesian updates piecemeal, by breaking up our data set into smaller segments and creating a sequence of new posteriors. But when those data segments are no longer IID, our assumption breaks down, and we require a new approach.

But all is not lost. In this section, we propose a method for combatting such a scenario, which we name the *Bayesian Kalman filter*, in recognition of the classic Kalman filter used in engineering, which is the direct inspiration for our our approach. A *Kalman filter* is a technique in dynamical systems theory that recursively estimates the state of a system over time, given measurements subject to uncertainty. Typically, one tracks both position and velocity, and the uncertainty of each (in the form of a covariance matrix), and further models the relation between true state and measurement, accounting for measurement error. The Kalman filter applies the linear dynamics to the prior system state to estimate a new system state, and then combines information from the new observed values and the intrinsic measurement uncertainty to provide an updated, superior estimation for the actual system state. For a details on Kalman filters in the context of general stochastic processes, see, for example, Hajek [2015]. The Kalman filter technique has been previously applied credibility theory, which constitutes a system of linear Bayes estimators, in Bühlmann and Gisler [2005].

To proceed, let us suppose we are given a sequence of data that are collected over time: $\mathcal{D}_1, \ldots, \mathcal{D}_m$, such that each set of data is collected sequentially at a distinct point in time (e.g., day-by-day). Now, a pure Bayesian update would result in the posterior

$$p(\theta|\mathcal{D}_1, \ldots, \mathcal{D}_m) = \left(\prod_{i=1}^{m} p(\mathcal{D}_i|\theta)\right) p(\theta),$$

treating each set of data as equivalent. This approach fundamentally fails to take into account the factor of time and the possibility of temporal changes to the distribution of data (e.g., switchpoints or trends).

To resolve this, we go back to Proposition 10.2, and recall that a *prior distribution* is a representation of our *belief* in the state of a system, and if we believe that the state of the system is, or can be, changing, then our prior distribution for day $m$ should not simply equal the posterior distribution of day $(m-1)$; i.e., we must explicitly break our tacit assumption that $\mathcal{D}_i$ and $\mathcal{D}_j$ are IID samples from the same distribution. In particular, we should explicitly counteract the variance shrinkage with a controlled *variance expansion*, which should be greater the more $\mathcal{D}_m$ differs from our prior day's posterior $p_{m-1}(\theta|\mathcal{D}_1, \ldots, \mathcal{D}_{m-1})$.

## Kalman Interpretation of Bayesian Update

Before proceeding, it is interesting to note that the standard Bayes' update rule is, essentially, a form of Kalman filter, with stationary (in-place) dynamics. To see this, let our state be defined as $\mu = \mathbb{E}[Y]$, and define the subscript notation as follows

$$\mu_{t|t-1} = \mathbb{E}[\mu|\mathcal{D}_1, \ldots, \mathcal{D}_{t-1}],$$
$$\mu_{t|t} = \mathbb{E}[\mu|\mathcal{D}_1, \ldots, \mathcal{D}_t].$$

The classic Kalman filter consists of two steps: prediction and update. For prediction, under a time-static model, we have

$$\mu_{t|t-1} = \mu_{t-1|t-1} \tag{10.23}$$

$$P_{t|t-1} = \mathbb{V}(\mu_{t|t-1}) = \mathbb{V}(\mathbb{E}[X|\Theta]) = \tau^2. \tag{10.24}$$

Next, we incorporate the new data set $\mathcal{D}_t$, using the summary statistics $n_t = |\mathcal{D}_t|$ and $\bar{y}_t = \frac{1}{n_t} \sum_{i=1}^{n_t} y_i$, to compute the *innovation* and its variance:

$$\delta_t = \bar{y}_t - \mu_{t|t-1},$$

$$S_t = \mathbb{V}(\delta_t) = \mathbb{V}(\bar{y}_t) = \frac{\sigma^2}{n} + \tau^2.$$

The *Kalman gain* is then computed as the ratio

$$K_t = PS^{-1} = \frac{\tau^2}{\tau^2 + \sigma^2/n_t} = \frac{n_t}{n_t + \sigma^2/\tau^2} = \lambda_t,$$

which is equivalent to the credibility factor Equation (10.14) of our Bayesian update. Finally, the Kalman update is given by

$$\mu_{t|t} = \mu_{n_t|n_{t-1}} + K_t(\bar{y}_t - \mu_{n_t|n_{t-1}})$$

$$= K_t\bar{y}_t + (1 - K_t)\mu_{t|t-1} \tag{10.25}$$

$$P_{n|n} = (1 - K_t)P_{n|n-1} \tag{10.26}$$

This is virtually synonymous with the Bayesian update Equations (10.12) and (10.13), except that the credibility factor is applied to $\tau^2$, as opposed to $\psi$. However, this is okay, as $\sigma^2$ approaches a limiting constant, whereas $\tau^2$ approaches zero, in the limit of infinite IID data.

Bühlmann and Gisler [2005] provides two extensions to this model, in the framework of credibility theory: the case of *orthogonal increments*, where $\mu(\Theta_t)$ is orthogonal to $\mu(\Theta_{t-1})$, in which case the update is still given by Equation (10.23), and the case of linear transformations (drift), where the update rule Equation (10.23) is replaced by a linear transformation

$$\mu_{t|t-1} = a\mu_{t-1|t-1} + b.$$

They further update the parameter $\tau^2$ using the equation

$$\tau_t^2 = \tau_{t-1}^2 + \mathbb{V}(\mu(\Theta_t) - \mu(\Theta_{t-1})),$$

which provides what we refer to as a *variance expansion*, which can be viewed as a counter-lever to the collapsing variance of subsequent Bayesian updates. We turn to our take on this rule next.

### Bayes–Kalman Filter

In this section, we discuss our approach to a *Bayes–Kalman filter*, for time-series data $\{\mathcal{D}_1, \mathcal{D}_2, \ldots\}$, such that, for each $t \in \mathbb{Z}_+$, we have IID data

$Y_{ti} \sim \text{EDM}(Y; \Theta)$, for $i = 1, \ldots, n_t$. Moreover, we express the conjugate prior as $\Theta \sim p(\Theta; \mu, \psi)$, such that the classic Bayesian update is given by Equations (10.12) and (10.13).

Now, if we were to make the IID assumption over all data sets $\mathcal{D}_1, \mathcal{D}_2, \ldots$ in our sequence, we would *only* have to apply the classic Bayesian update. The problem with this approach, however, is that in real life there can always be some exogenous influence that shifts the distribution of our data, and the classic Bayesian update leads to a variance shrinkage, such that $\psi \to 0$ as $n \to \infty$. Thus, when the underlying distribution changes, it takes longer for the Bayesian update to react, the longer a history it has already captured. We therefore propose a second step, which we call *variance expansion*, that is designed to counteract the effect of variance decay when a sudden change in behavior occurs. Note that we still assume that $Y \sim \text{EDM}(Y; \Theta)$, we only expand the variance of the parameter $\Theta$ to allow the classic Bayesian update rule to hone in on the new normal.

Drawing our inspiration from Kalman filters and evolutionary credibility models (Bühlmann and Gisler [2005]), we remedy these issues with our proposed Algorithm 10.1. Our procedure may still be considered a *two-step* update rule, though we break it into three sections. The *predict step* of the classic Kalman filter is here replaced with a *variance expansion* step: essentially we are predicting no change in the mean, $\mu_{t|t-1} = \mu_{t-1|t-1}$, as we are not modeling any explicit time dynamics, but an expansion to the variance, to account for the fact that successive data may not be IID, as

---

**Algorithm 10.1:** *Bayes–Kalman filter* (or *recursive Bayes*) for Bayesian updates of time series data.

---

**Input:** data series $\mathcal{D}_1, \mathcal{D}_2, \ldots$;
  primordial priors $\mu_0, \psi_0$;
  exponential decay factor $\alpha \in (0, 1)$
**Output:** Updated hyperparameters $(\mu_t, \psi_t)$.

1  Set $S_0, Z_0 = 0, 0$
2  **for** $t$ in $1, 2, \ldots$ **do**
3  $\quad$ *Variance Expansion*
4  $\quad$ $\psi_t = \psi_{t-1} + \max(0, Z_{t-1}^2 - 1)$
5  $\quad$ *Bayesian Update*
6  $\quad$ Credibility factor: $\lambda_t = \dfrac{n_t}{n_t + \phi/\psi}$
7  $\quad$ $\mu_t = \lambda_t \overline{y}_t + (1 - \lambda_t)\mu_{t-1}$
8  $\quad$ $\psi_t = (1 - \lambda_t)\psi_t$
9  $\quad$ *EWMA of variance and surprise estimates*
10 $\quad$ $S_t = \alpha s_t^2 + (1 - \alpha)S_{t-1}$; where $s_t^2$ is sample variance of $\mathcal{D}_t$
11 $\quad$ $Z_t = \alpha \dfrac{\overline{y_t} - \mu_t}{\sqrt{S_t(\psi_t + 1/n_t)}} + (1 - \alpha)Z_{t-1}$
12 **end**
13 **return** $(\mu_t, \psi_t)$

---

required for the standard Bayesian update. Recall that the parameter $\psi$ represents the ratio

$$\psi = \frac{\phi \tau^2}{\sigma^2} = \frac{\phi \mathbb{V}(\mathbb{E}[Y|\Theta])}{\mathbb{E}[\mathbb{V}(Y|\Theta)]}.$$

Increasing $\psi$ may therefore be viewed as an increase in our belief of the spread of population means between successive data sets, normalized by the expected variance. If we allow $\psi$ to collapse to zero, it becomes difficult to adapt to changing conditions.

Next, we have our *update step*, which has a familiar piece—the standard Bayesian update rule given by Equations (10.12) and (10.13)—and a new piece, representing an *exponentially weighted moving average* (EWMA) for our estimate $S_t$ of the expected variance $\sigma^2$ and our estimate of "surprise" $Z_t$.

*Note 10.4 (Exponentially weighted moving average).* A quick aside on EWMA. Given a time series $\{X_t\}$ and a weight parameter $\alpha \in (0, 1)$, we may define the sequence of averages

$$A_t = \alpha X_t + (1 - \alpha) A_{t-1}.$$

By expanding this recursion relation, we obtain

$$A_t = \alpha \sum_{s=0}^{n} (1 - \alpha)^s A_{t-s},$$

assuming a history of $n$ recursions. Note that the weights sum to unity in the limit as $n \to \infty$, as this result constitutes a geometric series.    ▷

So the estimate $S_t$ as an exponentially weighted average variance should be clear. Naturally, since we are averaging together the sample variances, which estimate $\mathbb{V}(Y|\Theta_t)$, we may treat $S_t$ as an estimate for $\sigma^2$.

Finally, we introduce the *surprise* $Z_t$, defined on line 11. First, let's consider the random variable $\overline{Y} - \mathbb{E}[Y]$, which is estimated by our numerator $\overline{y}_t - \mu_t$. The variance in this estimate is given by

$$\mathbb{V}(\overline{Y} - \mu) = \tau^2 + \frac{\sigma^2}{n} = \sigma^2 \left( \psi + \frac{1}{n} \right).$$

Thus $Z_t$ is an estimate for the number of standard deviations ($z$-score) the sample mean of the current data set is from our expected population mean, accounting for both the variance in $\Theta$ (as captured by $\tau^2$) as well as the sample variance given $\Theta$ (as captured by $\sigma^2/n$).

There are several final points regarding a few decisions we made when crafting this algorithm. First, note that we are summing the differences $(\overline{y}_t - \mu_t)$ directly, and not their squares. When we are in steady state, and the successive data sets are IID, the expected value of this random variable is zero, and positive outliers will cancel negative ones. So if $\overline{Y}_t$ is fluctuating around the mean randomly, we allow those fluctuations to cancel. However, when there has been a shift, one way or the other, our surprise factor will then build over time, as the fluctuations will suddenly be directional. Second, we calculate the surprise at the end of the update step, so that when we use it in the next predict step (i.e., variance expansion), we are more closely resembling the Kalman filter definition of prediction: we are updating $\psi_t$ based on information from the *previous* observation, not the current. We find this to be beneficial when there is an outlier that expands the variance, following our procedure, the variance will be expanded for the observation *following* the outlier, which grants us added protection from the final estimates succumbing to noise. Finally, note that our variance expansion step is only implemented when the surprise exceeds 1 (though, there is nothing that says that threshold cannot be adjusted). This prevents our algorithm from being *too reactive* to random outliers.

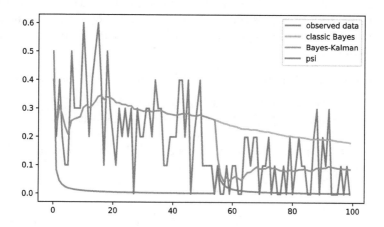

Fig. 10.4: Comparison of classic Bayes vs. the Bayes–Kalman filter. Exponential decay factor $\alpha = 0.2$.

To illustrate this algorithm, consider a time series consisting of ten draws from a Bernoulli random variable: $Y_{ti} \sim \text{Bern}(p)$, for $i = 1, \ldots, 10$. Moreover, suppose that the true value of $p$ is given by $p = 0.3$, for $t < 50$, and $p = 0.1$ for $t \geq 50$, constituting a *switchpoint*. Since on day 50, we have already performed 50 Bayesian updates, with total weight 500, it will require an additional 50 days before the two behaviors have a 50/50 blend. That is, with the class Bayesian update, our estimated probability will still be 20%, even on day 100! To contrast this, our Bayes–Kalman filter quickly locks onto the change, as shown in Figure 10.4.

## 10.2 Mixtures and Hierarchical Models

In this section, we will consider more advanced forms of Bayesian models. We begin by using a simple mixture to model population heterogeneity. We then discuss application of this technique to customer lifetime value models. Next, we extend our model to multi-level hierarchical models. Next, we discuss a certain linear Bayes estimator known as the credibility model, which is used extensively in the insurance industry. Finally, we resolve the conflict between time aspects of Bayesian data and variance reduction by devising a simple Bayesian Kalman filter.

### 10.2.1 Mixture Models

#### Finite Mixture Models

In this section, we will review two closely related topics: mixture models and hierarchical models. We begin with mixture models, which is a type of predictive model that is used to represent the presence of subpopulations within a given population, such that individual observations are not required to identify the component to which they belong. Formally, we have the following.

**Definition 10.4.** *A finite mixture model of dimension* $k$ *consists of a*

1. *a parametric sampling distribution* $F_\theta = p(Y|\theta)$,
2. *a set* $\mathcal{Z} = \{1, \ldots, k\}$ *of subpopulations or components,*
3. *a probability vector* $\pi \in \Delta^{k-1}$, *and*
4. *a set of* $k$ *distinct parameter values* $\theta_1, \ldots, \theta_k$ *corresponding to the* $k$ *components,*

*such that each observation is obtained by*

$$Z \sim \text{MultiBern}(\pi), \qquad (10.27)$$
$$Y|Z \sim F_{\theta_z}. \qquad (10.28)$$

Typically, one assumes the component value $Z$ for each observation to be unknown. In this case, the probability distribution for $Y$ is given by

$$p(Y) = \sum_{i=1}^{k} \pi_i p(Y|\theta_i). \qquad (10.29)$$

This is very similar to the predictive distribution from the Bayesian framework, when one views the probability vector $\pi$ as a discrete prior distribution over the parameter $\theta$. However, in this context, the probabilities $\pi$ are more suitably viewed as a measure of the variation in $\theta$ across different components of the population.

## Continuous Mixture Models and Population Heterogeneity

The definition of a mixture model can be extended to a continuous mixture as follows. In the continuous case, we consider, naturally, a continuum of subpopulations.

**Definition 10.5.** *A* continuous mixture model *consists of a*

1. *a parametric sampling distribution* $F_\theta = p(Y|\theta)$,
2. *a countable set* $\mathcal{Z} = \mathbb{N}$ *of individuals who, collectively, comprise the population, and*
3. *a distribution* $U$, *called the* mixing distribution *or* collective function,

*such that observations for each individual are obtained by*

$$\Theta \sim U, \tag{10.30}$$
$$Y|\Theta \sim F_\theta, \tag{10.31}$$

*where each individual has a unique value of* $\Theta$.

Naturally, the distribution for $Y$ is found by marginalizing over the random variable $\Theta$, according to

$$p(Y) = \int_{\mathcal{Z}} p(Y|\theta)U(\theta)\,d\theta. \tag{10.32}$$

Superficially, Equation (10.32) looks like the predictive distribution, if we view the collective function $U$ as a prior for the parameter $\theta$. However, there is one crucial philosophical difference, and that is that $U(\theta)$ no longer represents our *belief* about an unknown fixed parameter $\theta$, but rather it represents the *heterogeneity* of the population. In other words, each observation is a draw from $p(Y_i|\theta_i)$, where $\theta_i$ itself is drawn from $U(\theta)$.

Our reference to "individuals" in Definition 10.5 is not superfluous, as it allows the possibility of multiple observations from a given individual. In this case, we say that the $i$th individual has a parameter value drawn from the collective via the relation

$$\Theta_i \sim U.$$

Suppose, in particular, that individual $i$ has a parameter value $\Theta_i = \theta_i$. Next, we can consider multiple observations from that individual, each independent and identically distributed according to

$$Y_{ij}|\Theta_i = \theta_i \sim p(Y|\theta_i),$$

for, say, $j = 1, \ldots, r$.

We already saw from Equation (10.9) that the act of adding data to compute the posterior distribution inevitably results in a loss of variance in our belief over the parameter $\theta$. This is distinguished from a mixture model,

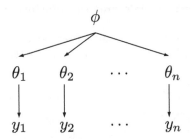

Fig. 10.5: Continuous mixture model.

for which the purpose of data is not to "narrow down" a single value for $\theta$, but to better refine the overall distribution of $\theta$ over the population. This is illustrated in Figure 10.5.

To add a Bayesian layer, we might consider the case in which the mixing distribution $U(\theta)$ is itself represented parametrically, with an unobserved hyperparameter $\phi$. The distribution for the random variable $Y$ can then be obtained by marginalizing over both $\theta$ and $\phi$, yielding

$$p(Y) = \iint p(Y|\theta)p(\theta|\phi)p(\phi) \, d\theta \, d\phi. \tag{10.33}$$

In this model, we have a mixing distribution $p(\theta|\phi)$, which could be considered as a secondary sampling distribution, and a prior distribution $p(\phi)$ over the hyperparameter $\phi$. The corresponding posterior distribution over $\phi$ follows from Bayes' law:

$$p(\phi|\mathcal{D}) \propto p(\phi) \underbrace{\int p(\mathcal{D}|\theta)p(\theta|\phi) \, d\theta}_{p(\mathcal{D}|\phi)}. \tag{10.34}$$

As the variance in our belief $p(\phi|\mathcal{D})$ vanishes with increasing data, we end up with, in the limit, a single value for the hyperparameter $\phi = \phi_0$, which determines the population distribution $p(\theta|\phi_0)$, the variance of which approaches a limit and does not shrink to zero. That limit represents the variance in $\theta$ across the population.

### 10.2.2 Hierarchical Models

A Bayesian mixture model is a special case of a more general structure called a hierarchical model.

**Definition 10.6.** *A Bayesian hierarchical model is a rooted tree (Definition 8.6), such that*

- *each node has an associated value;*

- *each leaf node has the same depth;*
- *the values of the leaf nodes are observed data;*
- *the value $\theta$ of each branch node $b$ is an unobserved parameter that determines a probability distribution $p(X|\theta)$, such that* child$(b) \sim p(X|\theta)$.

*The number of levels of a hierarchical model is the height of its associated tree graph. The level of a node is that node's height.*

A schematic for a simple two-level hierarchical model is shown in Figure 10.6. This is similar to a finite mixture model, except that the "buckets" are assumed to be known in this context. Continuous mixture models can be viewed as a special case of hierarchical models. Similarly, hierarchical models can be viewed as a cascade of nested mixture models. Hierarchical models are useful when the data are collected in a hierarchical fashion, or are themselves endowed with a hierarchical feature set; e.g., colleges within universities within states within regions within countries.

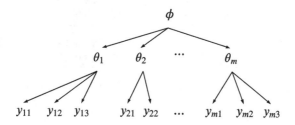

Fig. 10.6: Two-level hierarchical model.

## Empirical Bayes

The simplest approach to estimate the hyperparameters of a hierarchical model is using an empirical estimate. Though not technically Bayesian, this approach is given the name *empirical Bayes.*

**Definition 10.7.** *An* empirical Bayes hierarchical model *is a Bayesian hierarchical model for which the root-level parameter $\phi$ is estimated empirically; i.e., directly from the data. A complete Bayesian analysis involves prescribing a prior distribution $p(\phi)$ over the root-level parameter and performing a full Bayesian update with the data.*

*Example 10.7.* Ten laboratories conducted an identical experiment with a binary outcome, each with a random number of treatment units. Due to natural variability between the labs (different protocols, procedures, respondents for the experiment, etc.), we anticipate a natural variability in the average effect.

We can model this as a two-level hierarchical model, similar to Figure 10.6. In particular, the observed data are modeled by $Y_i \sim \text{Binom}(n_i, \theta_i)$, where $n_i$, the number of experimental units for laboratory $i$, is known, and where $\theta_i \sim \text{Beta}(\alpha, \beta)$ is distributed relative to an unknown beta distribution.

In order to construct a simulation, let us suppose that the true values of the hyperparameters are given by $\alpha = 2$ and $\beta = 8$. We can simulate a data set, as shown in lines 2–6 of Code Block 10.1. The results of such a simulation are given in the first three columns of Table 10.5.

| Group | $n_i$ | $y_i$ | $\hat{\alpha}_i'$ | $\hat{\beta}_i'$ |
|-------|-------|-------|---------|---------|
| 0 | 29 | 9 | 10.5327 | 24.9779 |
| 1 | 10 | 0 | 1.5327 | 14.9779 |
| 2 | 14 | 4 | 5.5327 | 14.9779 |
| 3 | 16 | 2 | 3.5327 | 18.9779 |
| 4 | 15 | 3 | 4.5327 | 16.9779 |
| 5 | 21 | 3 | 4.5327 | 22.9779 |
| 6 | 20 | 9 | 10.5327 | 15.9779 |
| 7 | 25 | 3 | 4.5327 | 26.9779 |
| 8 | 11 | 2 | 3.5327 | 13.9779 |
| 9 | 13 | 7 | 8.5327 | 10.9779 |

Table 10.5: Data and posterior estimates for Example 10.7.

Simulated data in hand, we may now proceed to use empirical Bayes to estimate the population parameters $\alpha, \beta$. Empirical Bayes follows a *bottom–up*, *top–down* approach, meaning we first estimate the hyper parameters, bottom–up, and then perform a Bayesian update for each treatment group, top–down.

Each individual $\theta_i$ can be estimated using that group's data, as $\hat{\theta}_i = y_i/n_i$. Given the set of estimates $\hat{\theta}_0, \ldots, \hat{\theta}_9$, we next compute the empirical mean and variance. Assuming these hidden parameters are distributed as a $\text{Beta}(\alpha, \beta)$ distribution, we can invert the relations for the mean and variance (Equations (2.50) and (2.51)) to obtain

$$\hat{k} = \frac{\mu(1-\mu)}{\sigma^2} - 1$$
$$\hat{\alpha} = \hat{k}\hat{\mu}$$
$$\hat{\beta} = \hat{k} - \hat{\alpha},$$

where we have defined $k = \alpha + \beta$, forming invertible relationships between $(\alpha, \beta)$, $(\mu, \sigma^2)$, and $(\mu, k)$. (Note that the variance of the beta distribution is equivalent to $\mu(1-\mu)/(k+1)$.)

The empirical estimates for $\hat{\alpha}$ and $\hat{\beta}$ (lines 13–14) yield $\hat{\alpha} = 1.5327$ and $\hat{\beta} = 4.9779$. Using these values a prior, we can next perform a Bayesian

```
1   # Simulate Data
2   alpha, beta = 2, 8
3   m = 10
4   theta_true = np.random.beta(alpha, beta, size=m) # hidden
5   n_i = np.random.randint(10, 30, size=m) #observed, given
6   y_i = np.random.binomial(n_i, theta_true) #observed
7
8   theta_hat = y_i / n_i
9   mu_hat = theta_hat.mean()
10  var_hat = theta_hat.var()
11
12  k_hat = mu_hat * (1-mu_hat) / var_hat - 1
13  alpha_hat = k_hat * mu_hat
14  beta_hat = k_hat - alpha_hat
15
16  df = DataFrame({'n':n_i, 'y':y_i, 'alpha_hat':alpha_hat,
        'beta_hat':beta_hat})
17  # Bayesian update for each group
18  df.alpha_hat += y_i
19  df.beta_hat += n_i - y_i
```

Code Block 10.1: Empirical Bayes simulation.

update for each group, yielding $\hat{\alpha}_i = \hat{\alpha} + y_i$ and $\hat{\beta}_i = \hat{\beta} + n_i - y_i$, completing the results in Table 10.5.                                                                                ▷

## 10.3 Modeling Customer Lifetime Value

We next turn to an important marketing application, and that is predicting a customer's lifetime value (defined below). This is relevant across a broad range of businesses, from coffee shops to software subscriptions to freemium apps. A *freemium product* is a product that is free to use, but that also offers premium content available for purchase (e.g., many mobile games use a freemium model). In this context, customers are often referred to as *users*, though the term *user* typically implies any user of the product, not limited to paying users or customers. An accessible and comprehensive overview of freemium products from a marketing perspective is found in Seufert [2014].

### 10.3.1 Frequency, Recency, and Monetization

We begin with a few basic definitions.

**Definition 10.8.** *The* (Customer) lifetime value (CLTV or LTV) *of a given user is the present value of all that customer's future purchases for a given product.*

When a customer makes a purchase, this amount is usually referred to as *revenue*, as it represents revenue for the company that produces the product. Since the future is, by definition, unknown, customer lifetime value is invariably described by some predictive model.

Additionally, there are two primary business models that are relevant when modeling customer lifetimevalue.

**Definition 10.9.** *A* contractual (or subscription) *product or service is a product in which payments are automatically recurring (e.g., monthly) at a fixed amount until the customer cancels.*

*A product that is not contractual—i.e., one in which customers may choose to make purchases of any amount at any time—is called* noncontractual.

In addition, we distinguish a certain class of product that is prevalent in software and mobile apps.

**Definition 10.10.** *A* freemium *product is one whose basic functionality is available for free, but that also offers premium content available for a fee. The premium content can be contractual or noncontractual, or a mixture of both.*

*Example 10.8.* The popular mobile game *Clash of Clans* is a freemium app in which the basic gameplay is free, but that offers *gems*, a form of *in-app currency*, available for purchase (see Table 10.6).

| Description | Amount of Gems | Price |
|-------------|----------------|-------|
| Pocketful of Gems | 80 | $0.99 |
| Pile of Gems | 500 | $4.99 |
| Bag of Gems | 1200 | $9.99 |
| Sack of Gems | 2500 | $19.99 |
| Box of Gems | 6500 | $49.99 |
| Chest of Gems | 14000 | $99.99 |

Table 10.6: Cost of gems, available as in-app purchase, for Clash of Clans.

Gems can then be used in the game to purchase various items of value: troop training, building, or resource speed-ups, building upgrades, purchase of resources (gold or elixir). In addition, Clash of Clans features occasional sales, in which specific items are available as a bundle, for which customers can receive 3x, 4x, or 5x the value for the price. Notice the in-app store is noncontractual, as users can make a purchase in any amount (of the available six options) at any time.                                             ▷

Summary statistics for a particular user are typically measured using three quantities: recency, frequency, and monetization.

**Definition 10.11.** *A customer's* recency, frequency, *and* monetization (RFM) *are defined as*

- recency: *the amount of time since the customer's last purchase;*
- frequency: *the rate at which purchase transactions are made;*
- monetization: *the average purchase amount.*

In a subscription model, frequency and monetization are fixed per the contract (e.g., \$9.99 per month until cancel). For a noncontractual model, freemium or otherwise, all three components play an important role in modeling customer lifetime value. For a freemium product with a fixed in-app store (see Example 10.8), the monetization can be described as a multinomial distribution over the available purchase amounts. For a nonfreemium, noncontractual product (e.g., Starbuck's Coffee), purchases cannot be so cleanly described, and so the total amount is usually recorded, as customers are able to buy any combination of products from the menu.

### 10.3.2 Pareto/NBD and Related Models

In this section, we introduce the classic Pareto/NBD model and review a number of variations that have been constructed through the years.

### Pareto/NBD Model

The Pareto/NBD model is a popular approach to modeling customer lifetime value in the context of noncontractual revenue streams; it was first introduced in Schmittlein, *et al.* [1987]. The basic model is defined as follows

**Definition 10.12.** *The* Pareto/NBD model *is a predictive customer lifetime value model built around the following key assumptions:*

1. *At any given time, a customer can be in one of two states: each customer is* alive *for a period of time, during which they may make a purchase at any time, until the customer becomes* dead *(or churned), at which point the customer permanently disengages with the product;*
2. *the frequency of purchases for any customer while alive follows a Poisson process; i.e., the number of transactions $N_t$ on the interval $[0, t]$ is represented as $N_t \sim \text{Poiss}(\lambda t)$, with rate parameter $\lambda$;*
3. *heterogeneity of transaction rates is described by a gamma distribution: $\lambda \sim \text{Gamma}(\alpha, \beta)$, with shape parameter $\alpha$ and scale parameter $\beta$;*
4. *the (unobserved) customer lifetime $\tau$ follows an exponential distribution with rate parameter $\mu$ (the churn rate);*
5. *heterogeneity in churn rate follows a gamma distribution: $\mu \sim \text{Gamma}(\gamma, \delta)$, with shape parameter $\gamma$ and scale parameter $\delta$;*
6. *the transaction rate $\lambda$ and churn rate $\mu$ vary independently across customers.*

We can use the results from Tables 10.2 and 10.3 to express the distributions of the number of transactions $X_t$ and customer lifetime $\tau$ in terms of their respective hyperparameters.

**Proposition 10.5.** *The marginal distributions for the count of transactions $X_t$ over a period $[0, t]$, for which a customer is alive, and customer lifetime $\tau$ in a Pareto/NBD model are given by*

$$p(X_t = x|t < \tau, \alpha, \beta) = \text{NBD}(x; \alpha, (1 + \beta t)^{-1}) \qquad (10.35)$$
$$p(\tau = t|\gamma, \delta) = \text{Lomax}(t; \gamma, \delta^{-1}). \qquad (10.36)$$

*respectively.*

*Proof.* Form the modeling assumptions of Definition 10.12, we have

$$p(X_t = x|t < \tau, \lambda) = \text{Poiss}(x; \lambda t) = \frac{e^{-\lambda t}(\lambda t)^x}{x!},$$

$$p(\lambda|\alpha, \beta) = \text{Gamma}(\lambda; \alpha, \beta) = \frac{\lambda^{\alpha-1}e^{-\lambda/\beta}}{\Gamma(\alpha)\beta^\alpha}$$

$$p(\tau = t|\mu) = \text{Exp}(t; \mu t) = \mu e^{-\mu t}$$

$$p(\mu|\gamma, \delta) = \text{Gamma}(\mu; \gamma, \delta) = \frac{\mu^{\gamma-1}e^{-\mu/\delta}}{\Gamma(\gamma)\delta^\gamma}.$$

Marginalizing, via

$$p(X_t = x|t < \tau, \alpha, \beta) = \int_0^\infty p(X_t = x|t < \tau, \lambda)p(\lambda|\alpha, \beta)\, d\lambda$$

$$p(\tau = t|\gamma, \delta) = \int_0^\infty p(\tau = t|\mu)p(\mu|\gamma, \delta)\, d\mu,$$

we obtain our results.                                                   $\square$

As a result of Proposition 10.5, we have the following quantities, which are useful in interpreting the practical implications of various configurations of hyperparameters.

**Corollary 10.1.** *For a customer in the Pareto/NBD model, with hyperparameters $(\alpha, \beta, \gamma, \delta)$, the expected lifetime, survival function, and expected number of transactions within a horizon are given by*

$$\mathbb{E}[\tau|\gamma, \delta] = \frac{\delta^{-1}}{\gamma - 1}, \ \textit{for } \gamma > 1 \qquad (10.37)$$

$$S(t|\gamma, \delta) = (1 + \delta t)^{-\gamma} \qquad (10.38)$$

$$\mathbb{E}[X_t|t < \tau, \alpha, \beta] = \alpha\beta t. \qquad (10.39)$$

*Moreover, the expected total number of transactions in a customer's lifetime is given by*

$$\mathbb{E}[X_\tau | \alpha, \beta, \gamma, \delta] = \frac{\alpha\beta}{\delta(\gamma - 1)}, \tag{10.40}$$

*whenever $\gamma > 1$.*

*Proof.* These follow from properties of the NBD and Lomax distributions. The final equation follows from Equations (10.37) and (10.39) and the independence assumption between the transaction rate and customer lifetime. $\square$

*Example 10.9.* We can write a simulation for a random customer with hyperparameters $(\alpha, \beta, \gamma, \delta)$, as shown in Code Block 10.2. If we run

```python
def simPareto(n_users, alpha=0.5, beta=0.5, gamma=2, delta=0.01,
        cohort_age=365):

    df_users = DataFrame({
            'user_id': np.arange(n_users),
            'transaction_rate': random.gamma(alpha, scale=beta,
                size=n_users),
            'lifetime': random.pareto(gamma,size=n_users)/delta
            })

    df_transactions = DataFrame([], columns=['user_id',
        'transaction_time'])

    for i, row in df_users.iterrows():
        time = 1 + random.exponential(scale=1/row.transaction_rate)
        while time < min(row.lifetime, cohort_age):
            df_transactions =
                df_transactions.append(DataFrame({'user_id':
                row.user_id, 'transaction_time': time}, index=[0]),
                ignore_index=True)
            time += random.exponential(scale=1/row.transaction_rate)

    df_transactions.loc[:, 'n_transactions'] = 1
    df_transactions = df_transactions.astype(int)

    return df_users, df_transactions
```

Code Block 10.2: Simulation of Pareto/NBD model

the simulation with hyperparameters $\alpha = 0.5$, $\beta = 0.5$, $\gamma = 2$, and $\delta = 0.01$, we should expect the average lifetime to be $\tau = 100$, as given by Equation (10.37). Moreover, we expect a first-year survival rate of approximately $S(365) \approx 0.0462$. The expected transaction rate is given by

$\mathbb{E}[\lambda|\alpha, \beta] = \alpha\beta = 0.25$, so that the average user should make approximately 25 transactions.

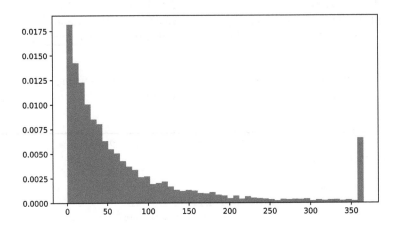

Fig. 10.7: Distribution of user lifetimes for Example 10.9.

The simulated distribution of user lifetimes is shown in Figure 10.7. The observed 1-year survival rate in our simulation was 4.71%, closely agreeing to the theoretical expected survival rate of 4.62%. Moreover, the average observed lifetime was 96.7 days, close to the predicted 100 day expected lifetime.

Finally, we note that the total number of transactions that resulted from our simulation was 19,233, or 19.233 transactions per user. The total expected lifetime transactions per user, however, as given by Equation (10.40), is 25. This discrepancy is due to the fact that there are a remaining 5% of users who will make transactions beyond the first year, ultimately brining the overall average up to 25.                                    ▷

In Example 10.9, we saw that the total number of transactions per user, for the first year, was 19.233, according to our simulation. This represents an approximately 20% haircut from the total expected *lifetime* transactions per user of 25. Naturally, we are interested in how to compute the expected number of transactions within a specified time horizon. This result is given as follows.

**Proposition 10.6.** *Let the random variable $X(H)$ represent the number of transactions during the time horizon $[0, H]$ for a customer in the Pareto/NBD model. Then the expected value of $X(H)$, for a customer with hyperparameters $\alpha, \beta, \gamma, \delta$, is given by*

$$\mathbb{E}[X(H)|\alpha,\beta,\gamma,\delta] = \frac{\alpha\beta}{\delta(\gamma-1)}\left[1 - \frac{1+\gamma\delta H}{(1+\delta H)^{\gamma}}\right] + \alpha\beta H(1+\delta H)^{-\gamma}. \quad (10.41)$$

*Proof.* In order to calculate this, we must divide the populations into two separate groups: those alive at time $H$ and those with a lifetime $\tau < H$:

$$\mathbb{E}[X(H)|\alpha,\beta,\gamma,\delta] = \mathbb{E}[X(H)|\tau \leq H, \alpha,\beta,\gamma,\delta]\mathbb{P}(\tau \leq H)$$
$$+ \mathbb{E}[X(H)|\tau > H, \alpha,\beta,\gamma,\delta]\mathbb{P}(\tau > H). \quad (10.42)$$

First, let us consider the group of customers alive at time $H$. From Equation (10.38), we know that this accounts for

$$\mathbb{P}(\tau > H) = S(H) = (1 + \delta H)^{-\gamma}$$

of all users. Since we are *only* interested in transactions *up to time* $H$, we can use Equation (10.39) to write

$$\mathbb{E}[X(H)|\tau > H, \alpha, \beta] = \alpha\beta H$$

for the customers who are alive at time $H$. Combining the previous two equations, we find that the contribution to $X(H)$ from users alive at time $H$ should be given by

$$\alpha\beta H(1 + \delta H)^{-\gamma},$$

the final term in Equations (10.41) and (10.42).

Second, let us consider users with a lifetime $\tau < H$. We can adjust the density for a users lifetime using

$$p(\tau|\tau \leq H) = \begin{cases} \dfrac{p(\tau)}{\mathbb{P}(\tau \leq H)} & \text{for } \tau \leq H \\ 0 & \text{for } \tau > H \end{cases}.$$

The expected lifetime for a user whose lifetime is less than $H$ is therefore given by

$$\mathbb{E}[\tau|\tau \leq H] = \int_0^H \frac{\tau p(\tau)}{\mathbb{P}(\tau \leq H)}\, d\tau.$$

Therefore, Equation (10.42) reduces to

$$\mathbb{E}[X(H)|\alpha,\beta,\gamma,\delta] = \alpha\beta \int_0^H \tau p(\tau|\alpha,\beta,\gamma,\delta)\, d\tau + \alpha\beta H(1 + \delta H)^{-\gamma}.$$

Next, one can show that

$$\int_0^H \tau p(\tau|\alpha,\beta,\gamma,\delta)\, d\tau = \int_0^H \gamma\delta x(1 + \delta x)^{-(\gamma+1)}\, dx$$

$$= \frac{1}{\delta(\gamma-1)}\left[1 - \frac{1+\gamma\delta H}{(1+\delta H)^{\gamma}}\right].$$

This completes the result. □

*Note 10.5.* For the choice of hyperparameters $\alpha = 0.5$, $\beta = 0.5$, $\gamma = 2$, $\delta = 0.01$, the expected number of transactions, as given by Equation (10.41), for the first year ($H = 365$) works out to be $\mathbb{E}[X(365)] = 19.62$. This closely matches the observed result (19.233) obtained from the simulation in Example 10.9.                                                                     ▷

### Variations: BG/NBD, BG, and BG/BB Models

Since the original publication of the Pareto/NBD model, a number of variations have been made that have offered substantial improvements to the model, especially in terms of computational ease of use. These include the BG/NBD model (Fader, *et al.* [2005]), the BG model for contractual products (Fader and Hardie [2007]), and the BG/BB model, which can be used for both contractual and noncontractual (Fader, *et al.* [2010]). A summary of the variations between these models is shown in Table 10.7. We note that the BG model is a degenerate case of the BG/BB model, for which the transaction probability of a customer while alive is 100%.

|  | | model | | | |
| --- | --- | --- | --- | --- | --- |
| component | | Pareto/NBD | BG/NBD | BG | BG/BB |
| transactions | rate | Poisson | Poisson | – | Bernoulli |
| | heterogeneity | gamma | gamma | – | beta |
| | predictive | NBD | NBD | – | BB |
| lifetime | rate | exponential | geometric | geometric | geometric |
| | heterogeneity | gamma | beta | beta | beta |
| | predictive | Pareto | BG | BG | BG |

Table 10.7: Development of LTV models.

The key conceptual leaps in these various models are the following:

- *replacing the exponential–gamma mixture with the beta–geometric mixture*: instead of allowing a user's lifetime to be unobserved, occurring at any point in time, we instead model churn as occurring only immediately following a transaction, with a constant probability for each customer;
- *replacing the Poisson–gamma mixture with the beta–Bernoulli mixture*: instead of modeling transactions as a Poisson process, we instead discretize time, creating a sequence of *transaction opportunities*, for which a customer that is alive makes a purchase with a constant probability.

### 10.3.3 The BG/BB Model

#### The Model

As we saw in Table 10.7, multiple variations of the classic Pareto/NBD model ultimately led to a highly flexible model that is computationally more tractable than the original Pareto/NBD and that captures a contractual model (BG model) as a special case. We therefore take a closer look at this final model, following the work presented in Fader, *et al.* [2010].

**Definition 10.13.** *The* BG/BB model *is a predictive customer lifetime value model built around the following key assumptions:*

1. *At any given time, a customer can be in one of two states:* alive *or* dead (churned); *a customer that is alive can make a purchase at any time, whereas a customer that has churned has permanently disengaged with the product;*
2. *time is characterized by a sequence of* transaction opportunities; *a customer that is alive will make a purchase during a given transaction opportunity with a probability p;*
3. *heterogeneity of the transaction probability follows a beta distribution:* $p \sim \text{Beta}(\alpha, \beta);$
4. *a customer that is alive will churn prior to any transaction opportunity with a probability $\theta$, such that each customer's unobserved lifetime follows a geometric distribution $\tau \sim \text{Geom}(\theta);$*
5. *heterogeneity in the churn rate follows a beta distribution: $\theta \sim \text{Beta}(\gamma, \delta);$*
6. *the transaction probability p and churn rate $\theta$ vary independently across customers.*

*The predictive distribution for the transaction probability follows a beta–Bernoulli distribution; the lifetime follows a beta–geometric distribution.*

The BG/BB model differs from the Pareto/NBD by a slight modification in the behavioral story, the key difference being that the BG/BB is a discrete version of the Pareto/NBD model. Instead of transactions occurring in continuous time, time is discretized into a sequence of transactional opportunities, for which a customer may either be active or inactive. Similarly, a user's lifetime is no longer continuous, but rather it is represented by an integer number of transaction opportunities; i.e., a user can no longer churn, as per the model, at an arbitrary moment in time, but each user can only churn following the completion of a transaction opportunity.

As we shall see, summary statistics for each user consist of the number $X$ of active transaction opportunities (which we call *actions*[1]), the time of last purchase $T_x$, and the time of the last transaction opportunity $N$. (Time, in this context, is the *cohort time*, or time since first purchase, as defined for online processes in Definition 5.20.) For a fixed number of transaction opportunities $n$, the number of distinct frequency–recency combinations therefore scales like $O(n^2)$, compared with the $O(2^n)$ possible full binary

---

[1] A *transaction* is itself a discrete event; when discretizing time, it is possible that a customer make multiple transactions within any fixed transaction opportunity. We therefore distinguish the atomic transactions from active transaction opportunities by using the term *action* to denote the latter.

vectors capturing the full (discretized) transaction history. This makes the BG/BB model highly scalable, as we only have to track the count of users with a particular frequency–recency combination.

To proceed, we first define our precise set of random variables.

**Definition 10.14.** *For a particular user in the BG/BB model, we define the following random variables*

- *$N$ represents the* cohort age *of the user, in the context of Definition 5.20;*
- *$Y_t$ represents the user being active during the tth transaction opportunity, for $t = 1, \ldots, n$;*
- *$X = \sum_{t=1}^{n}$ represents the number of active transaction opportunities (i.e., actions) observed;*
- *$T_x$ represents the time of the latest observed transaction opportunity;*
- *$\tau$ represents the (unobserved) lifetime, defined as the index of the final transaction opportunity for which a user was alive;*
- *$A_t = \mathbb{I}[t \leq \tau]$ represents the event in which the user is still alive at the tth transaction opportunity, for $t = 1, \ldots, n$.*

Note that

$$\mathbb{P}(\tau = t) = \text{Geom}(t; \theta) = \theta(1 - \theta)^t, \tag{10.43}$$

for $t = 0, 1, 2, \ldots$. Moreover, the survival function at $t_0$ is given by

$$S(t) = \mathbb{P}(\tau > t) = 1 - \mathbb{P}(\tau \leq t) = (1 - \theta)^{t+1}. \tag{10.44}$$

Consequently, $\mathbb{P}(A_t) = \mathbb{P}(\tau \geq t) = (1 - \theta)^t$.

*Note 10.6.* In general, a user does not become a customer until he or she makes an actual purchase. We can think of the variables defined in Definition 10.14 as occurring *subsequent* to the initial purchase transaction; i.e., we can augment the model with $A_0 = 1$ and $Y_0 = 1$ for all users. With this adjustment, the actual lifetime of a customer is $\tau + 1$, and the actual number of actions is $X + 1$. The definition of a churn opportunity occurring immediately *prior* to any transaction opportunity ($t = 1, 2, \ldots$) make more sense in this context, as all customers start off with an initial transaction at $t = 0$.                                                                              ▷

## Computation of Summary Statistics

The summary statistics for a user in the BG/BB model consist of the random variables in the triple $(X, T_X, N)$, as outlined in the previous paragraph. Moreover, time in the BG/BB model is discretized, so that we must also define which transaction opportunity each atomic transaction belongs to. In general, as inputs we will take a user table and a transaction table, similar to the outputs of the simulation in Code Block 10.2.

In Code Block 10.3 we have a function, `getBgbbSummary`, that takes a user and transactional dataframe as inputs, along with the length of the

```python
def getBgbbSummary(df_users, df_transactions, opp_window=7):

    df_transactions.transaction_time /= opp_window
    df_transactions.transaction_time =
        np.ceil(df_transactions.transaction_time).astype(int)
    df_reduced = df_transactions.groupby(['user_id',
        'transaction_time'], as_index=False)['n_transactions'].sum()
    df_reduced.loc[:, 'x'] = 1


    df_reduced = df_reduced.groupby('user_id',
        as_index=False).agg({'transaction_time':max, 'x':sum})
    df_reduced = df_reduced.rename(columns={'transaction_time':'tx'})
    n = df_reduced.tx.max()
    df_reduced.loc[:, 'n'] = n
    df_reduced.loc[:, 'freq'] = 1

    # add users without transaction record (x=0)
    x0_users = df_users[~df_users.user_id.isin(
        df_reduced.user_id.unique())].user_id.unique()
    df_reduced = df_reduced.append(DataFrame({'user_id':x0_users,
        'tx':0, 'x': 0, 'n':52, 'freq':1}), ignore_index=True)
    df_sum = df_reduced.groupby(['tx', 'x', 'n'],
        as_index=False)['freq'].sum()

    return df_reduced, df_sum
```

Code Block 10.3: Computation of BG/BB summary statistics.

opportunity window (e.g., 7-days to define a weekly discretization), and returns a reduced and summary dataframe as outputs. The reduced dataframe is still at the individual customer level, and contains the summary statistics for each individual customer. As we shall see in Corollary 10.5, however, we only require a frequency count for the number of customers for each combination of permissible summary statistics; this count is captured in the output dataframe df_sum.

## Marginal Distributions

Similar to Proposition 10.5, we can write the marginal distributions for the average customer, with hyperparameters $(\alpha, \beta, \gamma, \delta)$, as follows.

**Proposition 10.7.** *The marginal distributions for the count of* actions *(i.e., active transaction opportunities)* $X_t$ *over a period* $\{1, \ldots, t\}$*, for which a customer is alive, and customer lifetime* $\tau$ *in a BG/BB model are given by*

$$p(X_t = x | t \leq \tau, \alpha, \beta) = \text{BetaBin}(x; t, \alpha, \beta) \qquad (10.45)$$
$$p(\tau = t | \gamma, \delta) = \text{BetaGeom}(t; \gamma, \delta), \qquad (10.46)$$

*respectively.*

*Proof.* From the modeling assumptions of Definition 10.14, we have

$$p(X_t = x | t < \tau, p) = \text{Binom}(x; t, p) = \binom{t}{x} p^x (1 - p)^{t-x}$$

$$p(p | \alpha, \beta) = \text{Beta}(p; \alpha, \beta) = \frac{p^{\alpha-1}(1-p)^{\beta-1}}{B(\alpha, \beta)}$$

$$p(\tau = t | \theta) = \text{Geom}(t; \theta) = \theta(1 - \theta)^t,$$

$$p(\theta | \gamma, \delta) = \text{Beta}(\theta; \gamma, \delta) = \frac{\theta^{\gamma-1}(1-\theta)^{\delta-1}}{B(\gamma, \delta)}.$$

Marginalizing, via

$$p(X_t = x | t \leq \tau, \alpha, \beta) = \int_0^1 p(X_t = x | t < \tau, p) p(p | \alpha, \beta) \, dp$$

$$p(\tau = t | \gamma, \delta) = \int_0^1 p(\tau = t | \theta) p(\theta | \gamma, \delta) \, d\theta,$$

we obtain our result. □

As a result of Proposition 10.7, we have the following quantities, which are useful in interpreting the practical implications of various configurations of hyperparameters.

**Corollary 10.2.** *For a customer in the BG/BB model, with hyperparameters $(\alpha, \beta, \gamma, \delta)$, the expected lifetime, survival function, and expected number of transactions within a horizon are given by*

$$\mathbb{E}[\tau | \gamma, \delta] = \frac{\delta}{\gamma - 1}, \quad \text{for } \gamma > 1 \qquad (10.47)$$

$$S(t | \gamma, \delta) = \frac{B(\gamma, \delta + t + 1)}{B(\gamma, \delta)} \qquad (10.48)$$

$$\mathbb{E}[X_t | t < \tau, \alpha, \beta] = \frac{\alpha t}{\alpha + \beta}. \qquad (10.49)$$

*Moreover, the expected total number of transactions in a customer's lifetime is given by*

$$\mathbb{E}[X_\tau | \alpha, \beta, \gamma, \delta] = \frac{\alpha \delta}{(\alpha + \beta)(\gamma - 1)}, \qquad (10.50)$$

*whenever $\gamma > 1$.*

*Proof.* These follow from properties of the beta-binomial and beta-geometric distributions. The survival function for the beta-geometric distribution, as it is reported in Table A.2, was computed using Wolfram Alpha [2018], by simplifying the output[2] to the query:

```
sum beta(a+1,b+x)/beta(a,b) from x=0 to T.
```

The final equation follows from Equations (10.47) and (10.49) and the independence assumption between the transaction rate and customer lifetime.

□

### The Likelihood

Given the preceding definitions, we have the following result, due to Fader, *et al.* [2010].

**Proposition 10.8.** *The likelihood for a customer in the BG/BB model with summary statistics* $(x, t_x, n)$ *is given by*

$$p(x, t_x, n | p, \theta) = \sum_{t=t_x}^{n-1} p^x (1-p)^{t-x} \theta (1-\theta)^t + p^x (1-p)^{n-x} (1-\theta)^n. \quad (10.51)$$

*Note 10.7.* The only difference between the sum and the final term in Equation (10.51) is the factor of $\theta$, so that Equation (10.51) may alternatively be expressed as

$$p(x, t_x, n | p, \theta) = \sum_{t=t_x}^{n} p^x (1-p)^{t-x} \theta^{\mathbb{I}[t \neq n]} (1-\theta)^t,$$

where $\theta^{\mathbb{I}[t \neq n]} = 1$ when $t = n$. ▷

*Proof.* When calculating the probability of observing a user with summary statistics $(x, t_x, n)$, we must account for the $n - t_x + 1$ possibilities for which the user might have churned (since churn itself is not directly observable). The only thing we know for certain, is that the user was alive at time $t_x$, so that $\tau \geq t_x$. The user may have churned following any of the $n - t_x$ opportunities $\tau = t_x, t_x + 1, \ldots, n - 1$, or the user might still be alive on the $n$th opportunity, so that $\tau \geq n$. We therefore have the relation

$$p(x, t_x, n | p, \theta) = \sum_{t=t_x}^{n-1} \mathbb{P}(X = x | p, \tau = t) \mathbb{P}(\tau = t | \theta)$$
$$+ \mathbb{P}(X = x | p, \tau \geq n) \mathbb{P}(\tau \geq n | \theta).$$

---

[2] The output is also subtracted from 1, as the query corresponds to the cumulative distribution $C(T) = \sum_{x=0}^{T} \text{BetaGeom}(x; a, b)$.

Now,

$$\mathbb{P}(X = x|p, \tau = t) = p^x(1-p)^{t-x},$$
$$\mathbb{P}(X = x|p, \tau \geq n) = p^x(1-p)^{n-x},$$
$$\mathbb{P}(\tau = t|\theta) = \theta(1-\theta)^t,$$
$$\mathbb{P}(\tau \geq n|\theta) = (1-\theta)^n.$$

This completes the proof.                                                    □

Proposition 10.8 can be more readily understood in terms of an example. Consider a user with summary statistics $(x = 2, t_x = 3, n = 6)$; for example, as resulting from the binary vector 101000. There are $n - t_x + 1 = 4$ cases to consider, as shown in Table 10.8. The likelihood is obtained from this

| Case | 1 | 2 | 3 | 4 | 5 | 6 | $\mathbb{P}(X = 2|\text{case})$ | $\mathbb{P}(\text{case})$ |
|---|---|---|---|---|---|---|---|---|
|  | 1 | 0 | 1 | 0 | 0 | 0 | | |
| $\tau = 3$ | A | A | A | D | D | D | $p^2(1-p)$ | $\theta(1-\theta)^3$ |
| $\tau = 4$ | A | A | A | A | D | D | $p^2(1-p)^2$ | $\theta(1-\theta)^4$ |
| $\tau = 5$ | A | A | A | A | A | D | $p^2(1-p)^3$ | $\theta(1-\theta)^5$ |
| $\tau \geq 6$ | A | A | A | A | A | A | $p^2(1-p)^4$ | $(1-\theta)^6$ |

Table 10.8: Illustration of Proposition 10.8.

table by computing the sum

$$p(x = 2, t_x = 3, n = 6|p, \theta) = \sum \mathbb{P}(X = 2|\text{case})\mathbb{P}(\text{case}),$$

which is consistent with Equation (10.51).

**Proposition 10.9.** *The marginal likelihood, obtained by marginalizing the result from Proposition 10.8 over the respective mixing distributions, is given by*

$$L(\alpha, \beta, \gamma, \delta|x, t_x, n) = \sum_{t=t_x}^{n-1} \frac{B(\alpha + x, \beta + t - x)}{B(\alpha, \beta)} \frac{B(\gamma + 1, \delta + t)}{B(\gamma, \delta)}$$
$$+ \frac{B(\alpha + x, \beta + n - x)}{B(\alpha, \beta)} \frac{B(\gamma, \delta + n)}{B(\gamma, \delta)} \qquad (10.52)$$

*where $B(\cdot, \cdot)$ is the beta function, as defined in Equation (2.47). We call this expression the likelihood for the hyperparameters; i.e.,*

$$L(\alpha, \beta, \gamma, \delta|x, t_x, n) = p(x, t_x, n|\alpha, \beta, \gamma, \delta),$$

*where $L$ is viewed as a function of the hyperparameters.*

*Proof.* We may use Equation (10.51) to write the product of the likelihood with the mixing distributions as

$$p(x, t_x, n|p, \theta)p(p|\alpha, \beta)p(\theta|\gamma, \delta)$$

$$= \sum_{t=t_x}^{n-1} \frac{p^{\alpha+x-1}(1-p)^{\beta+t-x-1}}{B(\alpha, \beta)} \frac{\theta^\gamma(1-\theta)^{\delta+t-1}}{B(\gamma, \delta)}$$

$$+ \frac{p^{\alpha+x}(1-p)^{\beta+n-x-1}}{B(\alpha, \beta)} \frac{\theta^{\gamma-1}(1-\theta)^{\delta+n-1}}{B(\gamma, \delta)}$$

$$= \sum_{t=t_x}^{n-1} \left[ \frac{B(\alpha+x, \beta+t-x)}{B(\alpha, \beta)} \text{Beta}(p; \alpha+x, \beta+t-x) \right.$$

$$\left. \times \frac{B(\gamma+1, \delta+t)}{B(\gamma, \delta)} \text{Beta}(\theta; \gamma+1, \delta+t) \right]$$

$$+ \left[ \frac{B(\alpha+x, \beta+n-x)}{B(\alpha, \beta)} \text{Beta}(p; \alpha+x, \beta+n-x) \right.$$

$$\left. \times \frac{B(\gamma, \delta+n)}{B(\gamma, \delta)} \text{Beta}(\theta; \gamma, \delta+n) \right] \qquad (10.53)$$

Marginalizing over the hyperparameters, using

$$p(x, t_x, n|\alpha, \beta, \gamma, \delta) = \int_0^1 \int_0^1 p(x, t_x, n|p, \theta)p(p|\alpha, \beta)p(\theta|\gamma, \delta)\, dp\, d\theta,$$

yields our result, as expressed in Equation (10.52). □

**Corollary 10.3.** *The likelihood given by Equation* (10.52) *is equivalent to*

$$L(\alpha, \beta, \gamma, \delta|x, t_x, n) = \sum_{t=t_x}^{n} \frac{B(\alpha+x, \beta+t-x)}{B(\alpha, \beta)} \frac{B(\gamma+1, \delta+t)}{B(\gamma, \delta)}$$

$$+ \frac{B(\alpha+x, \beta+n-x)}{B(\alpha, \beta)} \frac{B(\gamma, \delta+n+1)}{B(\gamma, \delta)}. \quad (10.54)$$

*Proof.* This follows by direct comparison of Equations (10.52) and (10.54) and the identity

$$B(\gamma+1, \delta+n) + B(\gamma, \delta+n+1) = B(\gamma, \delta+n);$$

see Exercise 10.4.

Though this completes the proof, this distinction can be understood more intuitively by framing Equations (10.52) and (10.54) as

$$L = \sum_{t=t_x}^{n-1} \text{BetaBin}(x; t, \alpha, \beta)p(\tau = t|\gamma, \delta) + \text{BetaBin}(x; n, \alpha, \beta)p(\tau \geq n|\gamma, \delta)$$

$$= \sum_{t=t_x}^{n} \text{BetaBin}(x; t, \alpha, \beta)p(\tau = t|\gamma, \delta) + \text{BetaBin}(x; n, \alpha, \beta)p(\tau > n|\gamma, \delta),$$

respectively. Thus, we see that Equation (10.54) explicitly enumerates the possibility that a user's lifetime is exactly $\tau = n$ in the summation. This representation will be useful later on.                                                        □

In order to compute the gradient of the likelihood function Equation (10.52), we first define the quantity

$$\rho(\alpha, \beta, x, y) = \frac{B(\alpha + x, \beta + y)}{B(\alpha, \beta)}, \tag{10.55}$$

so that the likelihood may be expressed as

$$L = \sum_{t=t_x}^{n} \rho(\alpha, \beta, x, t - x)\rho(\gamma, \delta, u_t, t), \tag{10.56}$$

where we define $u_t = \mathbb{I}[t \neq n]$, as in Note 10.7.

Now, the derivatives of Equation (10.55), with respect to $\alpha$ and $\beta$, are given by

$$\frac{\partial \rho}{\partial \alpha} = \rho \left[ \psi(\alpha + x) - \psi(\alpha) + \psi(\alpha + \beta) - \psi(\alpha + \beta + x + y) \right], \tag{10.57}$$

$$\frac{\partial \rho}{\partial \beta} = \rho \left[ \psi(\beta + y) - \psi(\beta) + \psi(\alpha + \beta) - \psi(\alpha + \beta + x + y) \right], \tag{10.58}$$

where $\rho = \rho(\alpha, \beta, x, y)$ and $\psi(x)$ is the *digamma function*. (See Exercise 10.7.)

Given Equations (10.57) and (10.58), the gradient of the likelihood function Equation (10.56) can be expressed as

$$\frac{\partial L}{\partial \alpha} = \sum_{t=t_x}^{n} \frac{\partial \rho(\alpha, \beta, x, t - x)}{\partial \alpha} \rho(\gamma, \delta, u_t, t) \tag{10.59}$$

$$\frac{\partial L}{\partial \beta} = \sum_{t=t_x}^{n} \frac{\partial \rho(\alpha, \beta, x, t - x)}{\partial \beta} \rho(\gamma, \delta, u_t, t) \tag{10.60}$$

$$\frac{\partial L}{\partial \gamma} = \sum_{t=t_x}^{n} \rho(\alpha, \beta, x, t - x) \frac{\partial \rho(\gamma, \delta, u_t, t)}{\partial \gamma} \tag{10.61}$$

$$\frac{\partial L}{\partial \delta} = \sum_{t=t_x}^{n} \rho(\alpha, \beta, x, t - x) \frac{\partial \rho(\gamma, \delta, u_t, t)}{\partial \delta}. \tag{10.62}$$

**Corollary 10.4.** *The probabilities that a user in the BG/BB model, with hyperparameters $\alpha, \beta, \gamma, \delta$ and transactional summary statistics $(x, t_x, n)$, is still alive at times $t = n$ and $t = n + 1$ are given by*

$$p(A_n | \alpha, \beta, \gamma, \delta, x, t_x, n) = \frac{B(\alpha + x, \beta + n - x)}{B(\alpha, \beta)} \frac{B(\gamma, \delta + n)}{B(\gamma, \delta)} \cdot L^{-1} \tag{10.63}$$

$$p(A_{n+1} | \alpha, \beta, \gamma, \delta, x, t_x, n) = \frac{B(\alpha + x, \beta + n - x)}{B(\alpha, \beta)} \frac{B(\gamma, \delta + n + 1)}{B(\gamma, \delta)} \cdot L^{-1},$$

*respectively, where $L = L(\alpha, \beta, \gamma, \delta | x, t_x, n)$ is given by Equation (10.54).*

*More generally, the probability that such a customer is alive at some point in time in the future, $t = n + m$, for $m \geq 0$, is given by*

$$p(A_{n+m} | \alpha, \beta, \gamma, \delta, x, t_x, n) = \frac{B(\alpha + x, \beta + n - x)}{B(\alpha, \beta)} \frac{B(\gamma, \delta + n + m)}{B(\gamma, \delta)} \cdot L^{-1}.$$

(10.64)

*Proof.* The likelihood defined in Equation (10.52) defines a probability mass function over the $n - t_x + 1$ possibilities that a user's lifetime is $\tau = t_x, \ldots, n - 1$, or that the user is still alive at time $t = n$; i.e., that $\tau \geq n$. For example, the probability that the user's lifetime was exactly $\tau = t$, for $t = t_x, \ldots, n - 1$, is given by the likelihood of that event,

$$\frac{B(\alpha + x, \beta + t - x)}{B(\alpha, \beta)} \frac{B(\gamma + 1, \delta + t)}{B(\gamma, \delta)},$$

divided by the total likelihood of all the possible events, as computed in Equation (10.52). Similarly, the probability that a user is still alive at time $t = n$ is the final term divided by the total likelihood, which yields Equation (10.63).

Similarly, the probability $p(A_{n+1} | \alpha, \beta, \gamma, \delta, x, t_x, n)$ is obtained by applying similar logic to the equivalent expression Equation (10.54).

The proof of Equation (10.64) likewise follows along similar lines. We can see this result by expanding the numerator of the second factor of the final term of Equation (10.52) using the identity

$$B(\gamma, \delta + n) = \sum_{t=0}^{m-1} B(\gamma + 1, \delta + n + t) + B(\gamma, \delta + n + m).$$

This identity can be interpreted by equating the probability that the customer is alive at time $t = n$ with the sum of the probabilities of demise at times $t = n, \ldots, n + m - 1$ plus the probability that the customer is alive at time $t = n + m$. The result follows.  □

An illustration of the probability function Equation (10.63) is shown in Figure 10.8, for values $\alpha = 0.7$, $\beta = 0.5$, $\gamma = 1.9$, and $\delta = 11.6$, for users with various values of $(x, t_x)$ and $n = 12$ held fixed. Note that if $t_x = 12$, we have $p(A_{12} | t_x = 12) = 1$. Also note the general trend that the larger the difference $n - t_x$, the less likely it is that the customer is still alive at time $t = n$.

Now, for a fixed $n$, there are only $J_n = n(n+1)/2 + 1$ possible recency / frequency patterns. This follows since the random variable $T_x$ ranges from $T_x = 1, 2, \ldots, n$; given $T_x = t_x$, the number of transactions then ranges from $X = 1, \ldots, t_x$. (Since $T_x > 0$, this implies there must be at least one transaction.) In addition, we have the separate case for which $T_x = X = 0$.

As an immediate corollary to Proposition 10.8 and 10.9, we have the following expression for the log-likelihood of a *set* of customers.

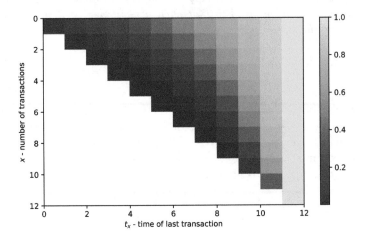

Fig. 10.8: Probability that a customer with cohort age $n = 12$ is alive at time $t = 12$, as a function of $x$ and $t_x$.

**Corollary 10.5.** *In the BG/BB model, consider a set of users $\mathcal{D} = \{(x, t_x, n)_i\}$. If we group the users into the $J = n(n+1)/2+1$ possible recency/frequency patterns, such that pattern $j$ consists of $f_j$ customers, then the log-likelihood is given by*

$$\ell(\alpha, \beta, \gamma, \delta) = \sum_{j=1}^{J_n} f_j \ln \left[ L(\alpha, \beta, \gamma, \delta | x_j, t_{x_j}, n) \right], \qquad (10.65)$$

*where the marginal likelihood function $L$ is given by Equation (10.52).*

Given this expression for the log-likelihood, we can now use the maximum likelihood estimate to approximate the hyperparameters $(\alpha, \beta, \gamma, \delta)$ for a given set of users. Oftentimes, numerical optimizers require an initial guess for the parameter set. A quick initial guess can be computed using

$$\hat{\theta} = \frac{\sum_{j=1}^{J} f_j x_j}{\sum_{j=1}^{J} f_j t_{x_j}} \qquad \text{and} \qquad \hat{\tau} = \frac{\sum_{j=1}^{J} f_j t_{x_j}}{\sum_{j=1}^{J} f_j},$$

and then using

$$\langle \alpha_0, \beta_0, \gamma_0, \delta_0 \rangle = \langle \hat{\theta}, 1 - \hat{\theta}, 2, \hat{\tau} \rangle.$$

This is, of course, not a very good approximation to the hyperparameters, but should be good enough to put the numerical solver on the right track.

The likelihood function given in Equation (10.65) can be easily coded, as shown in Code Block 10.4. The function `likelihood_kernel` computes Equation (10.52), the likelihood for a given customer with summary statistics $(x, t_x, n)$. The function `likelihood` then computes Equation (10.65)

```
1   from scipy.special import beta as BETA
2   from scipy.special import gamma as GAMMA
3
4   def likelihood_kernel(theta, x, tx, n):
5       alpha, beta, gamma, delta = theta
6       L = BETA(alpha+x, beta+n-x) * BETA(gamma, delta+n)
7       for t in range(tx, n):
8           L += BETA(alpha+x, beta+t-x) * BETA(gamma+1, delta+t)
9
10      return L / (BETA(alpha, beta) * BETA(gamma, delta))
11
12  def likelihood(theta, df_sum):
13      S = 0
14      for i, row in df_sum.iterrows():
15          L = likelihood_kernel(theta, int(row.x), int(row.tx),
                  int(row.n))
16          S += row.freq * np.log(L)
17
18      return -S
```

Code Block 10.4: Likelihood function for BG/BB Model.

from an input dataframe df_sum, formatted similarly to the output of getBgbbSummary in Code Block 10.3.

Next, we can simulate a set of data from the Pareto/NBD model, using the function defined in Code Block 10.2, and then convert the resulting data into the BG/BB format using Code Block 10.3. We then determine the maximum likelihood for the BG/BB model using Code Block 10.5.

```
1   alpha, beta, gamma, delta = 0.5, 0.5, 2, 0.01 # Pareto/NBD
2   df_users, df_transactions = simPareto(1000, alpha=alpha, beta=beta,
        gamma=gamma, delta=delta)
3   df_reduced, df_sum = getBgbbSummary(df_users, df_transactions)
4   p_hat = np.sum(df_sum.x * df_sum.freq) / np.sum(df_sum.tx *
        df_sum.freq)
5   T_hat = np.sum(df_sum.tx * df_sum.freq) / np.sum(df_sum.freq)
6
7   theta_0 = [p_hat, 1-p_hat, 2, T_hat]
8   bounds = [(0.01,100), (0.01,100), (0.01,100), (0.01,300)]
9   results = scipy.optimize.minimize(likelihood, theta_0,
        bounds=bounds, args=df_sum)
10  alpha, beta, gamma, delta = results['x'] # BG/BB
```

Code Block 10.5: MLE for simulated BG/BB data.

As an alternative, we can use the likelihood function in the form of Equation (10.56) and its corresponding gradient Equations (10.59)–(10.62) to subclass the `ObjectiveFunction` class from Code Blocks 5.7 and 5.8. This is shown in Code Block 10.7. For this approach, the `data` input argument to the constructor should be the `np.array` given by `X_in=df_reduced[['tx', 'x', 'n']].values`.

### Distribution of Horizon Actions

Once the maximum likelihood estimate has been fit, several quantities of managerial interest can readily be derived, involving the number of actions over a fixed time horizon $n$.

**Proposition 10.10.** *Let $X(n) = \sum_{t=1}^{n} Y_t$ be the number of active transaction periods from the first $n$ transaction opportunities. Given the hyperparameters of the BG/BB model, the PMF for the random variable $X(n)$ is given by*

$$
p(X(n) = x | \alpha, \beta, \gamma, \delta) = \sum_{t=x}^{n-1} \binom{t}{x} \frac{B(\alpha + x, \beta + t - x)}{B(\alpha, \beta)} \frac{B(\gamma + 1, \delta + t)}{B(\gamma, \delta)}
$$
$$
+ \binom{n}{x} \frac{B(\alpha + x, \beta + n - x)}{B(\alpha, \beta)} \frac{B(\gamma, \delta + n)}{B(\gamma, \delta)}. \quad (10.66)
$$

*Proof.* An individual that makes $X(n) = x$ actions must be alive for at least $x$ transaction opportunities. Supposing, then, that the customer's lifetime is $\tau = t$, the probability of making $x$ actions is given by the binomial distribution

$$
p(X(n) = x | \tau = t, p) = \mathrm{Binom}(x; t, p) = \binom{t}{x} p^x (1-p)^{t-x},
$$

for $x = t, \ldots, n$. Factoring in the distribution of lifetimes, we next write

$$
p(X(n) = x | p, \theta) = \sum_{t=x}^{n-1} p(X(n) = x | p, \tau = t) p(\tau = t | \theta)
$$
$$
+ p(X(n) = x | p, \tau \geq n) p(\tau \geq n | \theta)
$$
$$
= \sum_{t=x}^{n-1} \binom{t}{x} p^x (1-p)^{t-x} \theta (1-\theta)^t
$$
$$
+ \binom{n}{x} p^x (1-p)^{n-x} (1-\theta)^n.
$$

Our final result is obtained by marginalizing

$$
p(X(n) = x | \alpha, \beta, \gamma, \delta) = \int_0^1 \int_0^1 p(X(n) = x | p, \theta) p(p | \alpha, \beta) p(\theta | \gamma, \delta) \, dp \, d\theta,
$$

in the usual way. $\qquad \square$

```
 1   class BgbbLikelihood(ObjectiveFunction):
 2
 3       def _rho(self, alpha, beta, x, y):
 4           return BETA(alpha + x, beta + y) / BETA(alpha, beta)
 5
 6       def _rho_alpha(self, alpha, beta, x, y):
 7           rho = self._rho(alpha, beta, x, y)
 8           x = psi(alpha + x) - psi(alpha) + psi(alpha + beta) -
                   psi(alpha + beta + x + y)
 9           return rho * x
10
11       def _rho_beta(self, alpha, beta, x, y):
12           rho = self._rho(alpha, beta, x, y)
13           x = psi(beta + y) - psi(beta) + psi(alpha + beta) -
                   psi(alpha + beta + x + y)
14           return rho * x
15
16       def kernel(self, theta, x):
17           alpha, beta, gamma, delta = theta
18           l = np.zeros(len(x))
19           for i, row in enumerate(x):
20               tx, x, n = row
21               t = np.arange(tx, n+1)
22               u = (t != n).astype(int)
23               l[i] = np.sum(self._rho(alpha, beta, x, t-x) *
                       self._rho(gamma, delta, u, t))
24           return l
25
26       def kernelGrad(self, theta, x):
27           alpha, beta, gamma, delta = theta
28           l = np.zeros((4, len(x)))
29           for i, row in enumerate(x):
30               tx, x, n = row
31               t = np.arange(tx, n+1)
32               u = (t != n).astype(int)
33               l[0, i] = np.sum(self._rho_alpha(alpha, beta, x, t-x) *
                       self._rho(gamma, delta, u, t))
34               l[1, i] = np.sum(self._rho_beta(alpha, beta, x, t-x) *
                       self._rho(gamma, delta, u, t))
35               l[2, i] = np.sum(self._rho(alpha, beta, x, t-x) *
                       self._rho_alpha(gamma, delta, u, t))
36               l[3, i] = np.sum(self._rho(alpha, beta, x, t-x) *
                       self._rho_beta(gamma, delta, u, t))
37           return l
```

Code Block 10.6: BG/BB Likelihood and gradient; used with stochastic gradient descent; subclassed from Code Blocks 5.7 and 5.8.

Of course, we are often interested in the expected number of actions for a given customer within a given purchase horizon. This is given by our next result.

**Proposition 10.11.** *Let* $X(n) = \sum_{t=1}^{n} Y_t$ *be the number of active transaction periods from the first $n$ transaction opportunities. Given the hyperparameters of the BG/BB model, the expected number of active periods is given by*

$$\mathbb{E}[X(n)|\alpha, \beta, \gamma, \delta] = \frac{\alpha\delta}{(\alpha+\beta)(\gamma-1)}\left[1 - \frac{\Gamma(\gamma+\delta)}{\Gamma(\gamma+\delta+n)}\frac{\Gamma(1+\delta+n)}{\Gamma(1+\delta)}\right].$$

$$(10.67)$$

*Note 10.8.* When multiplied by the expected monetization rate per transaction opportunity, Equation (10.67) yields the expected LTV of a customer over the time horizon $n$.

One can further derive the precise probability mass function for $p(X(n) = x)$, as well as the expected number of active transactions over the period $(n, n+n^*)$; details of these calculations are found in Fader, *et al.* [2010]. ▷

*Proof.* We begin by computing the expected value of $X(n)$, given the transaction rate $p$ and churn rate $\theta$. Recall that the random variable $A_t$ represents the customer being alive at time $t$, we may write the conditional expectation as

$$\mathbb{E}[X(n)|p, \theta] = \sum_{t=1}^{n} \mathbb{P}(Y_t = 1|p, A_t)\mathbb{P}(A_t|\theta).$$

Now, according to our model,

$$\mathbb{P}(Y_t = 1|p, A_t) = p$$
$$\mathbb{P}(A_t|\theta) = \mathbb{P}(\tau \geq t|\theta) = (1-\theta)^t,$$

the final equality following from the fact that the user's lifetime is geometrically distributed. We therefore obtain

$$\mathbb{E}[X(n)|p, \theta] = p\sum_{t=1}^{n}(1-\theta)^n = \frac{p(1-\theta)}{\theta} - \frac{p(1-\theta)^{n+1}}{\theta}.$$

(See Exercise 10.5.)

Next, we marginalize the expectation over the mixing distributions to obtain

$$\mathbb{E}[X(n)|\alpha,\beta,\gamma,\delta] = \int_0^1\int_0^1 \mathbb{E}[X(n)|p,\theta]p(p|\alpha,\beta)p(\theta|\gamma,\delta)\,dp\,d\theta$$

$$= \int_0^1\int_0^1 \left[\frac{p^\alpha(1-p)^{\beta-1}\theta^{\gamma-2}(1-\theta)^\delta}{B(\alpha,\beta)B(\gamma,\delta)}\right.$$
$$\left. - \frac{p^\alpha(1-p)^{\beta-1}\theta^{\gamma-2}(1-\theta)^{\delta+n}}{B(\alpha,\beta)B(\gamma,\delta)}\right]\,dp\,d\theta$$

$$= \frac{B(\alpha+1,\beta)}{B(\alpha,\beta)}\left[\frac{B(\gamma-1,\delta+1)-B(\gamma-1,\delta+n+1)}{B(\gamma,\delta)}\right],$$

where the last equality follows by recognizing the penultimate integrand as the kernel of

$$\text{Beta}(p;\alpha+1,\beta)\left[\text{Beta}(\theta;\gamma-1,\delta+1)-\text{Beta}(\theta;\gamma-1,\delta+n+1)\right],$$

and then integrating. Further simplification, using the identity Equation (2.48), yields the result. □

For a given set of hyperparameters $(\alpha,\beta,\gamma,\delta)$, we can compute the expected lifetime (Equation (10.47)), the survival function (Equation (10.48)), the density function Equation (10.66), and the expected number of actions occurring within a horizon $(0,n)$ (Equation (10.67)), as shown in Code Block 10.7[3].

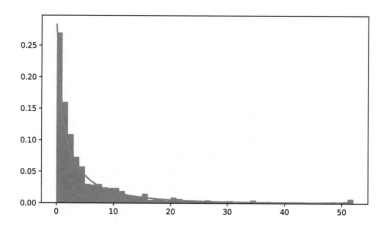

Fig. 10.9: Simulated and theoretical density function for $X(n)$.

Finally, we run the simulation of Code Block 10.5, by first simulating a cohort of users from the Pareto/NBD model and then converting the

---

[3] The function Xn relies on the combinatorial function comb from the math package, which computes $n$ *choose* $k$.

```
1   def ELT(gamma, delta, opp_window=7):
2       return opp_window * delta / (gamma-1)
3
4   def S(n, gamma, delta):
5       return GAMMA(delta + n + 1) / GAMMA(delta) * GAMMA(gamma+delta)
            / GAMMA(gamma+delta + n + 1)
6
7   def Xn(x, n, alpha, beta, gamma, delta):
8       S = comb(n, x) * BETA(alpha+x, beta+n-x) / BETA(alpha, beta) *
            BETA(gamma, delta+n) / BETA(gamma, delta)
9       for t in range(x, n):
10          S += comb(t, x) * BETA(alpha+x, beta+t-x) / BETA(alpha,
                beta) * BETA(gamma+1, delta+t) / BETA(gamma, delta)
11      return S
12
13  def EXn(n, alpha, beta, gamma, delta):
14      Xn = alpha * delta / (alpha + beta) / (gamma - 1)
15      R = GAMMA(gamma + delta) / GAMMA(gamma + delta + n)
16      S = GAMMA(1 + delta + n) / GAMMA(1 + delta)
17      Xn *= (1 - R * S)
18      return Xn
```

Code Block 10.7: Key quantities for BG/BB Model.

datasets to the BG/BB format. The expected one-year survival function for the Pareto/NBD, using the simulation parameters, is 4.62% (the same as in Example 10.9). The maximum likelihood for our simulation resulting in a set of hyperparameters for the BG/BB model, given by

$$\alpha = 0.72, \qquad \beta = 0.51, \qquad \gamma = 1.85, \qquad \delta = 11.60.$$

The BG/BB model parameters (for this particular simulation) yield a one-year ($n = 52$) survival function of 4.39% (Equation (10.48)). Moreover, the distribution of the number of actions (Equation (10.66)) is plotted against the histogram of actual frequency of actions in Figure 10.9. The observed mean number of actions was 5.734, which closely matches the expected number 5.97, given by Equation (10.67). This shows that even when data is generated from the Pareto/NBD model, the BG/BB model can be used to accurately describe purchase behavior.

### Bayesian Update for an Individual Customer

Oftentimes, we are interested in the incremental count of transactions occurring in the interval $(n, n+m]$. We can consider such a count for an arbitrary user with BG/BB hyperparameters $\alpha, \beta, \gamma, \delta$, or for a specific customer with given hyperparameters *and* transaction summary statistics $(x, t_x, n)$.

As it turns out, both cases can be captured under a single formulation. To see this, we first consider the following Bayesian update.

**Proposition 10.12.** *Given a lifetime mixture model*

$$\tau \sim \mathrm{Geom}(\theta),$$
$$\theta \sim \mathrm{Beta}(\gamma, \delta)$$

*and a censored observation that a user is still alive at time $t = n$, so that $\tau \geq n$, the posterior distribution is obtained by the Bayesian update*

$$\gamma' = \gamma$$
$$\delta' = \delta + n. \tag{10.68}$$

*Note 10.9.* The update rule of Proposition 10.12 is distinct from the one given in Table 10.2, which corresponds to a set of *observed* lifetimes as opposed to a single *censored* lifetime. (If the lifetime of a single user is *observed*, the update is $\gamma' = \gamma + 1$ and $\delta' = \delta + n - 1$.)    ▷

*Proof.* The likelihood that a customer is still alive at time $t = n$ is given by

$$p(A_n|\theta) = (1 - \theta)^n.$$

Therefore, by Bayes' rule, the posterior distribution is proportional to

$$p(\theta|A_n, \gamma, \delta) \propto p(A_n|\theta)p(\theta|\gamma, \delta) = (1 - \theta)^n \frac{\theta^{\gamma-1}(1 - \theta)^{\delta-1}}{B(\gamma, \delta)}.$$

The result given in Equation (10.68) follows.    □

The following gives us the posterior distribution for a customer in the BG/BB model with observed transaction history $(x, t_x, n)$, conditional on the customer being still alive at time $t = n$.

**Proposition 10.13.** *Given a user in the BG/BB model with a priori hyperparameters $\alpha, \beta, \gamma, \delta$ and observed transactional history $(x, t_x, n)$, the posterior distribution for the transaction and churn probabilities $p$ and $\theta$, assuming the customer is still alive at time $t = n$, is obtained by the Bayesian update*

$$\alpha' = \alpha + x \tag{10.69}$$
$$\beta' = \beta + n - x \tag{10.70}$$
$$\gamma' = \gamma \tag{10.71}$$
$$\delta' = \delta + n. \tag{10.72}$$

*Proof.* The likelihood of observing transaction history $(x, t_x, n)$ for a customer that is still alive at time $t = n$ is given by

$$p(x, t_x, n, A_n | p, \theta) = p^x (1-p)^{n-x} (1-\theta)^n.$$

Therefore, by Bayes' rule, the posterior distribution is proportional to

$$p(p, \theta | x, t_x, n, A_n, \alpha, \beta, \gamma, \delta) \propto p(x, t_x, n, A_n | p, \theta) p(p | \alpha, \beta) p(\theta | \gamma, \delta)$$

$$= p^x (1-p)^{n-x} (1-\theta)^n \frac{p^{\alpha-1}(1-p)^{\beta-1}}{B(\alpha, \beta)} \frac{\theta^{\gamma-1}(1-\theta)^{\delta-1}}{B(\gamma, \delta)}.$$

The result given in Equations (10.69)–(10.72) follows.    □

*Note 10.10.* In both Proposition 10.12 and 10.13, the Bayesian update of the lifetime hyperparameters is obtained by taking $\gamma' = \gamma$ and $\delta' = \delta + n$. Now, if a customer with transactional data $(x, t_x, n)$ has $t_x < n$, we have not truly observed that the customer is still alive at time $t = n$. However, in what follows, we will consider the number of *future* transactions, subsequent to the initial observed period $[0, n]$. If the cusomter is *not alive* at time $t = n$, the customer will make no further purchases. We are thus interested in conditioning on the case in which the customer is still alive, using the preceding Bayesian update rule, and then applying the law of total probability.    ▷

Next, we consider the general formula for the number of transactions of a user over a set of transaction opportunities subsequent to a particular point in time $t = n$. This result will be then combined with Proposition 10.12 and 10.13 to obtain two useful quantities.

**Proposition 10.14.** *Let $X(n, n+m) = \sum_{t=n+1}^{n+m} Y_t$ represent the number of transactions for a customer in the BG/BB model over the time frame $t \in (n, n+m]$. Suppose a customer in the BG/BB model with posterior transactional hyperparameters $\alpha', \beta'$ and lifetime hyperparameters $\gamma, \delta$ is alive at time $t = n$ with probability $\xi_n$, then the probability mass function for the random variable $X(n, n+m)$ is given by*

$$p(X(n, n+m) = y | \alpha', \beta', \gamma, \delta) = \mathbb{I}[y = 0] (1 - \xi_n)$$
$$+ \xi_n p(X(m) = y | \alpha', \beta', \gamma, \delta + n) \quad (10.73)$$

*where the function $p(X(n) = x | \alpha, \beta, \gamma, \delta)$ is defined by Equation (10.66).*
*Moreover, the expected number of transactions in this interval is given by*

$$\mathbb{E}[X(n, n+m) | \alpha', \beta', \gamma', \delta'] = \xi_n \mathbb{E}[X(m) | \alpha', \beta', \gamma', \delta'], \quad (10.74)$$

*where the function $\mathbb{E}[X(n) | \alpha, \beta, \gamma, \delta]$ is given by Equation (10.67).*

*Proof.* If the customer is not alive at time $n$, then the number of transactions over the interval $(n, n+m]$ must be zero. Thus, the general form of Equation (10.73),

$$p(X(n, n+m) \doteq y) = \mathbb{I}[y = 0](1 - \xi_n) + \xi_n p(X(n, n+m) = y|A_n),$$

follows. (We have suppressed the conditioning on $\alpha', \beta', \gamma', \delta'$ in this previous equation.)

But, assuming the customer is alive at time $t = n$, the number of transactions over the *next* $m$ transaction opportunities must be given by Equation (10.66), due to the memoryless nature of the churn process. Equation (10.73) follows.

Equation (10.74) follows since the term $\mathbb{I}[y = 0]\,(1 - \xi_n)$ does not contribute to the expectation. □

**Corollary 10.6.** *For a general customer in the BG/BB model with hyperparameters $\alpha, \beta, \gamma, \delta$, the probability mass function for $X(n, n+m)$ is given by*

$$p(X(n, n+m) = y \mid \alpha, \beta, \gamma, \delta) = \mathbb{I}[y = 0]\left[1 - \frac{B(\gamma, \delta + n)}{B(\gamma, \delta)}\right]$$

$$+ \sum_{t=y}^{m-1} \binom{t}{y} \frac{B(\alpha + y, \beta + t - y)}{B(\alpha, \beta)} \frac{B(\gamma + 1, \delta + n + t)}{B(\gamma, \delta)}$$

$$+ \binom{m}{y} \frac{B(\alpha + y, \beta + m - y)}{B(\alpha, \beta)} \frac{B(\gamma, \delta + n + m)}{B(\gamma, \delta)}, \quad (10.75)$$

*with expected value*

$$\mathbb{E}[X(n, n+m) \mid \alpha, \beta, \gamma, \delta] = \left(\frac{\alpha}{\alpha + \beta}\right)\left(\frac{\delta}{\gamma - 1}\right)\frac{\Gamma(\gamma + \delta)}{\Gamma(1 + \delta)}$$

$$\times \left[\frac{\Gamma(1 + \delta + n)}{\Gamma(\gamma + \delta + n)} - \frac{\Gamma(1 + \delta + n + m)}{\Gamma(\gamma + \delta + n + m)}\right]. \quad (10.76)$$

*Proof.* First, let us consider Equation (10.75). The probability that a generic customer is still alive at time $t = n$ is given by Equation (10.48) (evaluated at $n - 1$) as

$$\xi_n = p(A_n|\gamma, \delta) = \frac{B(\gamma, \delta + n)}{B(\gamma, \delta)}.$$

Next, we apply Equation (10.73), where the second term is made conditional upon the customer being alive at time $t = n$. However, from Equation (10.68), this is tantamount to the Bayesian update $\gamma' = \gamma$ and $\delta' = \delta + n$. Thus, the second term of Equation (10.73) is given by

$$\frac{B(\gamma, \delta + n)}{B(\gamma, \delta)}p(X(m) = y|\alpha, \beta, \gamma, \delta + n).$$

Upon substituting $n \to m$, $x \to y$, and $\delta \to \delta + n$ into Equation (10.66) and multiplying by $\xi_n$ (which has the effect of converting the denominator back to $B(\gamma, \delta)$), we obtain our first result.

Similarly, Equation (10.76) follows from Equation (10.74):

$$\mathbb{E}[X(n, n+m)|\alpha, \beta, \gamma, \delta] = \frac{B(\gamma, \delta+n)}{B(\gamma, \delta)} \mathbb{E}[X(m)|\alpha, \beta, \gamma, \delta+n].$$

Note that this second factor is obtained by replacing $n \to m$ and $\delta \to \delta + n$ in Equation (10.67). To simplify the result, it is useful to note that

$$\frac{B(\gamma, \delta+n)}{B(\gamma, \delta)} = \frac{\Gamma(\delta+n)\Gamma(\gamma+\delta)}{\Gamma(\delta)\Gamma(\gamma+\delta+n)},$$

and that

$$\frac{(\delta+n)\Gamma(\delta+n)}{\Gamma(\delta)} = \frac{\delta\Gamma(\delta+n+1)}{\Gamma(\delta+1)}.$$

The rest follows from algebra.                                    □

Naturally, Corollary 10.6 is not so useful, as it is addressing the behavior of a generic customer over the transaction intervals $(n, n+m]$. Typically, we do not consider a generic customer, but a specific customer with transaction history $(x, t_x, n)$, and ask how many future transactions that customer might make over the period $(n, n+m]$. This result is given as follows.

**Corollary 10.7.** *For a general customer in the BG/BB model with hyperparameters* $\alpha, \beta, \gamma, \delta$ *and observed transactional history* $(x, t_x, , n)$, *the probability mass function for* $X(n, n+m)$ *is given by*

$$p(X(n, n+m) = y \mid \alpha, \beta, \gamma, \delta, x, t_x, n) = \mathbb{I}[y = 0]\left[1 - \frac{B(\alpha', \beta')}{B(\alpha, \beta)} \frac{B(\gamma, \delta+n)}{B(\gamma, \delta)L}\right]$$

$$+ \sum_{t=y}^{m-1} \binom{t}{y} \frac{B(\alpha'+y, \beta'+t-y)}{B(\alpha, \beta)} \frac{B(\gamma+1, \delta+n+t)}{B(\gamma, \delta)L}$$

$$+ \binom{m}{y} \frac{B(\alpha'+y, \beta'+m-y)}{B(\alpha, \beta)} \frac{B(\gamma, \delta+n+m)}{B(\gamma, \delta)L}, \qquad (10.77)$$

*where* $\alpha' = \alpha + x$ *and* $\beta' = \beta + n - x$, *and* $L$ *is the likelihood given by Equation (10.52), with expected value*

$$\mathbb{E}[X(n, n+m) \mid \alpha, \beta, \gamma, \delta, x, t_x, n] = \frac{B(\alpha+x+1, \beta+n-x)}{B(\alpha, \beta)L}$$

$$\times \left(\frac{\delta}{\gamma-1}\right) \frac{\Gamma(\gamma+\delta)}{\Gamma(1+\delta)}$$

$$\times \left[\frac{\Gamma(1+\delta+n)}{\Gamma(\gamma+\delta+n)} - \frac{\Gamma(1+\delta+n+m)}{\Gamma(\gamma+\delta+n+m)}\right]. \qquad (10.78)$$

*Proof.* The result is obtained by applying Equations (10.73) and (10.74) using the probability that a customer is alive at time $t = n$:

$$\xi_n = \frac{B(\alpha + x, \beta + n - x)}{B(\alpha, \beta)} \frac{B(\gamma, \delta + n)}{B(\gamma, \delta)} L^{-1},$$

as given by Equation (10.63), and the Bayesian update found in Equations (10.69) and (10.72) in formulas Equations (10.66) and (10.67).

In the derivation for Equation (10.78), note that

$$B(\alpha + x, \beta + n - x) \left( \frac{\alpha + x}{\alpha + \beta + n} \right) = B(\alpha + x + 1, \beta + n - x).$$

The rest follows analogous to the proof of Corollary 10.6.                    □

*Note 10.11.* The results of Corollaries 10.6 and 10.7 are also derived in a different approach in Fader, *et al.* [2010]. There, the law of total probability is applied directly to the parameters $(p, \theta)$, and then the result is obtained by marginalizing over those parameters with respect to the two beta mixing distributions. The approach we presented, using Proposition 10.12 and 10.13 and Proposition 10.14 allows us to reveal the common structure between both expressions, as well as work directly in the hyperparameter space. Moreover, it is also more straightforward from a programming perspective to implement Equations (10.73) and (10.74) directly, as opposed to their more verbose counterparts.                    ▷

Finally, we can generalize the result of Proposition 10.14 one step further by applying it to an arbitrary future window. We state the result in the following theorem.

**Theorem 10.1.** *Let* $X(u, v) = \sum_{t=u+1}^{v} Y_t$ *represent the number of transactions for a customer in the BG/BB model over the time frame* $t \in (u, v]$ *for a customer in the BG/BB model with (prior) hyperparameters* $\alpha, \beta, \gamma, \delta$ *and behavioral data* $x, t_x, n$, *where* $n \leq u < v$. *If we define* $\xi_u$ *as the probability that the customer is alive at time* $t = u$, *i.e.,*

$$\xi_u = p(A_u | \alpha, \beta, \gamma, \delta, x, t_x, n),$$

*as given by Equation (10.64) (replacing* $n + m$ *with* $u$), *then the probability mass function for the random variable* $X(u, v)$ *is given by*

$$p(X(u, v) = y \mid \alpha, \beta, \gamma, \delta, x, t_x, n) = \mathbb{I}[y = 0] (1 - \xi_u)$$
$$+ \xi_u p(X(v - u) = y | \alpha + x, \beta + n - x, \gamma, \delta + u)\,(10.79)$$

*where the function* $p(X(n) = x | \alpha, \beta, \gamma, \delta)$ *is defined by Equation (10.66).*

*Moreover, the expected number of transactions in this interval is given by*

$$\mathbb{E}[X(u, v) | \alpha, \beta, \gamma, \delta, x, t_x, n] = \xi_u \mathbb{E}[X(v - u) | \alpha + x, \beta + n - x, \gamma, \delta + u], \quad (10.80)$$

*where the function* $\mathbb{E}[X(n) | \alpha, \beta, \gamma, \delta]$ *is given by Equation (10.67).*

*Proof.* This proof is a simple extension of Proposition 10.14 and Corollaries 10.6 and 10.7. The transactional data $(x, n, t_x)$ yield the updated hyperparameters $\alpha' = \alpha + x$ and $\beta' = \beta + n - x$. The probability that such a customer is alive at time $t = u \geq n$ is given by Equation (10.64). Moreover, conditioning on $A_u$ yields the updated hyperparameter $\delta' = \delta + u$. The result follows.    □

Note that Proposition 10.14 and Corollaries 10.6 and 10.7 are just special cases of Theorem 10.1. This general form is also useful for programmatic implementation, as shown in Code Blocks 10.8 and 10.9. In Code Block 10.8, we encode the likelihood function, as given in Equation (10.52), as the method `likelihood`. The probability that a customer is alive at time $t$, as given by Equation (10.64), is encoded as the function `probAlive`. The probability mass function Equation (10.66) and its expectation Equation (10.67) are encoded in Code Block 10.9 as `Xn` and `EXn`, respectively. Finally, the general results from Theorem 10.1 are encoded as `Xuv` and `EXuv`. Thus, probabilities and expectations for the number of future actions over arbitrary periods can be computed for any customer in the BG/BB model.

### The Ruse of Heterogeneity

The *ruse of heterogeneity*, as introduced in Vaupel and Yashin [1985] and described in the context of the BG model in Fader and Hardie [2007], is any apparent time dynamics in a mixture model that contains no explicit time-dynamic component. The observed time dynamics are therefore a specter in the data caused by a population's heterogeneity, and not any actual time-dependent behavior. As a simple example, we can restrict the BG/BB model to contractual users, thereby obtaining the BG model, where we can illustrate the concept with a user's retention rate.

In the limiting case of the BG/BB model as the hyperparameters $(\alpha, \beta) \to (\infty, 0)$, the transaction probability collapses to the particular value $p = 1$ and the model degenerates to the BG model, which is suitable for contractual customers, in the sense of Definition 10.9. This particular use case was introduced by Fader and Hardie [2007].

**Definition 10.15.** *The* retention rate $r_t$ *for a cohort at time t is defined by*

$$r_t = \frac{S(t)}{S(t-1)}.$$

*It represents the ratio of users who will survive beyond time t compared to those who survived until time t.*

In the BG model, we can compute the probability mass function for the user lifetime $\tau$ and its associated survival function by marginalizing Equations (10.43) and (10.44) over the churn parameter $\theta$.

```
 1   class Customer:
 2
 3       def __init__(self, alpha, beta, gamma, delta):
 4           self.alpha, self.beta, self.gamma, self.delta = alpha, beta,
                 gamma, delta
 5           self.x, self.tx, self.n = 0, 0, 0
 6
 7       def setData(self, x, tx, n):
 8           assert x <= tx, 'number of actions must be less than last
                 purchase time'
 9           assert tx <= n, 'last purchase time must be less than age'
10           self.x, self.tx, self.n = x, tx, n
11
12       def getGreeks(self, update=False):
13           if update:
14               return self.alpha + self.x, self.beta + self.n - self.x,
                     self.gamma, self.delta + self.n
15
16           return self.alpha, self.beta, self.gamma, self.delta
17
18       def likelihood(self):
19
20           alpha, beta, gamma, delta = self.getGreeks(update=False)
21           x, tx, n = self.x, self.tx, self.n
22           l = BETA(alpha+x, beta+n-x) / BETA(alpha, beta) *
                 BETA(gamma, delta+n) / BETA(gamma, delta)
23           for t in range(tx, n):
24               l += BETA(alpha+x, beta+t-x) / BETA(alpha, beta) *
                     BETA(gamma+1, delta+t) / BETA(gamma, delta)
25
26           return l
27
28       def probAlive(self, t):
29           assert t >= self.n, 'Alive probability for future time'
30
31           alpha, beta, gamma, delta = self.getGreeks(update=False)
32           x, n = self.x, self.n
33           l = self.likelihood()
34           p = BETA(alpha+x, beta+n-x) / BETA(alpha, beta) *
                 BETA(gamma, delta+t) / BETA(gamma, delta)
35
36           return p / l
```

Code Block 10.8: Customer-level predictions in BG/BB model.

```
1   def Xn(n, x, alpha, beta, gamma, delta):
2       assert x <= n, "number of actions cannot exceed horizon"
3
4       p = comb(n, x) * BETA(alpha+x, beta+n-x) / BETA(alpha, beta)
            * BETA(gamma, delta+n) / BETA(gamma, delta)
5       for t in range(x, n):
6           p += comb(t, x) * BETA(alpha+x, beta+t-x) / BETA(alpha,
                beta) * BETA(gamma+1, delta+t) / BETA(gamma, delta)
7
8       return p
9
10  def EXn(n, alpha, beta, gamma, delta):
11      x = alpha * delta / (alpha+beta) / (gamma-1)
12      x *= (1 - GAMMA(gamma+delta) / GAMMA(gamma+delta+n) *
            GAMMA(1+delta+n) / GAMMA(1 + delta))
13
14      return x
15
16  def Xuv(self, u, v, x):
17      assert u >= self.n
18      assert v >= u
19
20      alpha, beta, gamma, delta = self.getGreeks(update=True)
21      delta += u - self.n
22      xi = self.probAlive(u)
23      m = v - u
24      p = (1 - xi) if x == 0 else 0
25      p += xi * self.Xn(m, x, alpha, beta, gamma, delta)
26
27      return p
28
29  def EXuv(self, u, v):
30      assert u >= self.n
31      assert v >= u
32
33      alpha, beta, gamma, delta = self.getGreeks(update=True)
34      delta += u - self.n
35      xi = self.probAlive(u)
36      m = v - u
37
38      return xi * self.EXn(m, alpha, beta, gamma, delta)
```

Code Block 10.9: Customer-level predictions in BG/BB model (continued).

$$\mathbb{P}(\tau = t|\gamma, \delta) = \frac{B(\gamma + 1, \delta + t)}{B(\gamma, \delta)}, \tag{10.81}$$

$$S(t|\gamma, \delta) = \frac{B(\gamma, \delta + t + 1)}{B(\gamma, \delta)}. \tag{10.82}$$

(We leave the details as an exercise for the reader.) Given the survival function, we can now use Definition 10.15 to compute the retention rate.

$$r_t = \frac{B(\gamma, \delta + t + 1)}{B(\gamma, \delta + t)} = \frac{\delta + t}{\gamma + \delta + t}.$$

This result is described by Fader and Hardie [2007]:

> We immediately see that under the [BG] model, the retention rate is an increasing function of time, even though the underlying (unobserved) individual-level retention probability is constant. According to this model, there are no underlying time dynamics at the level of the individual customer; the observed phenomenon of retention rates increasing over time is simply due to heterogeneity (i.e., the high-churn customers drop out early in the observation period, with the remaining customers having lower churn probabilities).

Under the BG (or BG/BB) model, each individual has a constant, unobserved true churn rate, but the cohort consists of a mixture of users with varying churn rates described by a beta distribution. The ruse of heterogeneity is a case of burning the candle at both ends, albeit at different rates. As a cohort ages, users with a high churn rate are most likely to churn early, thus altering the mixture of the remaining users as the population matures.

### Monetization

So far, we have discussed numerous models that handle the recency and frequency components of a customer's lifetime value, vis-à-vis the transaction probability and unobserved customer lifetime. Naturally, such a picture is incomplete, as it has yet to include the important component of monetization. Two common approaches to modeling the monetization rate are the gamma–gamma mixture (Colombo and Jiang [1999]) and the normal–normal mixture (Schmittlein and Peterson [1994]).

## 10.4 Credibility Theory

Credibility theory is a mathematical framework for describing heterogeneous populations using certain linearity assumptions. Credibility theory arose in the insurance industry due to their need to understand and rate heterogeneous risk classes and to further understand the relationship between the individual and the collective. In particular, credibility theory

1. is used to model heterogeneous collectives (i.e., mixture models),
2. shows how we can combine individual and collective experience,
3. belongs to the field of Bayesian statistics,
4. represents a class of *linear* Bayes estimators,
5. has key advantages in terms of simplicity and structure,

It was developed in the seminal works of Bühlmann [1967] and Bühlmann and Straub [1970]. For an overview, see the texts Bühlmann [1970] and, for a more recent introduction, see Bühlmann and Gisler [2005], which also discusses hierarchical credibility models, multidimensional problems (regression), and certain evolutionary credibility models. We include a treatment of the theory here, as we believe the theory to have implications beyond rating insurance risks.

Bayesian models, other than a few cornerstone examples, are often unwieldy, complex, and analytically intractable, especially for multidimensional or hierarchical models, or when nonconjugate priors are prescribed. One popular remedy is the use of computer simulation and Markov-chain Monte Carlo methods, which we discuss in Section 10.5. Credibility theory, on the other hand, offers an analytic alternative with a number of key benefits:

1. *premiums*: estimates for the expected values $\mathbb{E}[X]$, known in credibility as *risk premiums*, are modeled directly;
2. *simplicity*: the formulas underlying credibility computations are simple and intuitive;
3. *structure*: the sampling distribution $p(X|\theta)$ and the mixing distribution $p(\theta)$ need not be specified, rather, everything we need from them is specified through their first few moments.
4. *extendability*: can easily be extended to hierarchical and multidimensional models.

### 10.4.1 Risks and Risk Premiums

Arising from the field of insurance, credibility theory has a few unique terms that are not standard elsewhere in statistics. We therefore begin our discussion by outlining several definitions that are useful in describing the mathematical problem that is addressed by credibility theory.

### Risk Premiums and Definitions

In credibility theory, the random variables under study are known as *risks*, as they, in their original application, represent insurance payouts and, thus, risks born by an insurance company. The basic idea underlying all of insurance is that a collective of individuals, each with an exposure to a particular risk (e.g., automobile accident, fire, theft, dropping your iPhone into a toilet, etc.), join together to form a "community-at-risk" that bears the risk in the form of a collective; i.e., by paying an insurance premium, individuals transfer their risk to an insurance company. Credibility calculations can be viewed from the perspective of an insurance company, that must understand how to quantify risk and, therefore, correctly assess their insurance premiums to all individuals insured. A priori, each individual is viewed as a random risk from the collective. However, as individuals amasses their respective claims histories, we may better assess the risk of each unique individual within the collective.

In particular, in the context of credibility, a random variable $X$ is referred to as a *risk*, whereas its expected value $\mathbb{E}[X]$ is referred to as a *risk premium*. For a given individual, we say that the claims history $X_1, \ldots, X_n \sim F_\theta$ is described by a certain *risk profile*, which is parameterized by a random variable $\Theta$. The risk profile $\Theta$ is itself distributed relative to a distribution $\Theta \sim U$, known as the *collective distribution*. From the perspective of the collective, we say that $F_\theta$ is a sampling distribution and $U$ is a mixing distribution, but from the perspective of each individual, we may view $U$ as a prior.

**Definition 10.16.** *Given a continuous mixture model, the* individual premium *is the random variable*

$$P^{\text{ind}} = \mathbb{E}[X|\Theta] = \mu(\Theta). \tag{10.83}$$

*Given the true parameter* $\Theta = \theta$ *of a particular individual, the* correct individual premium *is*

$$P^{\text{ind}}(\theta) = \mathbb{E}[X|\Theta = \theta] = \mu(\theta). \tag{10.84}$$

*The* collective premium *is the value*

$$P^{\text{coll}} = \mathbb{E}[P^{\text{ind}}] = \mu_0. \tag{10.85}$$

*And, finally, the* Bayes' premium *is the random variable*

$$P^{\text{Bayes}} = \tilde{\mu}(\Theta) = \mathbb{E}[P^{\text{ind}}|\mathcal{D}], \tag{10.86}$$

*where the expectation is with respect to the random variable* $\Theta$*, and where* $\mathcal{D}$ *represents an individual's observed history. The Bayes' premium is also referred to as the* experience premium.

Note that the individual premium is a random variable, as it is a function of the random variable $\Theta$. Similarly, the Bayes' premium is a random variable, as it is a function of the data $\mathcal{D}$. The collective premium, on the other hand, is a fixed value, since

$$\mu_0 = \mathbb{E}[\mathbb{E}[X|\Theta]] = \mathbb{E}[X].$$

Intuitively, the collective premium is the overall expected value of $X$ throughout the population. The individual premium is a random variable that represents the expected value of $X$ for individuals within the population.

These definitions are best illustrated by means of a few examples.

*Example 10.10 (Bernoulli–beta).* Consider the mixture model $X \sim \text{Bern}(\Theta)$, with $\Theta \sim \text{Beta}(\alpha, \beta)$. The individual premium is given by

$$P^{\text{ind}} = \mathbb{E}[X|\Theta] = \Theta,$$

the expected value of a Bernoulli random variable. Therefore, given the true probability $\Theta = \theta$ of a specific individual, the correct individual premium will be $P^{\text{ind}}(\theta) = \theta$.

Similarly, the collective premium is simply the expected value of $P^{\text{ind}}$, or

$$P^{\text{coll}} = \mathbb{E}[P^{\text{ind}}] = \mathbb{E}[\Theta] = \frac{\alpha}{\alpha + \beta}.$$

Next, let us suppose that we have a sequence of observations for a given individual in the population consisting of $n$ Bernoulli trials with a total of $k$ successes. the Bayes' premium (or experience premium) for that individual is obtained using our standard Bayesian update rules $\alpha' = \alpha + k$ and $\beta' = \beta + n - k$ for a binomial outcome, yielding

$$P^{\text{Bayes}} = \mathbb{E}[P^{\text{ind}}|\mathcal{D}] = \frac{\alpha + k}{\alpha + \beta + n}.$$

(Note that this equals the expected value of the posterior predictive distribution.)

Finally, we notice that, with a little bit of algebra, we can rewrite the experience premium in the form

$$P^{\text{Bayes}} = \lambda \overline{X} + (1 - \lambda) P^{\text{coll}},$$

where $\overline{X} = k/n$ and $\lambda$ is given by

$$\lambda = \frac{n}{n + \alpha + \beta}.$$

For this specific example, the Bayes' premium is a linear combination of the collective premium (which can be viewed as the prior expected value, which, for a random individual, is equal to the expected value of the population) and the observed data (the ratio of successes $\overline{X} = k/n$). Moreover, the particular weighting favors the experience as the number of observations $n$ increases.                                                                    ▷

*Example 10.11 (Poisson–gamma).* Consider the mixture model $X \sim \text{Poiss}(\Theta)$, with $\Theta \sim \text{Gamma}(\alpha, \beta)$. The individual premium is again given by

$$P^{\text{ind}} = \mathbb{E}[X|\Theta] = \Theta,$$

the expected value of a Poisson random variable. Therefore, given the true probability $\Theta = \theta$ of a specific individual, the correct individual premium will be $P^{\text{ind}}(\theta) = \theta$.

Similarly, the collective premium is simply the expected value of $P^{\text{ind}}$, or

$$P^{\text{coll}} = \mathbb{E}[P^{\text{ind}}] = \mathbb{E}[\Theta] = \alpha\beta.$$

Next, let us suppose that we have a sequence of $n$ observations of a IID Poisson random variable for a given individual with a total observed count of $X. = n\overline{X} = \sum_{i=1}^{n} X_i$. The Bayes' premium (or experience premium) for that individual is obtained using our standard Bayesian update rules for the Poisson–gamma mixture, or $\alpha' = \alpha + X.$ and $(1/\beta') = (1/\beta) + n$. This yields

$$P^{\text{Bayes}} = \mathbb{E}[P^{\text{ind}}|\mathcal{D}] = \left(\alpha + n\overline{X}\right)\frac{\beta}{1+n\beta} = \frac{\alpha\beta}{1+n\beta} + \frac{n\beta\overline{X}}{1+n\beta}.$$

(Note that this equals the expected value of the posterior predictive distribution.)

Finally, we notice that, with a little bit of algebra, we can rewrite the experience premium in the form

$$P^{\text{Bayes}} = \lambda\overline{X} + (1-\lambda)P^{\text{coll}},$$

where $\overline{X} = k/n$ and $\lambda$ is given by

$$\lambda = \frac{n}{n+1/\beta}.$$

Again we see that the Bayes' premium can be expressed as a linear combination of the collective premium and the observed data.    $\triangleright$

For a third example, consider the following.

*Example 10.12 (exponential–gamma).* Consider the mixture model $X \sim \text{Exp}(\Theta^{-1})$, and $\Theta \sim \text{Gamma}(\alpha, \beta)$, where $\Theta$ is the *rate* parameter (inverse scale parameter) of the exponential distribution. A simple comutation shows that

$$P^{\text{ind}} = \frac{1}{\Theta}$$

$$P^{\text{coll}} = \frac{1}{(\alpha-1)\beta}$$

$$P^{\text{Bayes}} = \frac{1/\beta + \sum_{j=1}^{r} X_j}{\alpha+n-1}.$$

Note that $P^{\text{coll}}$ is the expected value of a $\text{Lomax}(\alpha, \beta^{-1})$ distribution, which is the predictive distribution associated with the exponential–gamma mixture. The Bayes premium can again be expressed as a linear combination

$$P^{\text{Bayes}} = \lambda \overline{X}. + (1 - \lambda)P^{\text{coll}},$$

where

$$\lambda = \frac{n}{n + \alpha - 1}.$$

Note that if we instead specified the mixture model as $X \sim \text{Exp}(\Theta)$ and $\Theta \sim \text{Gamma}(\alpha, \beta)$, using the scale parameter as $\Theta$, the posterior distribution on $\theta$, given a set of data $X_1, \ldots, X_n$, would be proportional to

$$p(\Theta|\mathcal{D}) \propto \frac{e^{-x/\theta} \theta^{\alpha-2} e^{-\theta/\beta}}{\Gamma(\alpha)\beta^{\alpha}}.$$

The constant of proportionality must be found by integrating the right-hand side over $\theta$, which does not have a simple analytical expression. As a result, it is clear the Bayes' premium will not have a simple expression.  ▷

In each of the previous examples, we saw how the Bayes' premium can be expressed as a linear combination of the collective premium (i.e., the *prior*) and experience (i.e., an individual user's transactional history). In both of these cases, we were able to express the Bayes' premium as

$$P^{\text{Bayes}} = \lambda \overline{X} + (1 - \lambda)P^{\text{coll}},$$
$$\lambda = \frac{n}{n + \kappa},$$

for an appropriate choice of the constant $\kappa$. It turns out that the Bayes' premium follows this format whenever the sampling distribution is within the exponential family and the collective distribution is its conjugate prior. When that is not true, however, this simple linear combination fails to hold. In practice we are often interested in cases in which one has no right to assume that the collective distribution to be the conjugate prior of the sampling distribution. In such cases the Bayes' premium can quickly become difficult to express analytically. This problem is solved within the elegant field of credibility theory.

### 10.4.2 Bühlmann–Straub Model

In order to compute the Bayes' premium, once must specify both the sampling distribution as well as the mixing distribution. These distributions are often difficult to infer in practice. Even if a suitable class of distributions can be worked out, the resulting formula are often complex and unwieldy. The idea behind credibility theory is to determine a suitable estimator that is simple and intuitive, and that makes no suppositions on the form of the sampling or collective distributions. The requirement of simplicity is forced by restricting the class of estimators to those estimators that are linear in the data. The result is known as the *credibility estimator*, and it is essentially a linear Bayes estimator.

## Simple Credibility Model

Credibility theory was developed with two requirements in mind: simplicity and structure. (Bühlmann and Gisler [2005]). The requirement of simplicity means it should be straightforward for practitioners to apply and easy to interpret. The requirement of structure means that the model should not require prior knowledge of the specific form of the sampling distribution or the collective distribution. Bühlmann and Gisler [2005] states that specification of these models is, in practice, either artificial (if the distributional families are too large) or not helpful (if they are too narrow). Thus the model was developed without the requirement of specifying the specific distributional forms, instead relying on their first few moments. In order to describe the second moments mathematically, we require one additional definition before introducing the simple credibility model.

**Definition 10.17.** *Given a mixture model and the risk premiums defined in Definition 10.16, we define the individual risk variance, expected individual variance, and collective variance as*

$$\sigma^2(\Theta) = \mathbb{V}(X|\Theta) \tag{10.87}$$
$$\sigma^2 = \mathbb{E}[\sigma^2(\Theta)] \tag{10.88}$$
$$\tau^2 = \mathbb{V}(\mu(\Theta)), \tag{10.89}$$

*respectively. Moreover, the quotient $\kappa = \sigma^2/\tau^2$ is referred to as the credibility coefficient.*

For an individual with $\Theta = \theta$, the quantity $\sigma^2(\theta)$ represents the variance within that individual risk. The quantity $\sigma^2$ represents the average variance within an individual risk within the collective. And the quantity $\tau^2$ represents the variance between individual risk premiums. As an illustration, consider the case where $\tau^2$ is very large and $\sigma^2$ is very small. The average value for each individual is, in this case, spread out (large $\tau^2$), but the observations for a specific individual are tightly wound to each individual's average (small $\sigma^2$). In this case, therefore, it would be easy to identify individuals from the data.

*Example 10.13.* Consider again the Bernoulli–beta mixture from Example 10.10. We saw that the individual premium is given by

$$\mu(\theta) = \mathbb{E}[X|\Theta = \theta] = \theta.$$

Similarly, recall that

$$\sigma^2(\theta) = \mathbb{V}(X|\Theta = \theta) = \theta(1 - \theta),$$

so that we may compute

$$\sigma^2 = \mathbb{E}[\sigma^2(\Theta)] = \int_0^1 \theta(1-\theta)\mathrm{Beta}(\theta; \alpha, \beta)\, d\theta$$

$$= \frac{1}{B(\alpha, \beta)} \int_0^1 \theta^\alpha (1-\theta)^\beta\, d\theta$$

$$= \frac{B(\alpha+1, \beta+1)}{B(\alpha, \beta)}$$

$$= \frac{\alpha\beta}{(\alpha+\beta)(1+\alpha+\beta)}.$$

The last line follows from Exercise 10.10. It is interesting to compare this expression with

$$\sigma^2(\mathbb{E}[\Theta]) = \sigma^2 \left( \Theta = \frac{\alpha}{\alpha+\beta} \right) = \frac{\alpha\beta}{(\alpha+\beta)^2},$$

the variance in $X$ if $\theta = \mathbb{E}[\Theta] = \alpha/(\alpha+\beta)$. We see that the expected variance $\sigma^2$, marginalized over $\Theta$, is slightly less than the variance calculated at the expected value of $\Theta$.

Next, we find that

$$\tau = \mathbb{V}(\mu(\Theta)) = \mathbb{V}(\Theta) = \frac{\alpha\beta}{(\alpha+\beta)^2(1+\alpha+\beta)},$$

such that the credibility coefficient is given by

$$\kappa = \frac{\sigma^2}{\tau^2} = \alpha + \beta.$$

This makes sense in terms of the Bayesian interpretation of the hyperparameters of the Bernoulli–beta model: $\alpha$ represents the number of *virtual successes* and $\beta$ the number of *virtual failures*. The larger the sum $\alpha + \beta$, the more data are needed to overcome the prior. Moreover, when performing Bayesian updates, $\sigma^2 \to \sigma^2(\alpha/(\alpha+\beta))$ as $n \to \infty$, whereas $\tau \to 0$ as $n \to \infty$, again confirming the variance reduction aspects of a Bayesian update. In mixture models, e.g., credibility theory, the goal is not to shrink this variance, but to use the "prior" as a population distribution, which will instead converge to the variability of the population.                    ▷

*Example 10.14.* Consider the mixture model $X \sim \mathrm{Exp}(\Theta)$ and $\Theta \sim \mathrm{Gamma}(\alpha, \beta)$. *Note*: this is *not* the exponential–gamma mixture from Table 10.3, as this mixture is constructed using the *scale* parameter as opposed to the *rate* parameter. Nevertheless, we can compute the following quantities of interest:

$$P^{\mathrm{ind}} = \Theta$$
$$P^{\mathrm{coll}} = \alpha\beta$$
$$\sigma^2(\Theta) = \mathbb{V}(X|\Theta) = \Theta^2$$
$$\sigma^2 = \mathbb{E}[\Theta^2] = (\alpha+1)\alpha\beta^2$$
$$\tau^2 = \mathbb{V}(P^{\mathrm{ind}}) = \mathbb{V}(\Theta) = \mathbb{E}[\Theta^2] - \mathbb{E}[\Theta] = (\alpha+1)\alpha\beta^2 - \alpha^2\beta^2 = \alpha\beta^2.$$

In particular, the credibility coefficient is given by $\kappa = \sigma^2/\tau^2 = 1 + \alpha$. In particular, since $\kappa > 1$, the individual variation is always larger than the variation in the individual means.                                                                                    $\triangleright$

Before defining the credibility estimator, we need a basic result that relates the variance of a sample mean to the variances of Definition 10.17.

**Theorem 10.2.** *Let* $X_1, \ldots, X_n \sim \text{Mix}(F_\Theta, U)$ *be* IID *samples from the mixture distribution* $X_i \sim F_\Theta$, $\Theta \sim U$. *Then the variance of the sample mean is given by*

$$\mathbb{V}(\overline{X}) = \frac{\sigma^2}{n} + \tau^2, \tag{10.90}$$

*where* $\sigma^2$ *and* $\tau^2$ *are defined in Equations* (10.88) *and* (10.89), *respectively.*

*Proof.* From the law of total variance (Equation (2.58)), we have

$$\mathbb{V}(\overline{X}) = \mathbb{E}\left[\mathbb{V}(\overline{X}|\Theta)\right] + \mathbb{V}\left(\mathbb{E}[\overline{X}|\Theta]\right).$$

However, from Equation (1.23), we have

$$\mathbb{E}[\overline{X}|\Theta] = \mathbb{E}[X|\Theta] = \mu(\Theta)$$
$$\mathbb{V}(\overline{X}|\Theta) = \frac{\mathbb{V}(X|\Theta)}{n} = \frac{\sigma^2(\Theta)}{n}.$$

Therefore,

$$\mathbb{V}(\overline{X}) = \frac{\mathbb{E}\left[\sigma^2(\Theta)\right]}{n} + \mathbb{V}\left(\mu(\Theta)\right),$$

and the result follows.                                                                                    $\square$

Now we are in a position to define what we mean by the credibility estimator and prove an important result.

**Definition 10.18.** *Given a mixture model, the* credibility estimator $P^{\text{cred}}$ *or* $\hat{\mu}(\Theta)$ *is the unique estimator for the Bayes' premium that is linear in the observations*

$$P^{\text{cred}} = a_0 + \sum_{j=1}^{r} a_j X_j$$

*and that minimizes the expected quadratic loss*

$$\mathbb{E}\left[\left(\mu(\Theta) - P^{\text{cred}}\right)^2\right],$$

*where* $X_1, \ldots, X_r$ *constitute the observed values for a specific individual within the collective.*

   *The* homogeneous credibility estimator $P^{\text{hom}}$ *or* $\hat{\mu}^h(\Theta)$ *is the unique credibility estimator with the restriction that* $a_0 = 0$.

The credibility estimator can also be viewed as an orthogonal projection in the Hilbert space of square integrable functions (Bühlmann and Gisler [2005]), but for our purposes, it suffices to view it as the unique linear estimator that minimizes the mean-squared error. The homogeneous credibility estimator is more relevant in considering data from a collective, so we will defer discussion until the next section on the Bühlmann–Straub model. Definition 10.18 is, however, never applied directly, as we have the following fundamental result.

**Theorem 10.3.** *The credibility estimator for an individual risk premium in a mixture model is given by*

$$\hat{\mu}(\Theta) = \lambda \overline{X} + (1 - \lambda)\mu_0, \tag{10.91}$$

*where the* credibility weight $\lambda$ *is given by*

$$\lambda = \frac{n}{n + \sigma^2/\tau^2}. \tag{10.92}$$

*Moreover, the quotient* $\kappa = \sigma^2/\tau^2$ *is known as the* credibility coefficient.

*Proof.* Since the observed data $X_1, \ldots, X_r$ for an individual are independent and identically distributed, they are therefore invariant relative to permutations. Thus the coefficients $a_1, \ldots, a_r$ must be equal, and the linearity requirement of the credibility estimator can be reduced to

$$P^{\text{cred}} = a + b\overline{X}.$$

Next, let us consider the loss function

$$L(a, b) = \mathbb{E}\left[\left(\mu(\Theta) - a - b\overline{X}\right)^2\right].$$

Taking partial derivatives with respect to the coefficients $a$ and $b$ yields the equations

$$\mathbb{E}[\mu(\Theta) - a - b\overline{X}] = 0$$
$$\text{COV}(\overline{X}, \mu(\Theta)) - b\mathbb{V}(\overline{X}) = 0.$$

However, we have that

$$\text{COV}(\overline{X}, \mu(\Theta)) = \mathbb{V}(\mu(\Theta)) = \tau^2$$
$$\mathbb{V}(\overline{X}) = \frac{\sigma^2}{n} + \tau^2,$$

where the second equality follows from Theorem 10.2. Inserting these expressions into the partial derivative equations and solving yield the results

$$b = \frac{n}{n + \sigma^2/\tau^2}$$
$$a = (1 - b)\mu_0,$$

which proves the result. $\qquad\qquad\square$

*Note 10.12 (Credibility Rule of Thumb).* The credibility coefficient $\kappa$ thus has the following interpretation: when the sample size $n$ is equal to $\kappa$, the credibility is exactly 50%. Therefore, $\kappa$ represents the required sample size in order for the prior knowledge to balance with the experience in the credibility estimator. Similarly, when $n = 3\kappa$, the credibility is 75%.    ▷

*Note 10.13 (An Intuitive Principle).* The credibility estimator can also be understood intuitively as follows as a weighted average of the collective premium $P^{\mathrm{coll}} = \mu_0$ and the experience $\overline{X}$. In particular, the collective premium is the best *a priori* estimator, which has quadratic loss

$$\mathbb{E}\left[(\mu_0 - \mu(\Theta))^2\right] = \mathbb{V}(\mu(\Theta)) = \tau^2.$$

Similarly, the experience $\overline{X}$ is the best possible linear and individually unbiased estimator based on an individual's data, and has quadratic loss

$$\mathbb{E}\left[(\overline{X} - \mu(\Theta))^2\right] = \mathbb{E}[\sigma^2(\Theta)/n] = \frac{\sigma^2}{n}.$$

Finally, the credibility estimator is a weighted mean of these, where the weights are proportional to the inverse quadratic loss (precision) associated with each of the two components. This last statement can be realized by rewriting the credibility weight in the equivalent form

$$\lambda = \frac{n/\sigma^2}{n/\sigma^2 + 1/\tau^2},$$

so that the weight on the experience is proportional to $n/\sigma^2$ and the weight on the prior knowledge is proportional to $1/\tau^2$.    ▷

*Example 10.15.* Recall that the mixture $X \sim \mathrm{Exp}(\Theta)$ and $\Theta \sim \mathrm{Gamma}(\alpha, \beta)$ does not possess a simple expression for the Bayes' premium. Nevertheless, the credibility estimator is given by

$$P^{\mathrm{cred}} = \lambda \overline{X} + (1 - \lambda) P^{\mathrm{coll}},$$

where

$$\lambda = \frac{n}{n + \alpha + 1},$$

as we have previously computed the credibility coefficient $\kappa$ in Example 10.14.    ▷

## The Bühlmann–Straub Model

So far, we have discussed the simple credibility estimator as applied to a specific individual. We next generalize this model to account for observations from multiple individuals from the collective.

**Definition 10.19 (Bühlmann–Straub Model).** *The* Bühlmann–Straub
model *is a continuous mixture model for which the data are collected relative to specific individuals with given weights; i.e., data are of the form*
$(X_{ij}, w_{ij})$, *for individual* $i = 1, \ldots, m$ *and observation* $j = 1, \ldots, r$, *where*
$X_{ij}$ *is the observed value and* $w_{ij}$ *is its associated weight (given).*

Thus, individual $i$ has a risk profile $\Theta_i \sim U$, drawn randomly from the
collective distribution. Conditional on the individual $\Theta_i$, we therefore have

$$\mathbb{E}[X_{ij}|\Theta_i] = \mu(\Theta_i)$$
$$\mathbb{V}(X_{ij}|\Theta_i) = \frac{\sigma^2(\Theta_i)}{w_{ij}}.$$

We will follow our standard dot-summation notation, that

$$w_{i\cdot} = \sum_{j=1}^{r} w_{ij} \quad \text{and} \quad w_{\cdot\cdot} = \sum_{i=1}^{m} \sum_{j=1}^{r} w_{ij}.$$

*Note 10.14.* The *Bühlmann model,* which predates the Bühlmann–Straub
model, is a special case of the latter, in which the weights $w_{ij}$ are set
to equal one; i.e., $w_{ij} = 1$. The Bühlmann model is treated directly in
Bühlmann and Gisler [2005].                                          ▷

**Credibility in the Bühlmann–Straub Model**

In the Bühlmann–Straub model, the conditional individual variances are
inversely proportional to the weights. It follows that the best linear estimator (unbiased estimator with the minimum conditional variance) that is
obtainable from the data alone is the weighted average of the observations,
i.e.,

$$\overline{X}_{i\cdot} = \frac{1}{w_{i\cdot}} \sum_{j=1}^{r} w_{ij} X_{ij}. \tag{10.93}$$

**Theorem 10.4.** *The credibility estimator for the* $i$th *individual in the*
*Bühlmann–Straub model is given by*

$$\hat{\mu}(\Theta_i) = \lambda_i \overline{X}_{i\cdot} + (1 - \lambda_i)\mu_0, \tag{10.94}$$
$$\lambda_i = \frac{w_{i\cdot}}{w_{i\cdot} + \sigma^2/\tau^2}, \tag{10.95}$$

*where* $\mu_0$, $\sigma^2$, *and* $\tau^2$ *are defined as before.*

*Moreover, the homogeneous credibility estimator in the Bühlmann–*
*Straub model is given by*

$$\hat{\mu}^h(\Theta_i) = \lambda_i \overline{X}_{i\cdot} + (1 - \lambda_i)\hat{\mu}_0, \tag{10.96}$$

*where $\hat{\mu}_0$ is given by*

$$\hat{\mu}_0 = \frac{1}{\lambda.} \sum_{i=1}^{m} \lambda_i X_i, \tag{10.97}$$

*where $\lambda. = \sum_{i=1}^{m} \lambda_i$, as usual.*

We leave the proof to Bühlmann and Gisler [2005].

Note that the homogeneous credibility estimator replaces the collective premium $\mu_0$ with an estimate $\hat{\mu}_0$ which is formed by taking the *credibility-weighted average* of the individual means $\overline{X}_{i.}$. Note that this is *not* the weighted average

$$\overline{X} = \frac{1}{w..} \sum_{i=1}^{m} w_{i.} \overline{X}_{i.}$$

of all the data, but the *credibility-weighted* average

$$\hat{\mu}_0 = \frac{1}{\lambda.} \sum_{i=1}^{m} \lambda_i \overline{X}_{i.},$$

an important distinction. However, using the correct estimate for the collective premium, the homogeneous credibility estimator in the Bühlmann–Straub model has the following *balance property*

$$\sum_{i=1}^{m} \sum_{j=1}^{r} w_{ij} \hat{\mu}^h(\Theta_i) = \sum_{i=1}^{m} \sum_{j=1}^{r} w_{ij} X_{ij}.$$

We leave the proof of this fact to the interested reader.

*Example 10.16.* Consider again the mixture model $X \sim \text{Exp}(\Theta)$, with $\Theta \sim \text{Gamma}(\alpha, \beta)$. Let us consider a collective of ten random individuals, each with an known true parameter value $\Theta_i \sim \text{Gamma}(\alpha, \beta)$. Suppose the population has $\alpha = 4$ and $\beta = 5$. For these values of the population parameters, we have $\mu_0 = 20$, $\sigma^2 = 500$, and $\tau^2 = 100$, with $\kappa = 5$. A random data set for ten individuals is simulated in Code Block 10.10; the result is shown in Figure 10.10. The simulated data corresponding to the bottom half of Figure 10.10 are given in Table 10.9. Notice that, in both cases, the average individual's data is more spread out than the collection of individual means.

Since all data are weighted equally, each individual has the same credibility rating

$$\lambda = \frac{10}{10 + 5} = \frac{2}{3}.$$

The credibility premium for the $i$th individual is therefore obtained by

$$P_i^{\text{cred}} = \frac{2}{3} \overline{X}_{i.} + \frac{20}{3}.$$

```
 1  alpha, beta = 4, 5
 2  m, r = 10, 10
 3  thetas = np.random.gamma(alpha, scale=beta, size=m)
 4  thetas.sort()
 5
 6  X = np.zeros((m, r))
 7  for i in range(m):
 8      X[i, :] = np.random.exponential(scale=thetas[i], size=r)
 9
10  plt.figure(figsize=(8, 9/2))
11  for i in range(m):
12      color = default_colors[i]
13      plt.plot(thetas[i], 1, 'o', color=color)
14      plt.plot(X[i, :], np.zeros(r), '.', color=color)
15      for j in range(r):
16          plt.plot([thetas[i], X[i, j]], [1, 0], color=color)
17
18  # Credibility premiums
19  mu_0 = 20 # Or X.mean() for hom. estimator, since equally weighted.
20  la = 2/3
21  X_bar = X.mean(axis=1)
22  P_cred = la * X_bar + (1-la) * mu_0
23  mu_0_hat = X.mean()
24  sigma_2_hat = np.sum((X - X.mean(axis=1).reshape((m, 1)))**2) /
        (m*(r-1))
25  tau_2_hat = np.sum((X.mean(axis=1) - X.mean())**2) / (m-1) -
        sigma_2_hat / r
```

Code Block 10.10: Simulated data for individuals in an exponential–gamma mixture.

| $i$ | $\theta_i$ | $X_{i0}$ | $X_{i1}$ | $X_{i2}$ | $X_{i3}$ | $X_{i4}$ | $X_{i5}$ | $X_{i6}$ | $X_{i7}$ | $X_{i8}$ | $X_{i9}$ |
|---|---|---|---|---|---|---|---|---|---|---|---|
| 0 | 10.65 | 13.78 | 10.21 | 2.74 | 4.2 | 0.25 | 23.94 | 27.33 | 5.95 | 0.86 | 8.46 |
| 1 | 14.13 | 10.09 | 16.83 | 0.5 | 2.19 | 5.12 | 2.91 | 58.91 | 1.64 | 11.48 | 17.04 |
| 2 | 14.2 | 13.27 | 1.99 | 4.45 | 68.11 | 24.8 | 20.32 | 4.67 | 4.01 | 7.0 | 0.77 |
| 3 | 14.64 | 11.58 | 2.9 | 7.48 | 1.54 | 9.57 | 1.83 | 5.69 | 18.09 | 9.68 | 9.04 |
| 4 | 15.15 | 12.06 | 12.8 | 6.88 | 8.23 | 21.01 | 4.71 | 8.29 | 3.57 | 21.49 | 29.26 |
| 5 | 16.08 | 30.76 | 30.68 | 8.32 | 5.3 | 10.31 | 24.19 | 11.06 | 18.16 | 15.88 | 17.44 |
| 6 | 18.5 | 22.4 | 33.95 | 13.11 | 30.67 | 34.67 | 5.08 | 67.74 | 30.31 | 5.95 | 3.03 |
| 7 | 19.35 | 8.08 | 4.71 | 39.65 | 15.05 | 10.81 | 17.09 | 40.32 | 17.95 | 0.43 | 3.13 |
| 8 | 26.22 | 33.7 | 91.02 | 62.51 | 27.28 | 0.59 | 12.56 | 27.77 | 25.62 | 1.94 | 113.92 |
| 9 | 48.25 | 51.98 | 129.2 | 29.31 | 77.86 | 57.83 | 25.19 | 20.06 | 7.07 | 23.29 | 78.78 |

Table 10.9: Simulated data from Code Block 10.10.

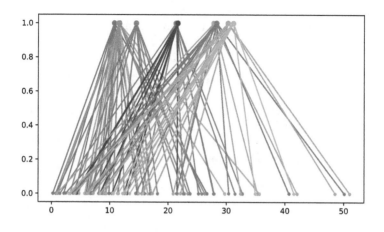

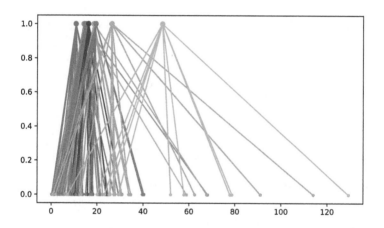

Fig. 10.10: Two visualizations of simulated data showing true individual risks $\theta_i$ (at $y = 1$) connected to each individual's simulated data (at $y = 0$). The bottom figure corresponds to the data shown in Figure 10.10.

| $i \rightarrow$ | 0 | 1 | 2 | 3 | 4 | 5 | 6 | 7 | 8 | 9 |
|---|---|---|---|---|---|---|---|---|---|---|
| $P_i^{\text{ind}}$ | 10.65 | 14.13 | 14.2 | 14.64 | 15.15 | 16.08 | 18.5 | 19.35 | 26.22 | 48.25 |
| $P_i^{\text{cred}}$ | 13.18 | 15.11 | 16.63 | 11.83 | 15.22 | 18.14 | 23.13 | 17.15 | 33.13 | 40.04 |

Table 10.10: True individual premium $P_i^{\text{ind}} = \theta_i$ and credibility estimate for each indivudal.

The true individual premium $\theta_i$ and credibility estimate are recorded in Table 10.10. If we did not know the collective premium $\mu_0$, we could estimate it as $\hat{\mu}_0 = 20.53$. Using this in the above computation would yield the homogeneous credibility estimate.                                                    $\triangleright$

**Estimates for Structural Parameters**

Naturally, it is often the case that the structural parameters $\sigma^2$ and $\tau^2$ are not known in advanced. The following result, originally due to Bühlmann and Straub [1970], but also discussed in Bühlmann and Gisler [2005], provides a natural estimator for these parameters.

**Theorem 10.5.** *The quantity*

$$\hat{\sigma}^2 = \frac{1}{m(r-1)} \sum_{i=1}^{m} \sum_{j=1}^{r} w_{ij} \left( X_{ij} - \overline{X}_{i\cdot} \right)^2 \tag{10.98}$$

*is an unbiased ($\mathbb{E}[\hat{\sigma}^2] = \sigma^2$) and consistent ($\hat{\sigma}^2$ converges in probability to $\sigma^2$ as $m \to \infty$) estimator for the structural parameter $\sigma^2$. Similarly, the quantity*

$$\hat{\tau}^2 = c \left[ \frac{m}{m-1} \sum_{i=1}^{m} \frac{w_{i\cdot}}{w_{\cdot\cdot}} (\overline{X}_{i\cdot} - \overline{X}_{\cdot\cdot})^2 - \frac{m\hat{\sigma}^2}{w_{\cdot\cdot}} \right], \tag{10.99}$$

*where*

$$c = \frac{m-1}{m} \left[ \sum_{i=1}^{m} \frac{w_{i\cdot}}{w_{\cdot\cdot}} \left( 1 - \frac{w_{i\cdot}}{w_{\cdot\cdot}} \right) \right]^{-1}, \tag{10.100}$$

*is an unbiased estimator for $\tau^2$, and consistent, as long as $\sum_{i=1}^{m} (w_{i\cdot}/w_{\cdot\cdot})^2 \to 0$ as $m \to \infty$; i.e., as long as no one individual is "dominating" as $m \to \infty$.*

*Note 10.15.* The means $\overline{X}_{i\cdot}$ and $\overline{X}_{\cdot\cdot}$ are both *weighted* means of the observed data.                                                    $\triangleright$

*Note 10.16.* For the case in which each individual has equal weight, i.e., $w_{1\cdot} = \cdots = w_{m\cdot}$, the constant $c = 1$.                                                    $\triangleright$

**Corollary 10.8.** *For the case in which each individual has equal weight, i.e., $w_{1\cdot} = \cdots = w_{m\cdot}$, the estimator $\hat{\tau}^2$ from Theorem 10.5 reduces to*

$$\hat{\tau}^2 = \frac{1}{m-1} \sum_{i=1}^{m} (\overline{X}_{i\cdot} - \overline{X}_{\cdot\cdot})^2 - \frac{\hat{\sigma}^2}{w_{i\cdot}}, \tag{10.101}$$

*where $w_{i\cdot} = w_{\cdot\cdot}/m$ is each individual's total weight.*

*Note 10.17.* The estimate $\hat{\sigma}^2$ given by Equation (10.98) should be compared with the mean within-group mean squares (Equations (3.66) and (3.68)). Similarly, the first term in the quantity $\hat{\tau}^2$ should be compared with the between-group mean squares (Equations (3.65) and (3.68)). The between-group mean squares MSW is an estimate for the variance of the population means. Thus, an estimate for $\sigma^2/n$ must be subtracted from it to obtain an estimate for $\tau^2$, due to Theorem 10.2.    ▷

*Note 10.18.* It is possible for the estimate $\hat{\tau}^2$ to be negative. Thus, it is better to replace the estimate $\hat{\tau}^2$ with

$$\hat{\tau}^2 \to \max(0, \hat{\tau}^2),$$

though this means that the estimate is no longer unbiased.    ▷

*Example 10.17.* Considering again the data from Table 10.9 in Example 10.16, we can compute the estimated structural parameters as $\hat{\sigma}^2 = 437$ and $\hat{\tau}^2 = 148$. These calculations are shown on the last two lines of Code Block 10.10. (Recall the true values are 500 and 100, respectively.) This results in the estimate $\hat{\kappa} = 2.95$, and would change the credibility weight to $\hat{\lambda} = 0.77$. Thus, for this specific example, the empirical credibility estimate would more closely favor the observed data.    ▷

## 10.5 Simulation

In practice, it is often difficult to analytically compute a posterior distribution for a given prior and a given set of data. Examples include non-conjugate priors or complex hierarchical models. Fortunately, there are a number of sophisticated techniques that allow us to sample the posterior distribution, providing robust estimates for many quantities of interest.

### 10.5.1 Markov Chain Monte Carlo Simulation

A *Monte Carlo method*, named for the Monte Carlo Casino in Monaco, is any computational algorithm that relies on random sampling to estimate a result. (For history of the early computational development of Monte Carlo methods in Los Alamos National Laboratory, see Metropolis [1987].) For example, to estimate the value of $\pi$, one could draw many random samples from the distribution Unif$([0,1] \times [0,1])$ and then compute the fraction of samples that satisfy $x^2 + y^2 < 1$. In the context of Bayesian statistics, a number of *Markov chain Monte Carlo methods* (MCMC) are often used. For such methods, the random samples are not independent, though they do satisfy the Markov property (Definition 5.12); i.e., any given random draw is a function of the past only through the immediately preceding instance.

## Gibbs Sampling

In many multidimensional Bayesian models, it is often difficult to sample from the joint posterior distribution directly, but not as difficult to sample from the conditional distribution for an individual or subset of parameters. Gibbs sampling is a particular MCMC algorithm designed for this situation. The Gibbs sampling algorithm is described in Algorithm 10.2.

---

**Algorithm 10.2:** Gibbs Sampling.

---

**Input:** A set of IID data $\mathcal{D} = Y_1, \ldots, Y_n \sim p(Y|\theta)$;
a partition $S$ of the parameter vector $\theta = (\theta_1, \ldots, \theta_d)$;
the conditional posterior distributions $p(\theta_i|\theta_{-i}, \mathcal{D})$, for
$i = 1, \ldots, d$;
an initial parameter value $\theta^0$;
number of desired samples $n$.

**Result:** Approximate sample $\theta^1, \ldots, \theta^n$ from $p(\theta|\mathcal{D})$.

1 **for** $t$ **from** $1$ **to** $n$ **do**
2     Generate a random permutation $\pi \in S_n$
3     **for** $j$ **from** $1$ **to** $d$ **do**
4         Draw a random
        $\theta^t_{\pi(j)} \sim p(\theta_{\pi(j)}|\theta^t_{\pi(1)}, \ldots, \theta^t_{\pi(j-1)}, \theta^{t-1}_{\pi(j+1)}, \ldots, \theta^{t-1}_{\pi(d)}, \mathcal{D})$
5     **end**
6 **end**
7 **return** $\theta^1, \ldots, \theta^n$.

---

For Gibbs sampling, a multidimensional parameter vector $\theta$ is broken into a set of $d$ subvectors, $\theta = (\theta_1, \ldots, \theta_d)$. Each sampling iteration is then carried out through a sequence of $d$ steps, where the $j$th subvector $\theta_j$ is sampled using the conditional posterior distribution, conditioned on the data and the most recent value of the remaining subvectors. Furthermore, the ordering is randomized at the beginning of each iteration. The final sample is not independent, as each sampled parameter $\theta^t$ depends on the previously sampled values, but only via the immediately preceding value $\theta^{t-1}$.

## Metropolis Algorithm

In considering Bayes' law in the form of Equation (10.2), the prior distribution $\mathbb{P}(\theta)$ and the likelihood $p(\mathcal{D}|\theta)$ are often easy enough to compute. The data distribution of the denominator,

$$p(\mathcal{D}) = \int p(\mathcal{D}|\theta)p(\theta)\, d\theta,$$

on the other hand, is often unwieldy and difficult to compute, especially analytically. Thus enters Metropolis. The Metropolis algorithm allows us to approximate a random sample from the posterior distribution $p(\theta|\mathcal{D})$

*without* the need for computing the marginal data distribution of the denominator. Before presenting the algorithm, we require the following definition.

**Definition 10.20.** *Given a parameter space $\Theta$, a* jumping distribution *(or* proposal distribution*) is any conditional distribution $J(\cdot|\cdot) : \Theta \times \Theta \to R_*$. A jumping distribution is said to be* symmetric *if $J(\theta|\phi) = J(\phi|\theta)$.*

Examples of symmetric jumping distributions include

$$J(\theta|\phi) = \text{Unif}([\phi - \delta, \phi + \delta])$$

and

$$J(\theta|\phi) = \text{N}(\phi, \delta^2),$$

where the parameter $\delta$ is a tuning parameter chosen so that the algorithm runs smoothly.

The Metropolis algorithm is a MCMC algorithm that determines a proposed parameter $\theta^*$ by sampling the jumping distribution and accepting or rejecting the proposed value based on the ratio of how likely the proposed value is compared with the likelihood of the current value. The details are presented in Algorithm 10.3.

---

**Algorithm 10.3:** Metropolis algorithm.

---

**Input:** A set of IID data $\mathcal{D} = Y_1, \ldots, Y_n \sim p(Y|\theta)$;
a prior distribution $p(\theta)$;
the likelihood function $p(\mathcal{D}|\theta)$;
a symmetric jumping distribution $J(\cdot|\cdot)$;
an initial parameter value $\theta^0$;
number of desired samples $n$.

**Result:** Approximate sample $\theta^1, \ldots, \theta^n$ from $p(\theta|\mathcal{D})$.

1 **for** $t$ **from** *1* **to** $n$ **do**

2     Sample $\theta^* \sim J(\theta|\theta^{t-1})$

3     Compute the acceptance ratio

$$r = \frac{p(\theta^*|\mathcal{D})}{p(\theta^{t-1}|\mathcal{D})} = \frac{p(\mathcal{D}|\theta^*)p(\theta^*)}{p(\mathcal{D}|\theta^{t-1})p(\theta^{t-1})} \tag{10.102}$$

    Set

$$\theta^t = \begin{cases} \theta^* & \text{with probability } \min(r, 1) \\ \theta^{t-1} & \text{with probability } 1 - \min(r, 1) \end{cases} \tag{10.103}$$

4 **end**

5 **return** $\theta^1, \ldots, \theta^n$.

---

## Metropolis–Hastings algorithm

The Metropolis–Hastings algorithm provides a subtle yet power generalization to the Metropolis algorithm, allowing for asymmetric jumping distributions. The asymmetry is corrected for in the acceptance ratio, as shown in Equation (10.105) of Algorithm 10.4.

---

**Algorithm 10.4:** Metropolis–Hastings algorithm.

**Input:**  A set of IID data $\mathcal{D} = Y_1, \ldots, Y_n \sim p(Y|\theta)$;
a prior distribution $p(\theta)$;
the likelihood function $p(\mathcal{D}|\theta)$;
a sequence of jumping distributions $J_t(\cdot|\cdot)$;
an initial parameter value $\theta^0$;
number of desired samples $n$.

**Result:**  Approximate sample $\theta^1, \ldots, \theta^n$ from $p(\theta|\mathcal{D})$.

1 **for** $t$ **from** $1$ **to** $n$ **do**
2     Sample $\theta^* \sim J(\theta|\theta^{t-1})$
3     Compute the acceptance ratio

$$r = \frac{p(\theta^*|\mathcal{D})/J_t(\theta^*|\theta^{t-1})}{p(\theta^{t-1}|\mathcal{D})/J_t(\theta^{t-1}|\theta^*)} \qquad (10.104)$$

    Set

$$\theta^t = \begin{cases} \theta^* & \text{with probability } \min(r,1) \\ \theta^{t-1} & \text{with probability } 1 - \min(r,1) \end{cases} . \qquad (10.105)$$

4 **end**
5 **return** $\theta^1, \ldots, \theta^n$.

---

It is interesting to note the Gibbs sampling method can also be derived as a special case of the Metropolis–Hastings algorithm.

### 10.5.2 PyMC3

The `pymc3` package (i.e., PyMC3) is an an advanced open-source Python package that implements many MCMC algorithms for us. This section is devoted to reviewing the package syntax and usage guide. An excellent resource on this package's predecessor, `pymc`, is Davidson-Pilon [2016]. Despite some changes in the syntax and usage of the software (PyMC3 was rebuilt from ground up), we still recommend this as an excellent resource for probabilisitic programming. A beginning guide for PyMC3 can be found in Salvatier, *et al.* [2016]. Additional documentation can be found on the developer's website `docs.pymc.io`.

*Note 10.19.* PyMC3 is built on top of Theano[4]. Variables in the context of PyMC3 are actually Theano variables. Theano, however, is no longer cur-

---

[4] `theano.readthedocs.io`.

rently maintained, so PyMC3 has imported much of Theano's functionality into its own libraries. As a result, many of the standard operators ($\leq$, $>$, if) do not work when defining new variables within PyMC3. Instead, the pymc3.math library should be used, which is based off the theano.tensor package. Where documentation in PyMC3 falls short, consult the Theano documentation.                                                                   ▷

### Variables and Models in PyMC3

PyMC3 has an extensive library of built-in random variables. The constructor for each consists of three primary components:

1. a name (a string), which is usually the same as the name of the Python variable where we are storing our object;
2. one or more (keyword) arguments that specify the distribution; these can be actual values or other random variables, in the case of hierarchical models; and
3. a number of optional keyword arguments (shape and observed; more on this later) that are universal across all PyMC3 variables.

Moreover, a PyMC3 variable cannot be defined outside of a *model context*. In other words, a PyMC3 model can be thought of as a collection of variables, and it is defined using a with statement. For example, in Code Block 10.11, we open a new model context, naming it model, and we define several random variables. Note that the first argument used to construct each variable is that variable's name.

```
import pymc3 as pm

with pm.Model() as model:
    x = pm.Normal("x", mu=0, std=2)
    y = pm.Normal("y", mu=3, std=2)
    z = pm.Deterministic("z", x+y)
```

Code Block 10.11: Constructing variables in PyMC3.

The variables x and y in Code Block 10.11 are normal random variables with different means (0 and 3), but the same standard deviation (2). The variable z is equivalent to the expression z = x + y, except we are tracking this variable in the context of our model.

In addition, there are two distinct types of PyMC3 variables:

1. *Stochastic*: Any variable that is not uniquely determined by its parents;
2. *Deterministic*: Any variable that is uniquely determined by its parents.

For example, the variables x and y in Code Block 10.11 are stochastic, because we do not know the next sample value, even such that we know mu and std. Similarly, the variable z is deterministic; if we know the values of both x and y, then we know exactly the value of the variable z.

*Note 10.20.* Both stochastic and deterministic variables are *random* variables!                                                                                    ▷

*Note 10.21.* At the present, PyMC3 does not support the observed keyword for deterministic variables. The Deterministic wrapper merely adds the variable to the final trace, so that we can view the posterior distribution of our deterministic random variables.

Therefore, *only stochastic variables can be observed*!

This means, in particular, that we cannot model our error term as a separate term, but must model it within the observed variable. For example, consider the two equivalent linear models

$$Y = a_1 X_1 + a_1 X_2 + a_3 X_3 + \epsilon,$$
$$\epsilon \sim N(0, \sigma^2),$$

and

$$Y \sim N(a_1 X_1 + a_1 X_2 + a_3 X_3, \sigma^2).$$

PyMC3 supports the latter, not the former. If we want to examine the posterior of our error term, we should, instead, define the error term as a deterministic variable

$$\epsilon = Y - a_1 X_1 + a_1 X_2 + a_3 X_3,$$

so that we can view its posterior trace.                                            ▷

In our definition of a causal structural model, the endogenous variables were deterministic, whereas the exogenous variables were (typically) stochastic. Thus a causal model can only be captured in PyMC3 with an appropriate modification, so that the endogenous variables are stochastic.

Finally, there are two optional keywords that can be used to construct PyMC3 variables. The first is shape (int), which can be used if we want to specify a random *vector* instead of a random variable. The second is observed (array), which takes as input any data set of actual observations. By passing a set of data through the observed keyword argument, we are implicitly defining our likelihood function, and fixing those values as actual observations of the model.

The cool thing is that a model, in PyMC3, is just a bag of variables, and the model's graph is inferred by the variable definition. This allows us to define complex Bayesian models with some ease.

*Example 10.18.* Let us consider normally distributed data with unknown mean $\mu$ and variance $\sigma^2$. We simulate such a data set on lines 2–4 of Code Block 10.12, from a population with true mean 12 and true standard deviation 4. The sample mean of our simulated data is 11.38, with sample standard deviation 4.55.

Aside from the sampling statistics, we have a prior belief in the mean and variance of the population. Suppose we have prior reason to suspect the mean to be around 8, and that we are 95% confident the mean is actually between 0 and 16. Additionally, we know that the standard deviation is less than ten, but don't have any feeling for whether it is on the low or high side. We can therefore construct a Bayesian hierarchical model as

$$X \sim N(\mu, \sigma^2),$$
$$\mu \sim N(8, 16),$$
$$\sigma \sim \text{Unif}(0, 10),$$

where we have used a normal and uniform prior for $\mu$ and $\sigma$, respectively.

```
1  # Simulate data.
2  mu, sigma = 12, 4
3  n = 20
4  data = np.random.normal(mu, sigma, size=n) # 11.38, 4.55
5
6  # Build a PyMC3 model
7  with pm.Model() as normal_model:
8      mu = pm.Normal("mu", mu=8, sd=4)
9      sigma = pm.Uniform("sigma", lower=0, upper=10)
10
11     x = pm.Normal("x", mu=mu, sd=sigma, observed=data)
```

Code Block 10.12: Simulation and PyMC3 model for Example 10.18.

Our model is easily encoded in lines 7–11 of Code Block 10.12. Note that mu, sigma, and x are each PyMC3 stochastic variables, as each have a random component that prevents one from deterministically specifying the outputs. In addition, the priors mu and sigma are the parents of the variable x, which, since we pass our data set into the constructor using the observed keyword, defines our likelihood function.    ▷

## Models and Sampling in PyMC3

*Example 10.19.* We conclude Example 10.18 with an implementation of the MCMC algorithm. To do this, we reopen our normal_model context and call the PyMC3 sample method, as shown in Code Block 10.13.

```
1   with normal_model:
2       idata = pm.sample(2000, tune=1500, return_inferencedata=True)
3
4   pm.plot_trace(idata)
5
6   idata.posterior["mu"].shape
7   idata.posterior["sigma"].shape
```

Code Block 10.13: Deploying the MCMC algorithm for Examples 10.18 and 10.19.

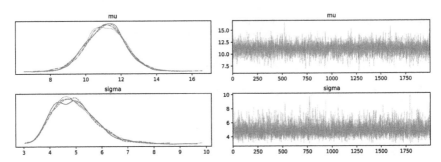

Fig. 10.11: Plot trace of posterior distributions for Example 10.19. Left: smoothed histogram (using kernel density estimation) for each of the four parallel simulations. Right: samples from MCMC plotted in sequential order.

PyMC3 also has a built-in method called plot_trace, which generates the plots shown in Figure 10.11. These contain a smoothed histogram for our posterior distributions for $\mu$ and $\sigma$ (left) and the individual samples plotted sequentially (right). The latter is a good visual check that our model has converged.    ▷

**A Bayesian Switchpoint Example**

To illustrate these ideas, we will follow an example from Davidson-Pilon [2016] (with a slightly modified story).

*Example 10.20.* Suppose we are running an email marketing campaign through an advertiser that sends out a certain number of emails per day, depending on our budget, bid, and market pressure. We have run our campaign for sixty days, and are analyzing the number of emails sent out per day, as shown in Figure 10.12. (We generated the data using Code Block 10.14.)

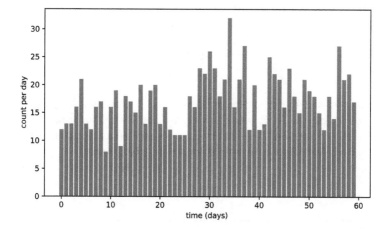

Fig. 10.12: Marketing Emails per day, in thousands.

```
lal, la2 = 15, 20
tau = 27
n = 60

data = np.zeros(n)
data[:tau] = random.poisson(la1, size=tau)
data[tau:] = random.poisson(la2, size=n-tau)
```

Code Block 10.14: Generating random marketing data for Example 10.20

From a visual inspection, it appears as though there was an increase in volume that occurred around day 28. We will use a Bayesian model to validate this observation and quantify the perceived lift.

We will model our data set using a *switchpoint model*; i.e., we will assume there is some time $\tau$ that divides our data into a before and after sets constituting time periods of different behavior. This can be captured by the following model:

$$X_t \sim \text{Poiss}(\lambda_t)$$
$$\lambda_t = \begin{cases} \lambda_1 & \text{if } t < \tau \\ \lambda_2 & \text{if } t \geq \tau \end{cases},$$

for some $\lambda_1$, $\lambda_2$, and $\tau$. Since we are modeling counts, we have used a Poisson random variable, with rate parameters $\lambda_1$ and $\lambda_2$. Of course, if the behavior is the same over the full period, we would have $\lambda_1 = \lambda_2$. The parameter $\tau$ represents the *switchpoint*, which divides the interval $[0, 60]$ into the two separate regions.

Since we do not know, a priori, the values of $\lambda_1$, $\lambda_2$, and $\tau$, we will specify the following prior distributions:

$$\lambda_1 \sim \text{Exp}(\alpha),$$
$$\lambda_2 \sim \text{Exp}(\alpha),$$
$$\tau \sim \text{Unif}(0, 60).$$

Finally, we set the hyperparameter $\alpha = \overline{X}_n$ to our sample mean, since $\mathbb{E}[\lambda|\alpha] = \alpha$. These are fairly weak priors, as they do not force too much information into our system. Moreover, since we are using numerical methods, we do not care whether or not these are conjugate priors.

We specify and train our model in Code Block 10.15 (using the data set generated from Code Block 10.14).

```
1   with pm.Model() as model:
2
3       lambda_1 = pm.Exponential("lambda_1", 1/data.mean())
4       lambda_2 = pm.Exponential("lambda_2", 1/data.mean())
5       tau = pm.DiscreteUniform("tau", lower=0, upper=n)
6
7       lambda_ = pm.math.switch(np.arange(n) < tau, lambda_1, lambda_2)
8       observation = pm.Poisson("observation", lambda_, observed=data)
9
10      step = pm.Metropolis()
11
12      trace = pm.sample(10000, tune=5000, step=step,
                return_inferencedata=True)
13
14  % mean of posterior samples
15  trace.posterior['lambda_1'].mean()
16  trace.posterior['lambda_2'].mean()
17  trace.posterior['tau'].mean()
```

Code Block 10.15: MCMC for switch point problem of Example 10.20.

The average values of the posterior samples from our simulation are $\lambda_1 = 14.8$, $\lambda_2 = 19.6$, and $\tau = 27.0$. We plot the stack trace from our simulation in Figure 10.13, which shows our posterior uncertainty distribution across each of the three hyperparameters. It is clear that the switch point occurs around $\tau = 27$, and that the Poisson rates before and after are closely grouped around 15 and 20, respectively. It is clear from our visual inspection that $\lambda_1$ and $\lambda_2$ have a negligible overlap.                    ▷

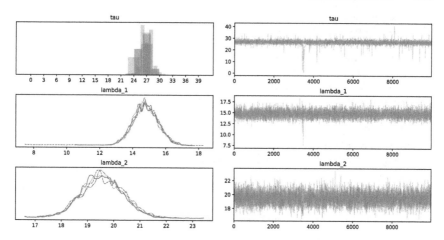

Fig. 10.13: Plot trace of posterior distributions for switchpoint Example 10.20. Left: smoothed histogram (using kernel density estimation) for each of the four parallel simulations. Right: samples from MCMC plotted in sequential order.

## A Bayesian Causality Example

*Example 10.21.* Let us return to the counterfactual calculation of the rain–sprinkler–wet pavement problem of Example 8.16. In this example, we computed the counterfactual

$$\mathbb{P}(W|\mathrm{do}(S)\mathrm{if}(R, \neg S, \neg W)),$$

which represents the probability of finding wet pavement in our counterfactual reality in which we intervened to ensure that the sprinkler was running, given that, in our actual reality it rained, the sprinkler was not running, and the pavement was not wet. In defining counterfactuals, we first implement our abduction (Definition 8.23), then our intervention, and then our prediction. We know see that our abduction is equivalent to a Bayesian update on our probability distribution $p(u)$ defined over the exogenous variables in $\mathcal{U}$. In our present example, we compute the above counterfactual quantity using PyMC3.

Our basic model is captured by Code Block 10.16. We begin by defining our three exogenous variables as uniform random variables over the interval $[0, 1]$. We then define our functional relationships $f_R$, $f_S$, and $f_W$. Now, ideally we could simply let these three variables be observed, and pass in the values 1, 0, and 0, for our if$(R, \neg S, \neg W)$ condition. The problem, however, is that these are deterministic variables, and therefore cannot be observed in PyMC3. To remedy this situation, we introduce a small error term, `err`, and create three Bernoulli random variables R, S, and W. Thus,

```
1   from pymc3.math import switch
2   with pm.Model() as cf_model:
3       UR = pm.Uniform("UR", 0, 1)
4       US = pm.Uniform("US", 0, 1)
5       UW = pm.Uniform("UW", 0, 1)
6       err = 0.001
7
8       pR = 0.3
9       fR = switch(UR<pR, 1-err, err)
10      R = pm.Bernoulli("R", p=fR, observed=[1])
11
12      pS = switch(eq(R,1), 0.1, 0.3)
13      fS = switch(US<pS, 1-err, err)
14      S = pm.Bernoulli("S", p=fS, observed=[0])
15
16      pW = switch(eq(R,1),
17               switch(eq(S,1), 0.9, 0.85),
18               switch(eq(S,1), 0.1, 0)
19               )
20      fW = switch(UW<pW, 1-err, err)
21      W = pm.Bernoulli("W", p=fW, observed=[0])
```

Code Block 10.16: PyMC3 encoding of the Rain–Sprinkler–Wet causal model of Example 10.21.

we allow for a small, but nonzero, fraction for which our causal model yields an incorrect result.

By conditioning on R, S, and W with the observed keyword, PyMC3's sample method will run an MCMC algorithm to simulate a sample for the posterior distributions of $U_R$, $U_S$, and $U_W$ under the abduction if $(R, \neg S, \neg W)$. Since the counterfactual reality springs from the posterior distribution over $\mathcal{U}$, we can copy and paste a counterfactual version of the model, and append it to the original, as shown in Code Block 10.17. It is also within this counterfactual reality that we make the intervention $do(S)$, which is captured on line 5. Note that our counterfactual random variables R_cf and W_cf are deterministic, so that PyMC3 records their trace under the posterior distribution on $\mathcal{U}$.

Finally, we run the model using the pm.sample method. The traceplot is shown in Figure 10.14. Note that the posterior distributions for $U_R$, $U_S$, and $U_W$ are consistent with the ones computed by hand in Example 8.16. Moreover, our simulation yields the final counterfactual probability of wet pavement very close to $1/3$, our prior result. This can be verified by accessing the posterior trace data for W_cf, as done in line 16, yielding a simulated result of 0.33275.                                              ▷

```
 1   with cf_model:
 2      R_cf = pm.Deterministic("R_cf", switch(UR<pR, 1, 0))
 3
 4      # Intervention
 5      S_cf = 1
 6
 7      pW_cf = switch(eq(R_cf,1),
 8                  switch(eq(S_cf,1), 0.9, 0.85),
 9                  switch(eq(S_cf,1), 0.1, 0)
10                  )
11      W_cf = pm.Deterministic("W_cf", switch(UW<pW_cf, 1, 0))
12
13   # Run the model.
14   with cf_model:
15      idata = pm.sample(2000, tune=1500, return_inferencedata=True)
16   idata.posterior["W_cf"].mean() # 0.33275
```

Code Block 10.17: PyMC3 encoding of the Rain–Sprinkler–Wet causal model of Example 10.21.

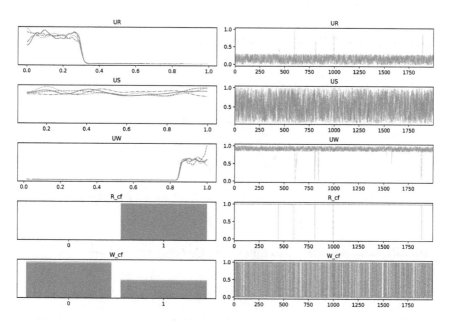

Fig. 10.14: The Rain–Sprinkler–Wet causal model of Example 10.21.

## Problems

**10.1.** Determine the probability that Joe is afflicted with the medical condition of Example 10.1 if he receives *two* positive test results.

**10.2.** For Example 10.6, show that the prior predictive distribution *for the number of successes K out of the next n results* is given by

$$p(K) = \text{BetaBin}(K; n, \alpha, \beta).$$

**10.3.** Let

$$X \sim \text{Poiss}(\lambda t),$$
$$\lambda \sim \text{Gamma}(\alpha, \beta).$$

Show that

$$p(X|\alpha, \beta) = \text{NBD}(x; \alpha, (1 + \beta t)^{-1}).$$

**10.4.** Prove that

$$B(\gamma + 1, \delta + n) + B(\gamma, \delta + n + 1) = B(\gamma, \delta + n)$$

**10.5.** In the proof of Proposition 10.11, show that

$$\mathbb{E}[X(n)|p, \theta] = p \sum_{t=1}^{n} (1 - \theta)^n = \frac{p(1 - \theta)}{\theta} - \frac{p(1 - \theta)^{n+1}}{\theta}.$$

*Hint*: recall the formula for a finite geometric series

$$\sum_{i=0}^{n-1} ar^i = \frac{a(1 - r^n)}{1 - r},$$

which, of course, goes to $a/(1 - r)$ in the limit as $n \to \infty$, for $|r| < 1$.

**10.6.** Prove Equations (10.81) and (10.82).

**10.7.** Prove Equations (10.57) and (10.58). *Hint*: Recall that

$$\frac{\partial B(x, y)}{\partial x} = B(x, y) \left( \frac{\Gamma'(x)}{\Gamma(x)} + \frac{\Gamma'(x + y)}{\Gamma(x + y)} \right),$$

and that $\psi(x) = \Gamma'(x)/\Gamma(x)$. *Hint*: The beta function is also symmetric.

**10.8.** Use Code Block 10.6, along with Code Block 5.7, to solve for the BG/BB parameters $\alpha, \beta, \gamma, \delta$ using stochastic gradient descent.

**10.9. (OPEN)** Is there a simple closed-form expression for $\mathbb{E}[\Theta]$ and $\mathbb{V}(\Theta)$ in terms of $\nu$, $\eta$, and $\phi$ for the conjugate prior Equation (10.15) of a general EDM?

**10.10.** Use the identity

$$B(\alpha + 1, \beta) = B(\alpha, \beta)\frac{\alpha}{\alpha + \beta}$$

and the symmetry of the beta function to show that

$$\frac{B(\alpha + 1, \beta + 1)}{B(\alpha, \beta)} = \frac{\alpha\beta}{(\alpha + \beta)(1 + \alpha + \beta)}.$$

# A

# Distributions

## A.1 Set Notation

We use the following conventions for standard sets:

$$\mathbb{Z} = \text{set of integers}$$
$$\mathbb{R} = \text{set of real numbers}$$
$$\mathbb{N} = \mathbb{Z}_* = \{n \in \mathbb{Z} : n \geq 0\}$$
$$\mathbb{Z}^+ = \{n \in \mathbb{Z} : n \geq 1\}$$
$$\mathbb{R}_* = \{x \in \mathbb{R} : x \geq 0\}$$
$$\mathbb{R}^+ = \{x \in \mathbb{R} : x > 0\}.$$

## A.2 Simulation

With few exceptions, random samples are computed using the `np.random` package and distributions are computed using the `scipy.stats` package.

```
from numpy import random
from scipy import stats

samples = np.random.exponential(0.2, size=n)
```

Code Block A.1: Example usage of random samples and distributions

## A.3 Discrete Distributions

## A.4 Continuous Distributions

| name | support | $p(x)$ | $S(x)$ | $\mathbb{E}[X]$ | $\mathbb{V}(X)$ | simulation |
|---|---|---|---|---|---|---|
| Bernoulli Bern($p$) | $\{0,1\}$ | $p^x(1-p)^{1-x}$ | — | $p$ | $p(1-p)$ | `random.binomial(1, p)` |
| binomial Binom($n,p$) | $\{0,\ldots,n\}$ | $\binom{n}{x}p^x(1-p)^{n-x}$ | | $np$ | $np(1-p)$ | `random.binomial(n,p)` `stats.binom(n,p)` |
| negative binomial NBD($r,p$) | $\mathbb{N}$ | $\dfrac{\Gamma(x+r)}{\Gamma(r)\Gamma(x+1)}p^r(1-p)^x$ | | $\dfrac{r(1-p)}{p}$ | $\dfrac{r(1-p)}{p^2}$ | `random.negative_binomial(r,p)` `stats.nbinom(r,p)` |
| shifted negative binomial sNBD($r,p$) | $\{r,r+1,\ldots\}$ | $\binom{x-1}{r-1}p^r(1-p)^{x-r}$ | | $\dfrac{r}{p}$ | $\dfrac{r(1-p)}{p^2}$ | `random.negative_binomial(r,p)+r` `stats.nbinom(r,p,loc=r)` |
| geometric Geom($p$) | $\mathbb{N}$ | $p(1-p)^x$ | $(1-p)^{x+1}$ | $\dfrac{1-p}{p}$ | $\dfrac{1-p}{p^2}$ | `random.geometric(p)-1` `stats.geom(p,loc=-1)` |
| shifted geometric sGeom($p$) | $\mathbb{Z}^+$ | $p(1-p)^{x-1}$ | $(1-p)^x$ | $\dfrac{1}{p}$ | $\dfrac{1-p}{p^2}$ | `random.geometric(p)` `stats.geom(p)` |

Table A.1: Catalogue of discrete distributions.

| name | support | $p(x)$ | $S(x)$ | $\mathbb{E}[X]$ | $\mathbb{V}(X)$ | simulation |
|------|---------|--------|--------|--------|--------|-----------|
| Poisson $\text{Poiss}(\lambda)$ | $\mathbb{N}$ | $\dfrac{e^{-\lambda}\lambda^x}{x!}$ | — | $\lambda$ | $\lambda$ | `random.poisson(lam)` `stats.poisson(lam)` |
| beta-binomial $\text{BetaBin}(n,\alpha,\beta)$ | $\{0,\ldots,n\}$ | $\dbinom{n}{x}\dfrac{B(\alpha+x,\beta+n-x)}{B(\alpha,\beta)}$ | — | $\dfrac{n\alpha}{\alpha+\beta}$ | $\dfrac{n\alpha\beta(\alpha+\beta+n)}{(\alpha+\beta)^2(\alpha+\beta+1)}$ | N/A `stats.betabinom(n,a,b)` |
| beta-geometric $\text{BetaGeom}(\alpha,\beta)$ | $\mathbb{N}$ | $\dfrac{B(\alpha+1,\beta+x)}{B(\alpha,\beta)}$ | $\dfrac{B(\alpha,\beta+x+1)}{B(\alpha,\beta)}$ | $\dfrac{\beta}{\alpha-1}^*$ | $\dfrac{\alpha\beta(\alpha+\beta-1)}{(\alpha-2)(\alpha-1)^2}^*$ | N/A N/A |

Table A.2: Catalogue of discrete distributions (continued); * defined only when denominator is positive.

| name | support | $p(x)$ | $S(x)$ | $\mathbb{E}[X]$ | $\mathbb{V}(X)$ | simulation |
|---|---|---|---|---|---|---|
| normal $N(\mu,\sigma^2)$ | $\mathbb{R}$ | $\dfrac{1}{\sigma\sqrt{2\pi}}e^{-(x-\mu)^2/(2\sigma^2)}$ | $\Psi(x)$ | $\mu$ | $\sigma^2$ | `random.randn()*sig+mu` `stats.norm(loc=mu,scale=sig)` |
| chi-squared $\chi_p^2$ | $\mathbb{R}^+$ | $\dfrac{x^{p/2-1}e^{-x/2}}{\Gamma(p/2)2^{p/2}}$ | | $p$ | $2p$ | `random.chisquare(p)` `stats.chi2(p)` |
| Student's T $t_p$ | $\mathbb{R}$ | $\dfrac{\Gamma((p+1)/2)}{\Gamma(p/2)\sqrt{p\pi}}(1+t^2/p)^{-(p+1)/2}$ | | $0$ | $\dfrac{p}{p-2}$* | `random.standard.t(p)` `stats.t(p)` |
| Snedecor's F $F_{p,q}$ | $\mathbb{R}^+$ | $\propto x^{p/2-2}\left[1+(p/q)x\right]^{-(p+q)/2}$ | | $\dfrac{q}{q-2}$* | | `random.f(p,q)` `stats.f(p,q)` |
| exponential $\mathrm{Exp}(\beta)$ | $\mathbb{R}_*$ | $\dfrac{1}{\beta}e^{-x/\beta}$ | $e^{-x/\beta}$ | $\beta$ | $\beta^2$ | `random.exponential(scale=b)` `stats.expon(scale=b)` |

Table A.3: Catalogue of continuous distributions; * defined only when denominator is positive.

| name | support | $p(x)$ | $S(x)$ | $\mathbb{E}[X]$ | $\mathbb{V}(X)$ | simulation |
|---|---|---|---|---|---|---|
| beta $\mathrm{Beta}(\alpha,\beta)$ | $(0,1)$ | $\dfrac{x^{\alpha-1}(1-x)^{\beta-1}}{B(\alpha,\beta)}$ | | $\dfrac{\alpha}{\alpha+\beta}$ | $\dfrac{\alpha\beta}{(\alpha+\beta)^2(\alpha+\beta+1)}$ | random.beta(a,b) stats.beta(a,b) |
| gamma $\mathrm{Gamma}(\alpha,\beta)$ | $\mathbb{R}^+$ | $\dfrac{x^{\alpha-1}e^{-x/\beta}}{\Gamma(\alpha)\beta^{\alpha}}$ | | $\alpha\beta$ | $\alpha\beta^2$ | random.gamma(a,scale=b) stats.gamma(a,scale=b) |
| Pareto $\mathrm{Pareto}(\alpha,\beta)$ | $[\beta,\infty)$ | $\dfrac{\alpha\beta^{\alpha}}{x^{\alpha+1}}$ | $\left(\dfrac{\beta}{x}\right)^{\alpha}$ | $\dfrac{\alpha\beta}{\alpha-1}^{*}$ | $\dfrac{\alpha\beta^2}{(\alpha-1)^2(\alpha-2)}^{*}$ | random.pareto(a)*b+b stats.pareto(a,scale=b) |
| Lomax $\mathrm{Lomax}(\alpha,\beta)$ | $\mathbb{R}_{*}$ | $\dfrac{\alpha}{\beta}\left[1+\dfrac{x}{\beta}\right]^{-(\alpha+1)}$ | $\left[1+\dfrac{x}{\beta}\right]^{-\alpha}$ | $\dfrac{\beta}{\alpha-1}^{*}$ | $\dfrac{\alpha\beta^2}{(\alpha-1)^2(\alpha-2)}^{*}$ | random.pareto(a)*b stats.lomax(a,scale=b) |
| Weibull $\mathrm{Weibull}(\alpha,\beta)$ | $\mathbb{R}_{*}$ | $\dfrac{\alpha}{\beta}\left(\dfrac{x}{\beta}\right)^{\alpha-1}e^{-(x/\beta)^{\alpha}}$ | $e^{-(x/\beta)^{\alpha}}$ | $\beta\Gamma(1+1/\alpha)$ | | random.weibull(a)*b stats.weibull_min(a,scale=b) |

Table A.4: Catalogue of continuous distributions (continued).

# References

Aalen, O.o., O. Borgan, H.K. Gjessing (2008) *Survival and Event History Analysis: A Process Point of View*, Springer.

Agresti, A. (2013) *Categorical Data Analysis*, 3rd ed., Wiley.

Agresti, A. (2015) *Foundations of Linear and Generalized Linear Models*, Wiley.

Agresti, A. (2019) *An Introduction to Categorical Data Analysis*, 3rd ed., Wiley.

Alpaydin, E. (2020) *Introduction to Machine Learning*, 4th ed., MIT Press.

Barber, D. (2012) *Bayesian Reasoning and Machine Learning*, Cambridge.

Beck, K., M. Beedle, A. van Bennekum, A. Cockburn, W. Cunningham, M. Fowler, J. Grenning, J. Highsmith, A. Hunt, R. Jeffries, J. Kern, B. Marick, R.C. Martin, S. Mellor, K. Schwaber, J. Sutherland, D. Thomas (2001) *The Agile Manifesto*, http://agilemanifesto.org.

Bentley, J.L. (1975) Multidimensional binary search trees used for associative search, *Communications of the ACM* **18**: 509–517.

Berger, P.D., R.E. Maurer, and G.B. Celli 2018 *Experimental Design: With Applications in Management, Engineering, and the Sciences*, 2nd ed., Springer.

Box, G.E.P., G.M. Jenkins, G.C. Reinsel, and G.M. Ljung (2016) *Time Series Analysis: Forecasting and Control*, 5th ed., Wiley.

Breiman, L. (1966) Bagging predictors, *Machine Learning* bf 26: 123–140.

Breiman, L. (2001) Random forests, *Machine Learning* bf 45: 5–32.

Brokwell, P.J. and R.A. Davis (1992) Time reversability, identifiability, and independence of innovations for stationary time series, *J. Time Series Analysis* **13**: 377–390.

Brokwell, P.J. and R.A. Davis (2016) *Introduction to Time Series and Forecasting*, 3rd ed., Springer.

Brown, J.W. and R.V. Churchill (2013) *Complex Variables and Applications*, 9th ed., McGraw–Hill.

Brier, G.W. (1950) Verification of forecasts expressed in terms of probability; *Monthly Weather Review* **78**(1)

Bühlmann, H. (1967) Experience rating and credibility, *ASTIN Bulletin* **4** 199–207.

Bühlmann, H. (1970) *Mathematical Methods in Risk Theory*, Springer.

Bühlmann, H. and A. Gisler (2005) *A Course in Credibility Theory and its Applications*, Springer.

Bühlmann, H. and E. Straub (1970) Glaubwürdigkeit für Schadensätze, *Bulletin of Swiss Ass. of Act.*, 111-133.

Buitinck, *et al.* (2013) API design for machine learning software: experiences from the scikit-learn project, in *ECML PKDD Workshop: Languages for Data Mining and Machine Learning*, p. 108–122.

Capinksi, M. and E. Kopp (2005) *Measure, Integral, and Probability*, 2nd ed., Springer.

Casella, G., and R.L. Berger (2002) *Statistical Inference*, 2nd ed., Brooks/-Cole.

Chawla, N.V., K.W. Bowyer, L.O. Hall and W.P. Kegelmeyer (2002) SMOTE: Synthetic Minority Over-sampling Technique, *J. Artificial Intelligence Research* **16**: p. 321–357.

Chen, T. and C. Guestrin (2016) XGBoost: A Scalable Tree Boosting System, *Proceedings of the 22nd ACM SIGKDD Int. Conf. on Knowledge Discovery and Data Mining* ACM 785–794.

Chong, E.K.P. and S.H. Zak (2008) *An Introduction to Optimization*, 3rd ed., John Wiley & Sons.

Colombo, R., W. Jiang (1999) A stochastic RFM model, *J. Interactive Marketing* **13**(3): 2–12.

Conway, J.B. (1978) *Functions of One Complex Variable I*, 2nd ed., Springer.

Cormon, T.H., C.E. Leiserson, R.L. Rivest, and C. Stein (2009) *Introduction to Algorithms*, 3rd ed., MIT Press.

Davidson-Pilon, C. (2016) *Bayesian Methods for Hackers: Probabilistic Programming and Bayesian Inference*, Addison Wesley. (See web for updated version https://camdavidsonpilon.github.io/Probabilistic-Programming-and-Bayesian-Methods-for-Hackers/.)

Devroye, L. (1986) *Non-Uniform Random Variate Generation*, Springer.

DiBenedetto, E. (2002) *Real Analysis*, Birkhäuser.

Dietterich, T. (2000) An experimental comparison of three methods for constructing ensembles of decision trees: bagging, boosting, and randomization, *Machine Learning* **40**: 139–158.

Dobson, A.J. and A.G. Barnett (2018) *An Introduction to Generalized Linear Models*, 4th ed., CRC Press.

Doerr, J. (2018) *Measure What Matters: How Google, Bono, and the Gates Foundation Rock the World with OKRs*, Portfolio.

Durbin, J. (1960) The fitting of time series models, *Review of the Institute of International Statistics* **28**: 233-244.

Dunn, P., and G.K. Smyth (2018) *Generalized Linear Models with Examples in R*, Springer.

Durrett, R. (2016) *Essentials of Stochastic Processes*, Springer.

Efron, B. (1971) Forcing a sequential experiment to be balanced, *Biometrika*, **58**, 403–417.

Fader, P., B. Hardie, and K.L. Lee (2005) Counting your customers the easy way: an alternative to the Pareto/NBD Model. *Mark. Sci.* **24**: 275–284.

Fader, P.S., B.G.S. Hardie (2007) How to project customer retention, *J. Interactive Marketing* **21**(Winter): 76–90.

Fader, P., B. Hardie, J. Shang (2010) Customer-base analysis in a discrete-time noncontractual setting. *Mark. Sci.* **29**: 1086–1108.

Fisher, R.A. (1935) *The Design of Experiments*, Oliver and Boyd.

Flach, P. (2012) *Machine Learning: The Art and Science of Algorithms that Make Sense of Data*, Cambridge.

Folland, G.B. (1999) *Real Analysis: Modern Techniques and their Applications*, 2nd ed., Wiley.

Freund, Y. and R.E. Schapire (1996) Experiments with a new boosting algorithm, In *The Thirteenth International Conference on Machine Learning*, ed. L. Saitta, 148–156, San Mateo, CA, Morgan Kaufmann.

Greedy function approximation: a gradient boosting machine, *Annals of Stats.* **29**(5): 1189–1232.

Stochastic gradient boosting, *Comp. Stats. and Data Analysis* **38**(4): 367–378.

Gelman, A., J.B. Carlin, H.S. Stern, D.B. Dunson, A. Vehtari, and D.B. Rubin (2014) *Bayesian Data Analysis*, 3rd ed., CRC Press.

Gèron, A. (2019) *Hands-on Machine Learning with Sci-Kit Learn, Keras, and TensorFlow: Concepts, Tools and Techniques to Build Intelligent Systems*, 2nd ed., O'Reilly.

Gill, J. and M. Torres (2020) *Generalized Linear Models: A Unified Approach*, 2nd ed., Sage Publishing.

Hajek, B. (2015) *Random Processes for Engineers*, Cambridge University Press.

Hamilton, J.D. (1994) *Time Series Analysis*, Princeton.

Härdle, W.K. and L. Simar (2019) *Applied Multivariate Statistical Analysis*, 5th ed., Springer.

Hastie, T, R. Tibshirani, and J. Friedman (2009) *The Elements of Statistical Learning: Data Mining, Inference, and Prediction*, Springer.

Ho, T.K. (1995) Random decision forests, *Proc. 3rd Int. Conf. Document Analysis and Recognition*, Montreal, QC, 278–282.

Ho, T.K. (1998) The random subspace method for construction decision forests, *IEEE Transactions on Pattern Analysis and Machine Intelligence* **13**: 340–354.

Hogg, R.V., E.A. Tanis, and D.L. Zimmerman (2015) *Probability and Statistical Inference*, 9th ed., Pearson.

Ishwaran, H., U.B. Kogalur, E.H. Blackstone, and M.S. Lauer (2008) Random survival forests, *Annals of App. Stats.* bf 2(3): 841–860.

Imbens, G.W., and D.B. Rubin (2015) *Causal Inference for Statistics, Social, and Biomedical Sciences: An Introduction*, Cambridge University Press.

James, G., D. Witten, T. Hastie, and R. Tibshirani (2013) *An Introduction to Statistical Learning: With Applications in R*, Springer.

Kahn, A.B. (1962) Topological sorting of large networks, *Communications of the ACM*, **5**(11): 558–562.

Kiefer, J. (1953) Sequential minimax search for a maximum, *Proceedings of the American Mathematical Society* **4**(3): 502–506.

Klein, J.P. and M.L. Moeschberger (2003) *Survival Analysis: Techniques for Censored and Truncated Data*, 2nd ed., Springer.

Kochenderfer, M.J. (2015) *Decision Making Under Uncertainty*, MIT Press.

Kohavi, R., D. Tang, and Y. Xu (2020) *Trustworthy Online Controlled Experiments: A Practical Guide to A/B Testing*, Cambridge University Press.

Kuhn, M. and K. Johnson (2013) *Applied Predictive Modeling*, Springer.

Lafore, R. (2003) *Data Structures and Algorithms in Java*, 2nd ed., SAMS.

Lambert, K.A. (2014) *Fundamentals of Python Data Structures*, Cengage.

Lee, K.D. and S. Hubbard (2015) *Data Structures and Algorithms with Python*, Springer.

Maruskin, J. (2010) Distance in the space of energetically bounded Keplerian orbits, *Celestial Mechanics and Dynamical Astronomy* **108**: 265–274.

Maruskin, J.M., D.J. Scheeres, and K.T. Alfriend (2012) Correlation of optical observations of objects in earth orbit, *Journal of Guidance Control and Dynamics* **32**: 194–209.

Maruskin, J. (2018) *Dynamical Systems and Geometric Mechanics: An Introduction*, 2nd ed., de Gruyter.

McElreath, R. (2016) *Statistical Rethinking: A Bayesian Course with Examples in R and Stan*, CRC Press.

Metropolis, N. (1987) The beginning of the Monte Carlo method, *Los Alamos Science* (1987 Special Issue dedicated to Stanislaw Ulam): 125–130.

Mood, A.M. (1950) *Introduction to the Theory of Statistics*. McGraw-Hill.

Morgan, S.L. and C. Winship (2015) *Counterfactuals and Causal Inference: Methods and Principles for Social Research*, 2nd ed., Cambridge University Press.

McKinney, W. (2017) *Python for Data Analysis*, 2nd ed., O'Reilly.

Newbold, P. (1974) The exact likelihood function for a mixed autoregressive moving average process, *Biometrika* **61**: 423–426.

Niculescu-Mizil and Caruana (2005) Predicting good probabilities with supervised learning, In ICML-05 *International Conference on Machine Learning*, Aug. 2005, 625–632; https://doi.org/10.1145/1102351.1102430.

Nielson, A. (2019) *Practical Time Series Analysis* O'Reilly.

Nyce, C.M. (2017) The winter getaway that turned the world upside down, *The Atlantic*.

Olive, D. (2017) *Linear Regression*, Springer.

Pearl, J. (2009) *Causality: Models, Reasoning, and Inference*, 2nd ed., Cambridge University Press.

Pearl, J., M. Glymour, and N.P. Jewell (2016) *Causal Inference in Statistics: A Primer*, Wiley.

Pedregosa, *et al.* (2011) Scikit-learn: Machine Learning in Python, *Journal of Machine Learning Research* **12**: p. 2825–2830.

Platt, J. (2000) Probabilistic outputs for support vector machines and comparison to regularized likelihood models; In Bartlett B., B. Schölkopf, D. Schuurmans, A. Smola (eds.) *Advances in Kernel Methods Support Vector Learning*, p. 61–74, MIT Press.

Reinhart, A. (2015) *Statistics Done Wrong*, no starch press.

Robbins, H. (1952) Some aspects of the sequential design of experiments, *Bulletin of the American Mathematical Society* **58**(5): 528–535.

Rosenbaum, P.R. (2002) *Observational Studies*, 2nd ed., Springer.

Ross, S. (2012) *A First Course in Probability*, 9th ed., Prentice Hall.

Salvatier, J., T.V. Wiecki, and C. Fonnesbeck (2016) Probabilistic programming in Python using PyMC3, *PeerJ Computer Science* 2:e55 DOI: 10.7717/peerj-cs.55.

Schapire, R.E. (1990) The strength of weak learnability, *Machine Learning* **5**: 197–227.

Schapire, R.E., Y. Freund, P. Bartlett, and W.S. Lee (1998) Boosting the margin: a new explanation for the effectiveness of voting methods, *Information Sciences* **179**: 1298–1318.

Schmittlein, D.C., D.G. Morrison, R. Colombo (1987) Counting your customers: who are they and what will they do next? *Manag. Sci.* **33**-1.

Schmittlein, D.C. and R.A. Peterson (1994) Customer base analysis: An industrial purchase process application. *Marketing Sci.* **13**(1): 41–67.

Seber, G.A.F. and A.J. Lee (2003) *Linear Regression Analysis* (2nd ed.), John Wiley.

Selvamuthu, D. and D. Das (2018) *Introduction to Statistical Methods, Design of Experiments, and Statistical Quality Control*, Springer.

Selvin, S. (1975a). A problem in probability (letter to the editor), *The American Statistician* **29**(1): 67–71.

Selvin, S. (1975b) On the Monty Hall problem (letter to the editor), *The American Statistician* **29**(3): 134.

Seufert, E.B. (2014) *Freemium Economics: Leveraging Analytics and User Segmentation to Drive Revenue*, Morgan Kaufmann, an imprint of Elsevier.

Shao, J. (2003) *Mathematical Statistics*, 2nd ed., Springer.

Shumway, R.H. and D.S. Stoffer (1999) *Time Series Analysis and Its Applications: With R Examples*, 4th ed., Springer.

Sutton, R.S. and A.G. Barto (2018) *Reinforcement Learning: An Introduction*, 2nd ed., MIT Press.

Takeuchi, H. and I. Nonaka (1986) The new new product development game, *Harvard Business Review*

Thorp, E.O. (1966) *Beat the Dealer: A Winning Strategy for the Game of Twenty-One*, Revised ed., Vintage.

Thorp, E.O. (2017) *A Man for all Markets: From Las Vegas to Wall Street, how I beat the dealer and the market*, Random House.

Trussell, J. and D.E. Bloom (1979) A model distribution of height or weight at a given age, *Human Biology* **51**: 523-536.

Vaupel, J.W. and A.I. Yashin (1985). Heterogeneity's ruses: some surprising effects of selection on population dynamics, *The American Statistician* **39**: 176–185.

Wasserman, L. (2004) *All of Statistics: A Concise Course in Statistical Inference*. Springer.

Wasserman, L. (2006) *All of Nonparametric Statistics*. Springer.

Wei, W.W.S. (2019) *Time Series Analysis: Univariate and Multivariate Models*, Pearson.

Withers, C.S. and S. Nadarajah (2014) The spectral decomposition and inverse of multinomial and negative multinomial covariances. *Brazilian Journal of Probability and Statistics*. Vol. 28, No.3, p. 376–380.

Wolfram Alpha LLC (2018) *Wolfram—Alpha*, wolframalpha.com.

Zou, H., T. Hastie, and R. Tibshirani (2007) On the "degrees of freedom" of the lasso, *The Annals of Statistics* **35**(5): 2173–2192.

# Index